Matthias Kolb

Strength of advanced ceramic breeder pebbles

Influence of fabrication methods, composition and annealing

Strength of advanced ceramic breeder pebbles

Influence of fabrication methods, composition and annealing

by
Matthias Kolb

KIT Scientific Publishing

Strength of advanced ceramic breeder pebbles
Influence of fabrication methods, composition and annealing

Zur Erlangung des akademischen Grades eines Doktor-Ingenieurs
von der KIT-Fakultät für Maschinenbau des
Karlsruher Instituts für Technologie (KIT) genehmigte Dissertation von

Dipl.-Ing. Holger Matthias Hartmut Kolb, geboren in Stuttgart

Tag der mündlichen Prüfung: 2. August 2018
Referenten: Prof. Dr. rer. nat. Michael J. Hoffmann, Prof. Dr.-Ing. Marc Kamlah

Impressum

KIT Scientific Publishing

Karlsruher Institut für Technologie (KIT)
KIT Scientific Publishing
Straße am Forum 2
D-76131 Karlsruhe

www.ksp.kit.edu

Print on Demand 2019 – Gedruckt auf FSC-zertifiziertem Papier

ISBN 978-3-7315-0899-1
DOI 10.5445/KSP/1000091030

Danksagung

Nun ist es also vollbracht. Die Dissertation ist geschrieben, in einem wissenschaftlichen Vortrag vorgestellt und verteidigt worden. Somit sind dies die letzten Zeilen die es zu schreiben gilt, und für gewöhnlich ist dies der persönlichste Teil der Arbeit. Ist es deshalb verwunderlich, dass man schon während der Phasen des Auswertens und des Schreibens immer wieder Sätze und Textfragmente im Kopf wälzt, wie die Danksagung im Kern auszusehen hätte? Sich Gedanken in den Vordergrund drängen, in welchem Stil man schreiben sollte? Kunstvoll oder sachlich? Knapp oder romanesk? Hymne oder Tirade? Ein Triumphzug, ein Fiebertraum? Es war vermutlich nicht unklug, zuerst die Arbeit fertigzustellen und sich mit etwas Abstand der Danksagung zu widmen.

So ist es nun an der Zeit zurückzublicken, und dieser Blick ist ja in vielen Fällen kein verlässlicher, mindestens jedoch ein anderer. Ob mir Vieles in Erinnerung geblieben ist oder bleiben wird? Bestimmt wird mit der Zeit einiges verklärt werden. Etwa Momente der Depression, in denen nichts ging: „Was bedeute ich, wenn ich nichts schreibe, was von Bedeutung ist? Dann bedeute ich nichts." Stunden, in denen man „mit Bleistift schreibt" und das Vorhaben zu promovieren als Ganzes in Frage stellt. Nur, um anschließend wie ein psychisch Kranker in Augenblicke großer Euphorie zu taumeln, wenn sich aus den Messwerten und Bildern ein Nugget an Erkentnis waschen ließ. Woran sich postwendend Tage ätzender Zweifel anschließen, die das Gewonnene zu zersetzen versuchen. Aber was soll ich klagen? Das ist die Essenz meines Berufes! Und ich werde diesem Fluß ins Herz der Finsternis abermals folgen.

Wer auch immer sich ebenfalls auf diese Reise begeben möchte, möge aus meinen Fehlern lernen und diesen einen Ratschlag beherzigen: „Was ihr auch macht, macht es nicht selbst. Ob ihr verblendet, oder erhellt: was ihr auch macht, macht es nicht selbst." Ich bin diesbezüglich vor Allem Christina Odemer zu großem Dank verpflichtet, die für mich sämtliche Schliffe präpariert, die Messungen per optischem Mikroskop durchgeführt, und die allgemeinen Abläufe im Labor geregelt hat. Ferner möchte ich Ulrike Maciejewski für die Bruchlastmessungen und für einige der REM-Bilder danken, sowie Margret Offermann für die Dichtemessungen. Thomas Bergfeld und seinen Mitarbeitern gebührt mein Dank für die Elementbestimmungen.

Natürlich bin ich mir bewusst, welches Privileg ich hatte, im Rahmen dieser Arbeit für drei Monate in Japan zu arbeiten. Nicht nur dafür, dass sie mir diesen Forschungsaufenthalt ermöglicht haben, möchte ich mich bei Regina Knitter, Tsuyoshi Hoshino und Masaru Nakamichi bedanken. Dankbar bin ich außerdem den Mitarbeitern bei JAEA/QST in Rokkasho, als da im Besonderen wären: Suzuki-san, Sekya-san und Jae-Hwan Kim, ohne die ich mit Sicherheit nicht so gut zurechtgekommen wäre und auch viele meiner Ausflüge die Planungsphase nicht überstanden hätten. Ferner komme ich nicht umhin, in meiner Danksagung dieser Doktorarbeit auf die Förderung des Auslandsaufenthaltes durch das Karlsruhe House of Young Scientists (KHYS) hinzuweisen.

„Alles was ich immer wollte, war Alles“ mit dieser Arbeit. Manch einer mag das anmaßend finden – vermutlich nicht zu Unrecht. In jedem Fall birgt es die reelle Gefahr, sich im Verfassen und Ausarbeiten seines Opus Magnum wortwörtlich zu verlieren. Dass mir dies nicht passiert ist, verdanke ich zu großen Teilen Olga und der wundervollen Zeit, die wir abseits des Promovierens zusammen erlebt haben. Auch meine Brüder haben ihren Anteil daran, dass mir immer bewusst blieb, was im Leben wirklich zählt. Meinen Eltern danke ich in erster Linie dafür, dass sie meiner Neugier in jedem Alter neue Nahrung gegeben haben und mich zu einem kritisch denkenden Menschen erzogen haben. Dadurch habe ich die Werkzeuge erhalten, die ich brauchte, um so weit zu kommen.

Zu guter Letzt möchte ich dieses Kapitel mit einer Konstatierung schließen, derer ich letztlich während des Promovierens immer wieder gewahr wurde.

„The stars are indifferent to astronomy.“

Kurzfassung

Um Fusionskraftwerke mit dem Treibstoff Tritium zu versorgen, werden Lithium-reiche Verbindungen um das Fusionsplasma herum angebracht, damit Neutronen aus der Fusionsreaktion darin eingefangen werden und somit Tritium "erbrütet" wird. Das europäische Helium-gekühlte Kugelschüttbett Blanket ist eine der vielversprechendsten technischen Umsetzungen dieses Konzepts und es verwendet keramische Kugeln als feste Brutmaterialien. Während des Betriebs eines Fusionskraftwerks müssen diese Brutkeramikkugeln sowohl hohen Temperaturen, als auch erheblichen mechanischen Spannungen standhalten. Fortschrittliche keramische Brutmaterialien die entweder aus den zwei Verbindungen Lithiumorthosilikat, Li_4SiO_4, und Lithiummetatitanat, Li_2TiO_3, oder aus reinem Lithiummetatitanat mit Lithium-Überschuss bestehen, werden gegenwärtig als für diese Anwendung am besten geeignete Materialien betrachtet. Generell ist eine hohe mechanische Festigkeit der Kugeln von hoher Bedeutung für das Funktionieren des Blankets, da nur eine unbedenklich kleine Menge an Kugeln toleriert werden kann die unter diesen Bedingungen zerbrechen. Es ist deshalb das Ziel dieser Dissertation die Versagensfestigkeit und die Versagensmechanismen dieser Kugeln zu untersuchen, damit diese für die Verwendung in einem Fusionskraftwerk approbiert werden können und um Ansatzpunkte für weitere Verbesserungen zu erarbeiten. Dafür sollen ebenfalls die wesentlichen Effekte die zur Schädigung der Materialien führen identifiziert werden. Damit die Brutkeramikkugeln sicher in einem Fusionskraftwerk eingesetzt werden können, ist es weiterhin unerlässlich dass diese eine ausreichend hohe Langzeitstabilität aufweisen. In dieser Arbeit soll deshalb darüber hinaus die Entwicklung der Festigkeit und der Versagensmechanismen als Funktion der Einsatzdauer untersucht werden.

Um die zweiphasigen Kugeln herzustellen, wird ein Schmelze-basierender Herstellprozess verwendet, während ein konventioneller Sinterprozess mit einem nasschemischen Formgebungsprozess für die Herstellung der Lithiummetatitanat-Kugeln mit Lithium-Überschuss verwendet wird. Da das relevante quasi-binäre System Li_4SiO_4-Li_2TiO_3 eutektisch ist, hat die Zusammensetzung einen entscheidenden Einfluss auf das Gefüge der Kugeln bei einer Schmelze-basierten Herstellung. Um diesem Umstand Rechnung zu tragen, wurden sowohl eine über- und als auch eine untereutektische Zusammensetzung für die Untersuchungen gewählt. Da die Zusammensetzung der Lithiummetatitanat-Kugeln mit Lithium-Überschuss feststeht, spielt der verwendete Herstellprozess eine wesentliche Rolle. Infolgedessen wird bei diesen Kugeln der Einfluss der Emulsionsmethode und der Sol-Gel-Methode auf die Kugeleigenschaften untersucht.

Um den Effekt der Zweitphase auf die Kugeleigenschaften über den gesamten Konzentrationsbereich zu untersuchen, ist der Schmelze-basierte Herstellprozess ungeeignet. Deshalb wurde im Zuge dieser Arbeit die Emulsionsmethode angepasst um zweiphasige Kugeln beliebiger Zusammensetzungen herstellen zu können. Zusätzlich wurden Proben mit der nachgewiesenermaßen steiferen Verbindung Lithiumaluminat anstelle von Lithiummetatitanat hergestellt, um den Einfluss einer weiteren Zweitphase zu untersuchen. Um die Langzeitstabilität der fortschrittlichen Brutkeramikkugeln zu analysieren, wurden die angeführten vier Sorten über 4 bis 128 Tage unter fusionsrelevanten Bedingungen ausgelagert. Jede Probe wurde eingehend charakterisiert im Hinblick auf die Elementzusammensetzung, die Phasenzusammensetzung und Eigenspannungen, sowie die Morphologie der Kugeln und ihrer Porosität. Die hierzu notwendigen Untersuchungen wurden hauptsächlich per ICP-OES, Röntgendiffraktometrie sowie Rasterelektronenmikroskopie, optischer Mikroskopie und Helium-Pyknometrie durchgeführt. Die mechanische Festigkeit der Kugeln wurde in uniaxialen Drucktests bei Raumtemperatur ermittelt. Um Rückschlüsse auf die Versagensmechanismen ziehen zu können, wurde für jede Probe eine Weibullanalyse durchgeführt, sowie die Bruchstücke von ausgewählten Proben untersucht.

Anhand der vorliegenden Ergebnisse dieser Arbeit ist es unzweifelhaft, dass die Festigkeit der fortschrittlichen Brutkeramikkugeln vom Herstellprozess, der Zusammensetzung und der Auslagerungsdauer, bzw. der Einsatzdauer, bestimmt wird. Deshalb ergeben sich die Festigkeit und die Versagenswahrscheinlichkeit einer einzelnen Kugel als Kombination all dieser Punkte.

Durch den Schmelze-basierten Herstellprozess werden Oberflächenrisse in die Kugeln eingebracht, die das Versagensverhalten der Mehrzahl der Kugeln bestimmen. Des Weiteren führt dieser Prozess zu erheblichen Eigenspannungen in den Kugeln nach der Kristallisation. Während des Auslagerns werden diese Spannungen abgebaut was einen deutlichen Einfluss auf die Festigkeit der Kugeln hat. Beide Sorten Lithiummetatitanat-Kugeln mit Lithium-Überschuss versagen aufgrund von Volumenrissen. Jedoch führen die unterschiedlichen Herstellmethoden zu merklichen Unterschieden im Versagen der Kugeln und der Kugelmorphologie. Trotz dieser Unterschiede wird die Festigkeit beider Sorten Kugeln nach der Auslagerung von der Korngröße der Kugeln bestimmt. Die Ergebnisse der zweiphasigen Kugeln die durch die Emulsionsmethode hergestellt wurden, weisen eindeutig daraufhin, dass die Bruchzähigkeit eine Funktion des Lithiummetatitanatgehalts ist. Die Resultate zeigen weiterhin, dass durch die Verwendung von Lithiumaluminat anstelle von Lithiummetatitanat die Bruchzähigkeit noch deutlich erhöht werden kann.

Abstract

In order to supply fusion power plants with tritium as fuel, lithium rich compounds are placed around the fusion plasma to capture impinging neutrons from the fusion reaction and thus 'breed' tritium. The European helium cooled pebble bed blanket is one of the most promising technological realizations of this concept and it applies ceramic breeder pebbles as solid breeder material. During the operation of the fusion power plant, these ceramic breeder pebbles have to withstand a combination of high temperatures and significant mechanical stress. Advanced ceramic breeder materials that either consist of the two compounds lithium orthosilicate, Li_4SiO_4, and lithium metatitanate, Li_2TiO_3, or purely of lithium excessive lithium metatitanate, are currently the most promising materials for this application. Generally, a high mechanical strength of the pebbles is of great importance for the functioning of the fusion blanket, because an unobjectionable fragmentation of the pebble bed during operation must be realized. It is thus the goal of the presented doctoral thesis to investigate the failure strength and mechanisms of these pebbles, in order to be able to qualify the pebbles for use and to provide starting points for future improvements. Therefore the determinant effects that generate the observed failure mechanisms shall be identified as well. For using breeder pebbles safely in a fusion power plant, it is furthermore imperative to provide a sufficient degree of long term stability. Therefore the evolution of the mechanical strength and the failure mechanisms shall be investigated in this work as a function of the duration of use.

For the fabrication of the two-phase pebbles, a melt-based process is used, whereas the lithium excessive lithium metatitanate pebbles are fabricated by a conventional sintering process after they were shaped by wet-chemical techniques. As the relevant quasi-binary system Li_4SiO_4-Li_2TiO_3 is eutectic, the composition of the pebbles will significantly influence the pebble microstructure in a melt-based fabrication process. In order to address this effect, one hypo- and one hypereutectic composition were chosen for this work. Since the composition of the lithium excessive lithium metatitanate pebbles is fixed, the fabrication technique plays a much more important role. Consequently, the effect of the emulsion method and the sol-gel method on the pebble properties is investigated.

In order to also determine the influence of the second phase on these pebble properties over the whole composition range, the melt-based fabrication is inapplicable. Consequently, the emulsion method was modified in this work and transferred to produce pebbles with lithium orthosilicate and lithium metatitanate contents at will. Additionally, pebbles with the reportedly stiffer compound lithium aluminate instead of lithium metatitanate as second phase were fabricated to test the effect of an other second phase on the mechanical strength. For investigating the long-term stability of the advanced ceramic breeder pebbles, the mentioned four grades were annealed in a fusion relevant environment for 4 to 128 days. All samples were thoroughly characterized in terms of their elemental composition, their constituting phases and residual stresses as well as their morphology and porosity, by mainly using ICP-OES, X-ray diffraction as well as scanning electron microscopy, optical microscopy and helium pycnometry.

The mechanical strength of the pebbles was investigated by uniaxial compression experiments at room temperature. To gain insight into the failure mechanisms of the pebbles, a Weibull analysis was performed for all samples as well as an analysis of the pebble fragments of selected samples. From the results of this work, it is unambiguously clear that the strength of advanced ceramic breeder pebbles is governed by the fabrication process, the material composition and the time for that these pebbles are annealed, i.e. the duration of usage. Consequently the strength and the probability of failure of an individual pebble is a combined result of all of these points.

The melt-based fabrication generally leads to the formation of surface cracks, which determine the failure behavior of the majority of pebbles. This fabrication method also leads to considerable residual stresses within the pebbles after crystallization. During the annealing, these stresses are removed, which greatly affects the mechanical strength of the samples. While both types of lithium excessive lithium metatitanate pebbles fail from volume defects, the different fabrication processes leads to a significant disparity of the failure behavior and the pebble morphology. In spite of these differences, the mechanical strength of these pebbles after annealing is equally governed by their grain size. The results of the biphasic emulsion method pebbles that contain lithium metatitanate clearly indicate that their fracture toughness is a function of the lithium metatitanate concentration. The results also show that by using lithium aluminate instead of lithium metatitanate, the fracture toughness of the pebbles can be increased even more.

Table of contents

1 Introduction

Today's power generation is facing an enormous challenge, not only within Germany or Europe but anywhere in the developed and developing world. Current studies estimate that the worldwide demand for electrical power will at least double or even quadruple within this century [1]. With dwindling fossil fuels in general and their limited availability for many nations, an alternative way of power generation must be established to ensure adequate power generation. The need for substantial progress in power generation is even more pressing if man-made climate change is taken into account. While the demand for affordable electrical power increases, the existing carbon dioxide emitting power generation holds no future prospect and has to be scaled back and replaced to keep and increase the level of power generation. Currently, short-run solutions to reduce the share of carbon dioxide emitting power plants like bio-energy consume significant portions of fertile soils, which are indispensable for providing nutrition to an ever-growing world population, and as biofuels are oftentimes fabricated from esculent plants, these plants may become prohibitive on a global scale for poor people, who make up a large proportion of the world population. Yet, bio-energy may still contribute to the energy generation in the future, if these issues are solved [2–4].

Nuclear fission power plants were once thought to be immune to such issues and provide electricity at low cost to any developed nation in abundance. However, severe accidents like the Three Mile Island incident [5], the Chernobyl catastrophe [6,7] and the Fukushima disaster [8] have led to a rejection of this technology by a growing proportion of the public throughout the world and especially within Germany [9–12]. As a consequence of the Fukushima disaster, the German government decided to refrain from nuclear fission energy and to shut down all fission power plants by the year 2022 [9,13]. On top of these safety issues, fission power plants face a number of other challenges. For one, the supply of uranium is very limited [14], which resembles the case of fossil fuels. Yet, new generations of nuclear power plants try to overcome this issue [15]. However, the probably most serious downside of fission power plants arises from the insecure disposal or storage of the generated long-living radioactive waste. As the waste has to be stored safely for tens of thousands of years, it is indeed questionable whether safe storage can be guaranteed. Also, the establishment of final storages reliably leads to resolute protests by the residents living nearby. To overcome the specific issues of fission power plants i.e. safety, large fuel supply and no need for a final storage of radioactive waste, fusion power plants are being developed. If these goals can be achieved, fusion power plants fulfill the technological requirements for being a highly attractive option for future power generation [1,16].

The core of a fusion power plant is its, usually circular, vacuum chamber in which a highly energetic hydrogen plasma is ignited (see Figure 1.1). Within the plasma, the fusion of two light atomic nuclei into one heavier nucleus can be realized, while a small amount of mass is 'lost'. The generated energy ΔE is proportional to the mass deficit Δm after the fusion reaction according to Albert Einstein's famous equation

$$\Delta E = \Delta mc^2 \tag{1.1}$$

where c represents the velocity of light in vacuum [17,18]. Currently fusion power plants are expected to use the hydrogen isotopes deuterium ^2H and tritium ^3H as fuel, since the only alternative fusion reaction, regarding a large enough reaction cross section, is that of ^2H and ^3He [19]. Yet, the latter reaction can probably not be realized on earth, because ^3He is an extremely rare isotope. The complete deuterium-tritium fusion reaction is given by

$$^2_1H + ^3_1H \rightarrow ^4_2He + ^1_0n + 17.6\ \text{MeV} \tag{1.2}$$

and generates the most abundant helium isotope ^4He as well as one neutron n. Whereas deuterium availability is not critical since it can be extracted from water and is a stable nuclide with a share of 0.015 %, tritium is scarce on earth, owing to its relatively short half-life of about 12.3 years. This circumstance makes tritium the key fuel for the operation of fusion power plants. In addition to that, tritium cannot be stored for long periods, owing to its decay

$$^3_1H \rightarrow ^3_2He + e^- + \bar{\nu}_e + 18.56\ \text{keV} \tag{1.3}$$

which shows a relatively short half-life [20]. However, in contrast to ^3He, tritium can be generated or 'bred' within the fusion power plant itself, which also eliminates the problem of tritium storage. The breeding exploits the transmutation of lithium to tritium and helium by neutron capture as

$$^6_3Li + ^1_0n \rightarrow ^3_1H + ^4_2He + 4.78\ \text{MeV} \tag{1.4}$$
$$^7_3Li + ^1_0n \rightarrow ^3_1H + ^4_2He + ^1_0n - 2.47\ \text{MeV}. \tag{1.5}$$

The two stable lithium isotopes ^6Li and ^7Li, occurring naturally by 7.42 % and 92.58 % respectively, show a significantly different cross section for the capturing of neutrons. Also from an energetic point of view, both reactions are different (see equations (1.4) & (1.5)) [18]. While the breeding reaction of ^7Li is endothermic, its ^6Li counterpart is significantly exothermic and therefore preferable [19]. Also, the less abundant lithium isotope's cross section for capturing a neutron is several orders of magnitude higher than for ^7Li in the relevant energy regime of the incident neutron. The required neutrons for the transmutation reaction to take place are produced in the deuterium-tritium fusion reaction itself. The lithium for breeding tritium is kept in the so-called breeder blankets, which are steel structures surrounding the plasma. Within the breeder blanket, the neutrons are moderated and multiplied, to increase the generation of tritium to sufficient levels. The simultaneous burn and production of tritium to fuel the power plant is referred to as a 'closed fuel cycle'. One of the breeder blanket concepts proposed by the European Union, the so-called helium cooled pebble bed blanket (HCPB), features pebble beds of lithium ceramics as lithium reservoir and beryllium as neutron multiplier [21]. It will be tested in the International Thermonuclear Experimental Reactor, ITER, which is currently under construction in Cadarache, France (see Figure 1.1). Currently, the size of the ceramic pebbles for the HCPB blanket lies within the range between 250 µm and 630 µm [22–24]. However, all other blanket concepts for ITER which use a solid breeder material, and which are thus called solid breeder blankets, will apply 1000 µm pebbles [24].

Apart from lithium, the elements, of which the ceramic compound is made of, have to qualify as so-called low activation elements. Such elements do not transmute into long-living radioactive nuclides by neutron capture. The eventual requirement that the breeder materials, made from those elements, have to fulfill is that they decay to hands-on-level activity (10 μSv/h) within 100 years after use. As a result no final storage will be required for the breeder ceramics as opposed to fuels of fission power plants. Elements like oxygen, silicon and titanium easily qualify as low-activation elements. An also important property of tritium breeder materials is given by the quotient of lithium and volume, the so-called 'lithium density'. An as high as possible lithium density is needed to keep the size of the blankets reasonably small and to maximize the chance of neutrons hitting a lithium nucleus instead of another element's one. The latter will not only increase the tritium generation, but also reduce the overall activation of the blanket during use and the necessary amount of beryllium for neutron multiplication. To further increase the generation of tritium to suitable levels, all currently discussed blanket concepts consider a significant enrichment of the tritium breeder material with ^6Li above the natural fraction of 7.42 %. As the necessary separation of the two lithium isotopes involves expensive processes, a high lithium density also reduces the costs of the breeder material, because of the higher natural amount of ^6Li that comes with a higher lithium density. On top of that, unburned ^6Li has to be recovered after use. Therefore breeder ceramics must be recyclable with relative ease. Moreover, remote recycling at activities below 10 mSv/h is highly anticipated. Recycling of the breeder ceramics not only reduces operation costs, but also reduces the generation of waste.

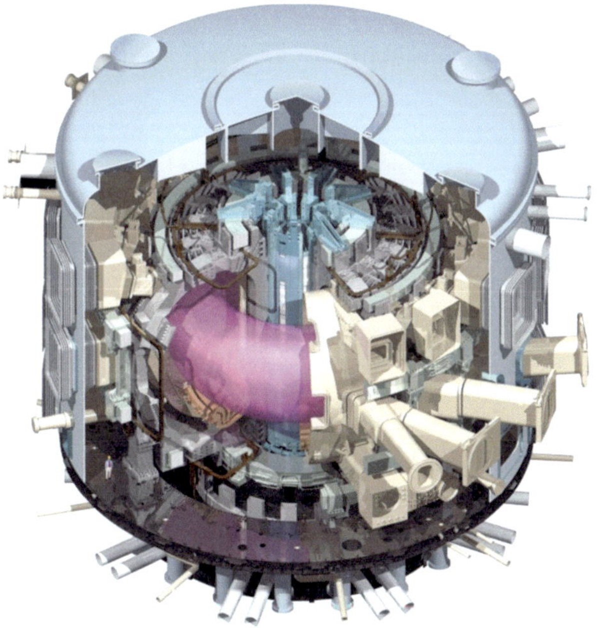

Figure 1.1: Drawing of the ITER tokamak, © ITER Organization, http://www.iter.org/ [25].

Apart from breeding tritium, another major task of the breeder material is the extraction of the generated fusion energy. About 80 % of the fusion energy is carried by the generated neutrons as kinetic energy. By inelastic collisions with the breeder material, the neutrons are decelerated and their kinetic energy is converted into heat. In addition to that, the exothermic transmutation of ^6Li sets free a high amount of energy. By both processes, the breeder material is significantly heated and thus has to withstand the resultant high temperatures. These high temperatures are also desired to maximize the thermodynamic efficiency of a fusion power plant. The so-generated heat has to be removed and in a second step it is converted into electricity.

The ceramic pebble beds, as well as the beryllium pebble beds, do not contribute to the structural rigidity of the blanket modules, and breeder materials are thus generally classified as functional materials. Yet, the pebbles will be exposed to various external forces during the operation of the blanket. To avoid or minimize fracture of pebbles and thus to ensure an adequately efficient fuel cycle, the mechanical rigidity of the pebbles has to be good enough to withstand these forces. Lithium orthosilicate, Li_4SiO_4, and lithium metatitanate, Li_2TiO_3, fulfill all requirements reasonably well and are usually considered as alternatives to each other for the testing in ITER. Yet, lithium orthosilicate based breeder materials exhibit comparably weak mechanical strength with regard to lithium metatitanate, whereas lithium metatitanate shows a comparably low lithium density. Also a comparably effective recycling process for lithium metatitanate based materials as for lithium orthosilicate based ones has not been demonstrated up to now. Recycling of breeder materials will be no issue for ITER. However, moving to a future demonstration power plant (DEMO) or even further to a commercial fusion power plant will significantly increase the severity of the requirements, thus the existing breeder materials have to be ameliorated. Within Europe, the development of lithium metatitanate has seized while the development of lithium orthosilicate based breeder materials is on-going at KIT. Currently Europe favors lithium orthosilicate based materials as solid breeders. Japan on the other hand favors lithium metatitanate and therefore leads the development of this compound for solid breeder application.

Since 2012 mixed breeder pebbles, consisting of both of the phases, are being developed at KIT in an effort to combine the positive aspects of both compounds in one material and thus leap forward towards a solid breeder material for DEMO. Moreover at QST, Japan, also newly developed pebbles on the basis of lithium metatitanate are available which feature a deliberate lithium excess. These pebbles are also developed in view of DEMO. Yet, both materials still have to prove their maturity for application.

The aim of this work is to characterize these new, so-called advanced solid breeder pebbles with a special regard to the mechanical strength and the failure mechanisms that determine the mechanical strength. In addition to that, the determinant effect that generate the observed failure mechanisms shall be identified as well. For the mixed breeder pebbles, also the influence of the second phase on the mechanical strength is of great interest. Since the breeder pebbles will be used for several years within a fusion blanket not only the performance of the pebbles in the as-fabricated state is important but also the evolution of the mechanical strength and the failure mechanisms with the duration of usage. Besides the irradiation defects that the materials will accumulate during use in a fusion blanket and the burning of lithium, the involved thermal annealing will be most responsible for changes of the material during its application. To investigate the effects of neutron irradiation on the pebbles, similar to what is expected within a fusion reactor, is very expensive and enormously time-consuming. In contrast, the thermal annealing of the advanced breeder pebbles in a relevant non-nuclear environment is far less intricate and suitable for analyzing this important aspect of the long-term stability of these advanced solid breeder pebbles. Therefore the investigation of the mechanical strength and the failure mechanisms of the pebbles as a function of thermal annealing are also one objective of this work. Finally, suggestions for future adaptions of the fabrication processes and potentially the solid breeder materials themselves shall be derived from this knowledge, so that the quality of the advanced solid breeder pebbles can be further increased in the future.

2 Literature review

In this chapter, the relevant literature for the conduction of this work is summarized and displayed. To give an overview of the European solid breeder blanket concept and the role that breeder ceramics play in it, a brief introduction into this field of fusion technology is given. Subsequently, the compounds that are considered as basis for breeder ceramics are presented with regard to their physical and chemical properties. Also the most relevant fabrication processes for these materials are detailed. As multi-phase materials provide an opportunity to improve the properties of a single phase material, the relevant, known phase equilibria for these compounds are summarized.

As the breeder materials are used as a filling of pebbles, the mechanical stresses that arise in samples of spherical geometry are important. For its simplicity, the mechanical testing of the materials is carried out by uniaxial compression tests, and for that reason, the stresses that go along with the uniaxial loading are given in this chapter. Also, as the temperature of the pebbles is changed during the fabrication processes, the thermal stresses that are generated as a result in spherical samples are outlined. With the data of the strength of the pebbles, a Weibull analysis can give additional insight into the underlying failure mechanism and provide useful information on the reliability of the pebbles when they are exposed to mechanical stresses. Therefore, the concepts used to perform the Weibull analysis are also introduced in this chapter.

2.1 The European solid breeder blanket concept for ITER

As scientists recognized that the realization of fusion power not only relies on solving fundamental issues, like keeping the deuterium-tritium plasma burning, but also building a machine that can technically realize the solutions to the fundamental issues, a more engineering oriented branch in fusion research, namely fusion technology, was established. The difficulties in building a fusion device increase significantly when a constantly operating power plant has to be realized, which is the ultimate goal. As a result, a number of topics surface which are otherwise of no concern. A list of all topics which are currently addressed by fusion technology would be interminable. One of these topics, however, is the blanket, which surrounds the plasma of a fusion power plant (see Figure 2.1). As the generation of tritium, the harvesting of the generated fusion power and the shielding of radiation sensitive components are all performed by the blanket, the design of this component is of utmost importance [26,27]. It was always recognized that using a solid material for breeding of tritium (thus called breeder material), the blanket is more easily realized in a technical way as if liquids were used. Within Europe, the development and design of solid breeder blanket concepts is today mainly performed at KIT and dates back to first research from the 1980s [29]. From these first concepts, the solid breeder blanket design evolved in several steps while a number of other concepts were proposed as well [30–35].

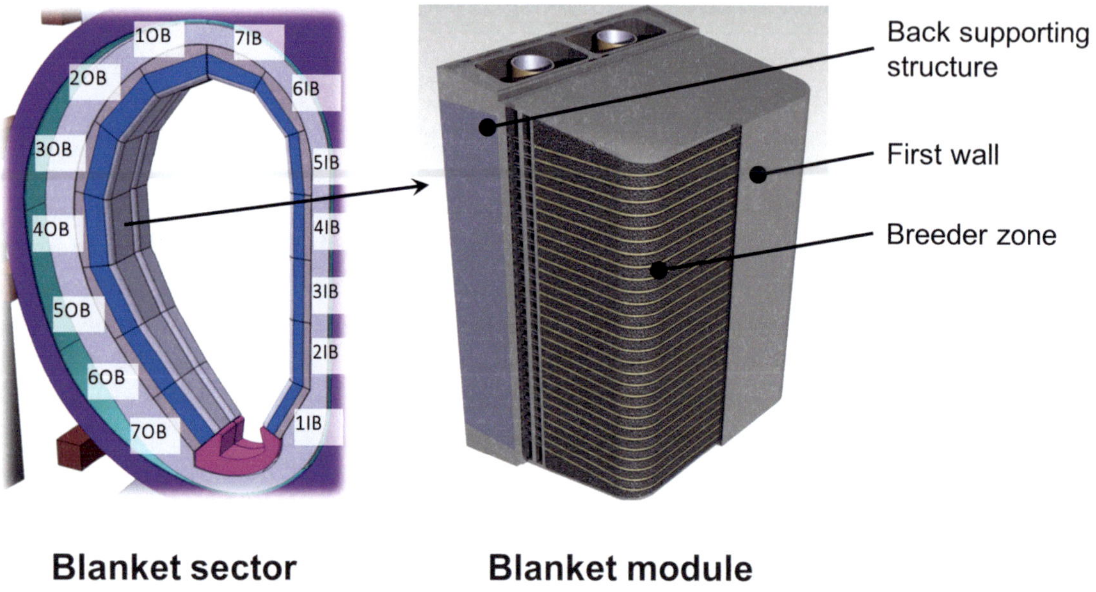

Figure 2.1: Illustration of a cross section of the 2016 design of the EU DEMO reactor, detailing the location of the blanket, adapted from [28].

In 1995, the European Union selected the two most promising blanket concepts for further development to eventually reach a demonstration power plant (DEMO) blanket design [36]. Along with a liquid breeder blanket which utilizes the low melting eutectic composition of the lithium-lead binary system, the so-called helium cooled pebble bed (HCPB) blanket concept was selected. The HCPB concept as well as the liquid breeder blanket rely on a structure of EUROFER, which is the EU reference reduced activation ferritic-martensitic steel. To cope with the deposited heat in the structural material, which is especially challenging in the plasma facing first wall, as it encounters direct bombardment of highly energetic particles, the steel structure of the HCPB blanket is cooled by a high pressure helium loop.

The HCPB concept features two functional materials, the breeder material and the neutron multiplier material, which are implemented as pebble beds to avoid excessive stresses during operation and to leave room for irradiation induced swelling. As already mentioned in section 1 the realization of fusion power on earth relies on the generation of tritium, which has to be bred from lithium. The use of pure lithium within a fusion power plant, however, leads to severe safety risks as lithium heavily and readily reacts with all major constituents of air. Also, the low melting point of pure lithium would require low temperatures within the blanket and thus a low thermodynamic efficiency of the whole power plant. In addition, low blanket temperatures are highly problematic for the designated structural material EUROFER since it does not provide sufficient ductility below 300 °C in a fusion environment [37]. As a result lithium compounds showing a significantly higher chemical and thermal stability are selected as breeder material.

From equations (1.4) and (1.5) it is evident that one neutron, which is generated by the fusion of one deuterium nucleus and one tritium nucleus (see equation (1.2)), is required to generate one tritium atom. Therefore in an ideal case, achieving tritium self-sufficiency would not require neutron multiplication, if each neutron generated in the D-T plasma was to generate one tritium atom by transmuting a lithium nucleus. Yet, there are numerous reasons for a less than ideal generation of tritium, with the most apparent being that the chances of a neutron hitting a lithium nucleus are significantly reduced due to the surrounding steel structure and the use of a lithium compound rather than pure lithium. Therefore the use of a neutron multiplier is unavoidable. Currently, beryllium is the most promising material for this purpose. The quotient of the generated amount of tritium and the theoretically necessary amount, i.e. no losses occur, is called tritium breeding ratio (TBR). For a reliable operation of a fusion power plant, the TBR must reach at least 1.15 [38].

In 2003 to 2005, the design of the HCPB concept was revised and finalized [39]. Up to this day, it still is the latest EU DEMO blanket design that is fully validated. The volume within the blanket module is subdivided into a number of individual units in which the functional material pebble beds are placed. Referring to their primary function, these units are called 'breeder units'. One reason for subdividing the blanket is the concern that the stiffness of the blanket module would be otherwise insufficient. Therefore a stiffening grid structure is introduced into the module with which the mechanical design requirements can be met. This structure also separates the individual breeder units from each other. Within the breeder units, alternating beds of breeder and neutron multiplier material are kept and separated by helium cooled steel plates. To remove the bred tritium from the breeder unit, the breeder pebble bed is purged by a low pressure helium flow with an addition of 0.1 % of hydrogen. This gas mixture is referred to as reference purge gas. The hydrogen is introduced into the purge gas in order to facilitate the release of tritium as exchange reactions are common [40]. Of course, also the beryllium beds are purged in this fashion to remove gaseous transmutation products.

In the EUROfusion research and development program, which was established in 2014, the further advancement of this DEMO blanket concept is one major topic. Whereas the design of a DEMO blanket has yet to complete the conceptual phase, the present construction of ITER in Cadarache, France, requires the actual fabrication of blanket modules with which the solid breeder blanket concept can be put to the test under relevant operating conditions, until ITER is commissioned. In order to be consistent with the DEMO design of the early 2000s, the test blanket module (TBM) for ITER was derived from this design in the years 2008 to 2010. Therefore no principal differences between these designs exist. Yet, the dimensions of the ITER port, into which the TBM will be mounted, are to be considered, which leads to a significant reduction of the size of the blanket. The blanket module's cuboid shaped casing consists of a first wall plate which is bent two times by 90° to seamlessly close the top and the bottom, while both sides are closed by cap plates (see Figure 2.2) [41]. The enclosed volume is structured by a stiffening grid which holds two columns of eight individual breeder units, which are inserted from the back. The back is closed by the manifold system which distributes the helium coolant into the cooling channels of the steel structure and the purge gas into the pebble beds and leads the reflux back into gas piping.

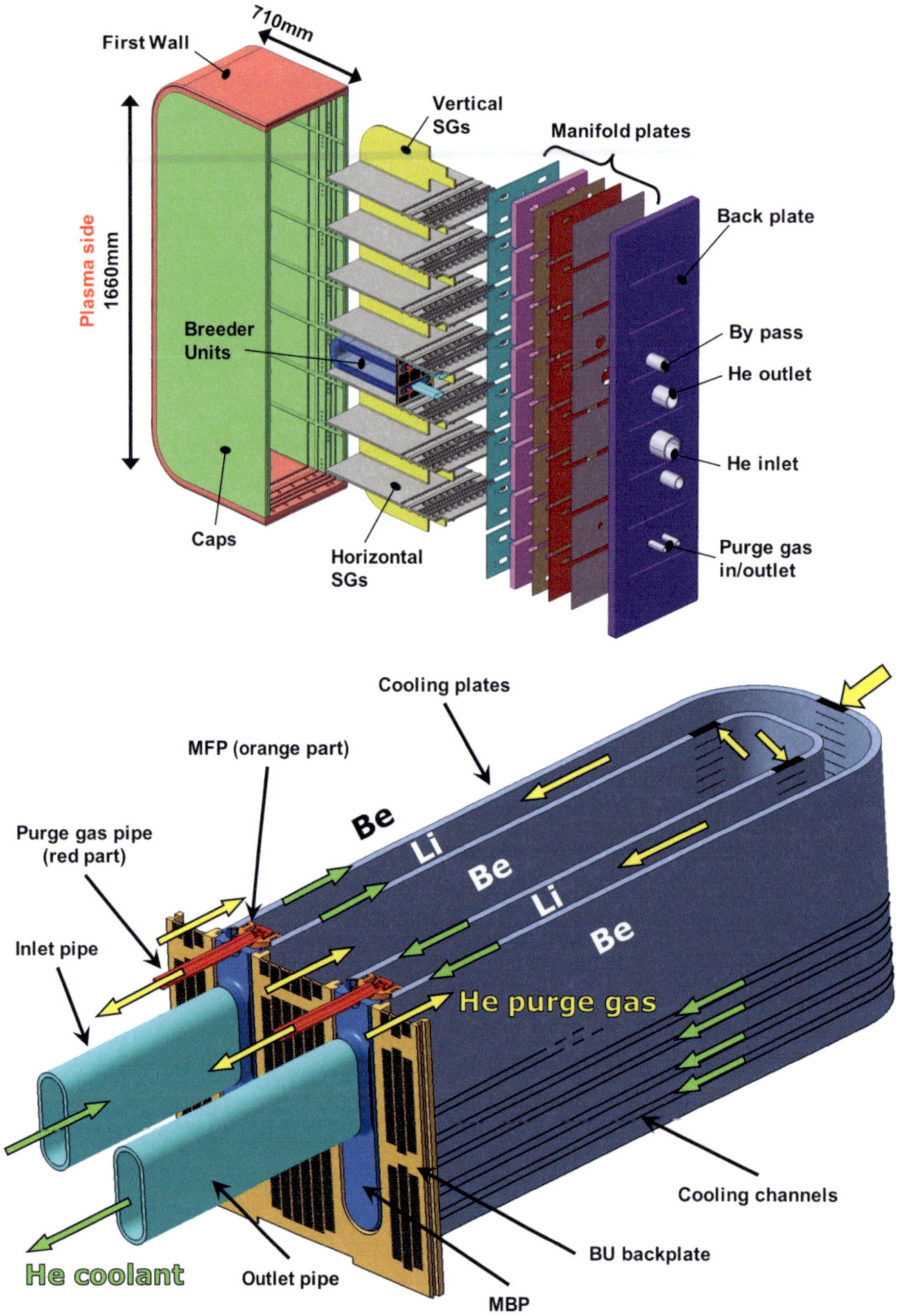

Figure 2.2: CAD visualization of the ITER TBM (top) and detail view of breeder unit (bottom), reproduced from [41].

A breeder unit is composed of two vertically aligned U-shaped cooling plates which are attached to the manifold system of the TBM in cantilever position. Three volumes are circumvented by the cooling plates. The U-shaped interior volume, also called breeder zone, is filled with the breeder pebble bed while the two remaining spaces of the breeder unit are filled with neutron multiplier pebble beds. A detailed CAD visualization of the breeder unit is given in Figure 2.2. The purge gas flow is first distributed into the neutron multiplier beds and subsequently enters the breeder pebble bed through micro-holes in the cooling plates. Otherwise the cooling plates hermetically separate the pebble beds. Each breeder unit also features a simple manifold system to handle the coolant and the purge gas flows appropriately and separately.

The breeder units will be filled with the pebbles in two separate steps. In a first step, the breeder pebbles will be filled into the U-shaped space and subsequently the manifold system of each breeder unit is welded onto the back to close the breeder zone. After finishing all other assembly steps and the post-welding heat treatment, the neutron multiplier is filled into the breeder units through holes in the caps, which will be closed in a final step. The use of pebble beds instead of a monolithic piece of ceramic enables the application of complex designs and geometries like for the ITER TBM. Because of the fluid-like behavior of the pebble beds, an excellent geometrical adaptability is achieved. It is moreover possible to fill the breeder unit at any convenient point during the fabrication process as the filling requires minimal intrusion into the TBM structure.

The design of the ITER TBM and the breeder units was carried out assuming that ceramic breeder pebbles consist of lithium orthosilicate, Li_4SiO_4, and 10 mol% of lithium metasilicate, Li_2SiO_3, with a ^6Li enrichment of 90 %. The size of these pebbles ranges from 250 µm to 630 µm. For the breeder pebble bed a packing factor, i.e. the occupied volume of the pebbles divided by the available volume, of 64 % was considered to be reasonably achievable. As neutron multiplier, beryllium pebble beds, with the individual pebble size being 1000 µm, are chosen and a packing factor of 63 % was anticipated. Both pebble beds are purged from the generated gaseous products by an 8 Nm^3/h gas flow of reference composition at 500 °C and a pressure of 0.4 MPa [40]. With the expected ITER operation parameters, a mean tritium production rate of about $108 \cdot 10^{12}$ T/cm^3s was ascertained by neutronic calculations [42].

The neutronic calculations also provide information about the volumetric heating of the pebble beds during operation. Cismondi et al. [43] modelled the breeder unit in its final design by 3D computational fluid dynamics taking the expected volumetric heating during an ITER pulse into account. Figure 2.3 shows the resultant expected temperature profiles within the functional material pebble beds of the ITER TBM breeder unit. It is obvious, that both pebble beds are not heated uniformly and that there are significant temperature gradients within them. A temperature window of 400 °C to 920 °C is considered for the lithium orthosilicate based breeder material [41], which is clearly not exceeded. Yet, with the volumetric heating of the breeder zone thermal stresses are acting on the pebble bed. Gan et al. [44] modelled these stresses for the EU reference material and a slice of the actual breeder zone by means of a discrete element method. For the relevant case of a polydisperse pebble bed, the calculated hydrostatic thermal stresses range from about 4 MPa to about 7 MPa, depending on the heating power and the packing of the pebble bed.

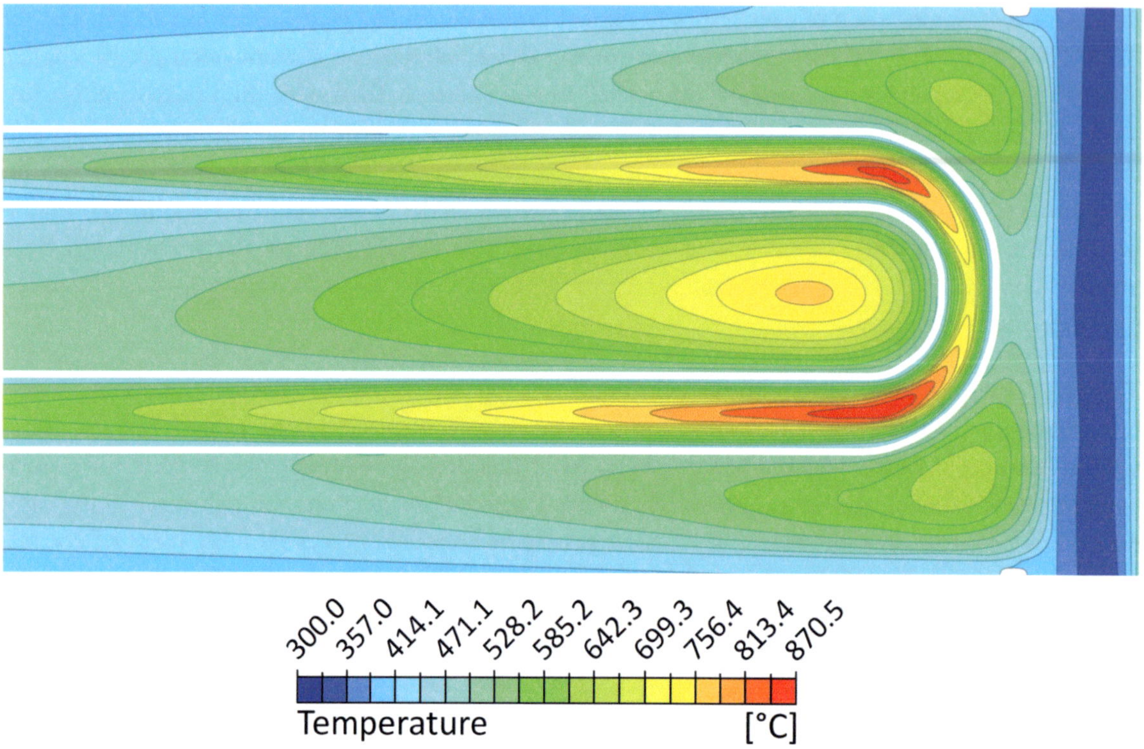

Figure 2.3: Heat profiles within the pebble beds of a breeder unit for the expected ITER operation parameters obtained by Cismondi et al., reproduced from [43]. The maximum temperature of the lithium orthosilicate bed is expected to lie at about 870 °C, which does not exceed the anticipated maximum temperature of 920 °C.

2.2 Breeder materials and relevant lithium compounds

As the use of lithium is mandatory in a fusion power plant to generate the needed tritium to ensure enduring operation, highly thermally stable lithium compounds were proposed as an attractive technological solution as they are obviously not prone to melting during operation. Especially dilithium oxide, Li_2O, and most ternary lithium oxide compounds show high melting points well above 1200 °C. Of course, oxygen saturated compounds are also attractive with respect to safety. The appreciable number of stable ternary lithium oxide compounds leads, however, not to an as large number of possible ceramic breeder materials. As explained in chapter 1 the breeder compounds must consist of low activating elements to obviate long-term or even final storage of fusion power plant components after use. The simplest lithium compound to fulfil the requirement of low activation is also dilithium oxide. Yet this compound may be problematic because of possibly large lithium vaporization [45]. Still, a number of other elements also qualify as low activation elements, among them silicon and titanium which form highly stable ternary compounds with lithium and oxygen. A detailed list of other low activation elements is given in by Cheng [46,47].

2.2.1 Universal requirements for breeder materials

As the example of Li_2O shows, the suitability of a breeder material for use in a fusion power plant depends on a number of properties. Although it is surely eligible to choose a material which delivers excellent performance in all regards, already the sheer number of requirements renders it difficult to find such a material. In addition, a number of properties may only be maximized while minimizing another. For instance a high specific surface area, which is beneficial for the release of tritium, will certainly reduce the mechanical strength of the pebbles since the increase in surface area involves an increase of the pebble porosity. Additionally, increased pebble porosity also reduces the overall amount of lithium within the pebble bed. Therefore it is aimed for an optimum compromise of all properties. In the following, the universal requirements, as discussed in literature, for any solid breeder material are listed from an engineering point of view.

- Low tritium inventory, which equals fast release of the generated tritium [32,48–50]: The need to keep the lithium inventory as low as possible originates from a number of reasons. If a high amount of tritium is permanently stored within the blanket, it might be released in case of an accident and therefore this is a safety relevant issue. Also the more tritium is stored in the blanket the more external tritium is needed in the start-up phase of a power plant until tritium self-sufficiency is reached.
- Low lithium partial pressure at high temperatures, resulting in low lithium transport [48,49]: Lithium transport and evaporation are also safety relevant topics. If lithium was transported with the purge gas stream into the ancillary systems connected to the blanket, it might lead to corrosion as these systems are not designed to cope with a lithium attack. Also, lithium evaporation is unwanted because the amount of lithium in the breeder zone is likely to be reduced and also the breeder pebble composition changes more significantly if evaporation was a factor on top of the anticipated burn-up.
- Large amount of lithium per volume, i.e. high lithium density [32]: Obviously, the more lithium is stored within the breeder zone of the blanket, the more tritium can be generated per volume. Therefore the expensive enrichment with ^6Li might be reduced or the blanket volume might be reduced, or both.
- Low neutron absorption cross section by the elements constituting the lithium compound [32]: Highly neutron absorbing elements like boron have to be avoided in fusion relevant material as a reduced number of free neutrons reduces the generation of tritium in return.
- Chemical compatibility with the structural material, the purge gas and the coolant in case of an accident at operating temperatures [32,49]: Naturally all materials which are in contact with each other during normal operation have to be compatible. Furthermore, all involved substances within the blanket must be compatible with each other also if it is not foreseen that they get in contact with each other. This may happen in case of an accident all too well for substances that are in close vicinity of each other.
- Mechanical and structural stability at high neutron fluence, high lithium burn-up and high temperatures, which means that excessive swelling and fragmentation cannot be tolerated [32]: During operation of the blanket, significant stresses are expected which must be withstood by

the breeder material although it does not have a structural function [49]. These stresses may originate from thermal expansion mismatch between the steel structure and/or swelling.

- Low neutron activation to minimize the need for storage of used material [32]: As pointed out in section 1, for fusion energy to be an attractive option, long-term storage of used material is not acceptable.
- Industrially producible at reasonable cost [32]: If a full size DEMO or fusion power plant is built, large amounts of breeder material will be necessary to fill the blankets of such a device. Consequently not only the ability to produce these amounts must be ensured, but also the production costs must be as low as possible to ensure competitive prices for fusion energy in the market.
- Constant heat transfer or thermal conductivity of the breeder material during operation to allow for good control of the breeder temperature [49]: To reliably predict the temperatures within the blanket and to ensure that the temperatures stay within acceptable levels without adapting the cooling parameters during operation, constant thermal conductivity of the pebble beds is a necessary requirement. This is mostly associated with the mechanical integrity of the majority of the pebbles within the pebble bed.
- Providing the possibility of easy reprocessing [49]: As ^6Li is an expensive and necessary isotope for tritium breeding, the possibility to reprocess used material and re-use the unburnt ^6Li will increase the overall cost efficiency of a power plant and significantly reduce the necessary space for medium term storage.

Presently, lithium orthosilicate and lithium metatitanate are considered to provide a good compromise of properties to meet all requirements. Also the fabrication routes of these materials are reasonably well established, to meet the production related requirements. Therefore all ITER parties chose either one of these two compounds as basis for their candidate breeder material.

2.2.2 Lithium orthosilicate

Lithium orthosilicate, Li_4SiO_4 or LOS, was selected as a candidate breeder material [51,52]. Although lithium orthosilicate features a comparably low silicon dioxide fraction, relations to the processing of glass and glass melts were drawn, so that glass-like processing was considered as an alternative to conventional shaping and sintering processes [52].

2.2.2.1 Physical and chemical properties of lithium orthosilicate

At room temperature, lithium orthosilicate crystallizes in monoclinic structure. Basically two monoclinic structures are mentioned in literature, which bear close similarity, since one is a 7x2 superstructure of the simpler monoclinic structure. Currently, the larger structure is considered the ordered room temperature modification. Yet also a triclinic structure, Γ-Li_4SiO_4, is reported by Hoppe et al. [53] as equilibrium room temperature modification, however, it does not seem to appear in practice. Both structures are composed of individual SiO_4 tetrahedra which are separated by lithium ions. At about 665 °C, lithium orthosilicate undergoes a reversible and displacive second order phase transformation

[54,55]. In literature, this phase transformation is not described comprehensively. However, it is reported, that there is little change in the unit cell volume and that the expansion of the lattice parameters shows no visible discontinuity. Also above 665 °C, at least one additional second-order phase transformation is reported by Kleykamp [54] at about 725 °C, however, there are no further reports to shed some light on these alleged phase transformations. It has to be concluded, that the equilibrium crystal structure at room temperature also represents the high temperature crystal structure reasonably well, if thermal expansion is taken into account. The basic structure properties of all known modifications are given in Table 2.1. The most recent determination of the crystal structure of α-Li_4SiO_4 at room temperature was carried out by Deng et al. [56]. Their results are in good accordance with the previously reported structure by de Jong et al. [57], which, however were obtained at 150 K. At a temperature of 1255±5 °C, lithium orthosilicate is reported to melt congruently.

Table 2.1: Basic crystallographic parameters of the reported modifications of lithium orthosilicate as well as their equilibrium temperature range of existence. (*: reported for 150 K)

Phase	Equil. temp. of existence	Space group	a [Å]	b [Å]	c [Å]	α [°]	β [°]	γ [°]	Ref.
α-Li_4SiO_4	< 665 °C	$P2_1/m$ (11)	11.5556	6.0986	16.7300	90	99.0580	90	[56]
α-Li_4SiO_4	< 665 °C	$P2_1/m$ (11)	11.532*	6.075*	16.678*	90	99.04*	90	[57]
Γ-Li_4SiO_4	< 665 °C	P-1 (2)	7.519	5.648	5.031	124.15	97.18	100.26	[53]

The fundamental mechanical properties of polycrystalline lithium orthosilicate were investigated from the late 1980s until the end of the 20th century. An overview of the obtained properties is given in Table 2.2. Mostly, the porosity p and the grain size d_g of the samples are used as microstructure parameters which define the mechanical properties.

In terms of chemical properties, Ortiz-Landeros et al. [62], Schauer and Schumacher [63] as well as Alcantar-Vázquez et al. [64] demonstrate the significant reactivity of lithium orthosilicate with humidity. Once the water molecules react with the surface, hydroxyl groups, bound to lithium or silicon, are formed. These lithium hydroxyl groups may also react with carbon dioxide, which is present in ambient atmosphere to form lithium carbonate. While this process already takes place at low temperatures, lithium orthosilicate is also investigated as potential high temperature carbon dioxide adsorbent [65]. The low chemical resistance of lithium orthosilicate also shows in dissolving experiments where boiling in dilute nitric acid is sufficient for an almost complete dissolution [66].

Table 2.2: Fundamental mechanical properties of polycrystalline lithium orthosilicate.

Property	Value	Validity range	Ref.
Young's modulus / GPa	$E = 110 \times (1-p)^3 \left(1 - 2.5 \times 10^{-4}(T-293)\right)$	$0.02 \leq p \leq 0.32$	[58,59]
Poisson's ratio	$\nu = 0.24$	$p = 0.1, T = 293\ K$	[59]
Tensile failure strength / MPa	$\sigma_{ft} = 150 \times d_g^{-0.5} \times \frac{1850}{T} e^{-4.3p}$	$4 \leq d_g \leq 50$ µm $0.02 \leq p \leq 0.32$	[60]
Compressive strength / MPa	$\sigma_{fc} = 940 \times d_g^{-0.5} e^{-7.5p} ln\left(\frac{1850}{T}\right)$	–	[60]
Ultimate bending strength / MPa	$\sigma_{ub} = 275 \times d_g^{-0.5} e^{-4.3p}$	$4 \leq d_g \leq 50$ µm $0.02 \leq p \leq 0.32$ $T = 293\ K$	[61]

2.2.2.2 Fabrication of lithium orthosilicate-based breeder pebbles by melt-based processes

Within Europe, two melt-based processes for the fabrication of lithium orthosilicate based breeder pebbles exist. These processes are detailed in the following, as pebbles fabricated by one of these processes are used in this study.

EU reference fabrication process

The current EU reference breeder material is fabricated by a melt spraying process. The production of the pebbles was carried out at Schott AG, Mainz. As detailed in literature [52,67,68], the fabrication setup, illustrated in Figure 2.4, consists of a vertically aligned tube furnace, which heats a platinum alloy crucible ① above the melting point of pure lithium orthosilicate. The crucible holds lithium hydroxide and silica as raw materials, which are co-melted to lithium orthosilicate melt ②. The composition of the melt is adjusted to achieve slightly over-stoichiometric lithium orthosilicate pebbles with 2.5 wt.% silicon dioxide excess to form 9.50 mol% lithium metasilicate, Li_2SiO_3, as second phase. Once a homogenous melt is established, the melt flows out of the crucible in a few millimeter wide melt jet and is sprayed to small droplets with air ③. The generated droplets solidify during their flight. These droplets are mostly found to crystallize by a size dependent combination of heterogeneous and homogeneous nucleation. During solidification, lithium orthodisilicate, $Li_6Si_2O_7$ forms peritectically at 1025 °C (see also section 2.2.5). Since the pebbles are cooled relatively quickly, this phase is retained within the pebbles in a metastable state instead of decomposing eutectoidally into the equilibrium phases at 1020 °C [67,69]. For very small pebbles with diameters below 50 µm, no crystalline phases are observable by X-ray diffraction [69]. After cooling to room temperature, the pebbles are collected and sieved.

The pebble size fraction of interest lies between 250 μm and 630 μm, which represents 50 wt.% of the produced pebbles [70]. It has been shown, that this so-called 'Schott process' can be used to effectively reprocess used material without changing the material properties perceivably [67,71]. Also, it is a rather facile batch process and with the current design the production capacity reaches 1 kg per batch [52]. Upscaling of this process is considered to be relatively unproblematic [52]. However, there are no published solutions for the advancement into a continuous fabrication process.

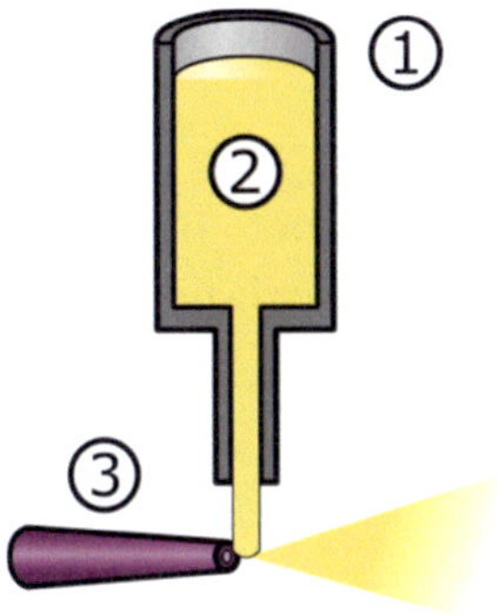

Figure 2.4: Illustration of the Schott melt spraying process. After Kolb et al. [23].

KALOS process

While featuring a lot of positive aspects, the Schott process atomizes the melt jet in a rather chaotic and harsh way [23]. This leads to a number of unwanted effects. First of all, the pebble size cannot be controlled effectively, resulting in a low yield. Additionally, the harsh way of atomization introduces large defects, i.e. mainly pores, into the pebbles. Consequently the mechanical rigidity and the pebble bed density are reduced. To overcome this issue an advanced process was developed on basis of the Schott process. For the advanced process, the way of droplet forming was changed to the simple decay of a jet of liquid. This is a well understood and described technique and can produce monodisperse droplets once the optimal process parameters are known and applied [72]. Thus this way of droplet forming has the potential to raise the yield significantly with regard to the Schott process. Another important advantage of this technique is that the gentle forming of the pebbles introduces fewer defects into them. Therefore the density of the pebbles will increase and also the mechanical rigidity should improve. In addition to that, more sophisticated cooling methods can be applied, as the flight path of the droplets is very narrow and the droplet velocity is also of a manageable scale, in contrast to the Schott process. The so-modified Schott process is named 'KALOS' (KArlsruhe Lithium OrthoSilicate) process and is established at KIT.

Since 2011, so-called modified or advanced ceramic breeder pebbles, consisting of lithium metatitanate as second phase instead of lithium metasilicate, are fabricated by the KALOS process [73]. The starting materials are commercially available lithium hydroxide monohydrate and silicon dioxide to form lithium orthosilicate as well as titanium dioxide to form lithium metatitanate as second phase. The powders are weighted in stoichiometrically according to

$$2(2 - x_{LMT})LiOH \cdot H_2O + (1 - x_{LMT})SiO_2 + x_{LMT}TiO_2 \xrightarrow{\Delta E} (1 - x_{LMT})Li_4SiO_4 + x_{LMT}Li_2TiO_3 + 3(2 - x_{LMT})H_2O \tag{2.1}$$

with x_{LMT} as fraction of Li_2TiO_3 in Li_4SiO_4. After several hours of blending, the starting powders are dried at 150 °C in vacuum to remove the hydrate water from the lithium hydroxide and also adhering water from the powders' surface. After drying, the powders are introduced into the crucible of the fabrication facility, which is subsequently heated to 500 °C and for 4 hours [74]. At this temperature, lithium hydroxide melts and reacts, in particular with the silicon dioxide powders, to precursor phases. During these reactions, the remaining water is set free. The water is condensed, and thus removed from the process, outside of the furnace.

After retrieving the water from the powder mixture, the crucible is heated to a temperature above 1300 °C to melt the powders as a whole and to reach the anticipated products. Principally lower temperatures would be sufficient to finish the reactions. However a reasonably low viscosity of the melt has to be reached to process it, thus higher temperatures are applied. As there are indications that the melt separates into two liquid phases at these temperatures as Knitter et al. report [75], it is quickly processed once the desired processing temperature is reached. The KALOS setup is schematically detailed in Figure 2.5.

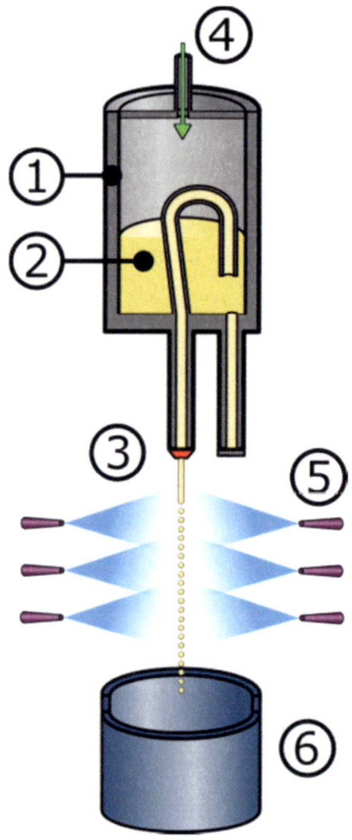

Figure 2.5: Illustration of the KALOS melt spraying process [76].

As for the Schott process, a platinum alloy crucible is used since platinum alloys were found to be reasonably resistant to lithium orthosilicate melts at high temperatures. The KALOS crucible consists of a body of dispersion hardened platinum-rhodium 90/10 because of its good high temperature strength, and an inner, U-shaped tube made of platinum-gold 95/5 through which the melt is driven to the outlet. The U-shaped tube is necessary to prevent an unintended outflowing of the melt. Compared to rhodium bearing platinum alloys, the platinum-gold alloys are wetted less by lithium orthosilicate melts, and as a result the melt is not driven through the outlet solely by capillary forces.

To heat the crucible, a two-zone resistance heater furnace is used. The bottom heating zone compensates for the heat losses due to the necessary opening for the melt jet [72]. Otherwise, an increase of the melt viscosity or even premature solidification of the melt at the outlet would be possible. A maximum temperature of 1600 °C can be achieved within the furnace, which limits the manageable lithium metatitanate content to approximately 40 mol%. Additionally, the platinum alloys become too soft at high temperatures to safely operate the facility. Therefore and to reduce corrosive attacks to the crucible materials as much as possible, the temperature is only set as high as necessary. As the melt viscosity highly depends on the composition of the melt, the processing temperature is also adjusted accordingly. However, experimental difficulties may also require adjustments of the processing temperature during the fabrication.

By applying a gauge pressure on the melt ④, the fabrication process can be started deliberately generating a melt jet. Ideally the droplets that form after breaking from the melt jet, and accordingly the pebbles that form from them upon cooling, are roughly double the size of the melt jet diameter. As pebbles of sizes between 250 µm and 1250 µm are to be predominantly produced by the process, the melt jet diameter is adjusted accordingly to a suitable diameter. A pin-hole nozzle of 300 µm diameter, also made of platinum-gold 95/5, but dispersion hardened to avoid deformation at high temperatures, is welded to the crucible outlet ③ for this reason. To adjust the velocity of the flowing melt, the gauge pressure can be varied up to 1 bar. However, the dynamics in the liquid jet are also altered with the gauge pressure and thus an optimal pressure exists. The standard gauge pressure lies at 400 mbar, which ensures an adequate yield in the desired pebble diameter range and a stable melt jet. Synthetic air is chosen as pressure gas to avoid reducing the melt. Especially the formation of Si^0 must be inhibited, because it readily reacts with the platinum crucible forming a low melting eutectic (see Tanner and Okamoto [77]).

As soon as the droplets break from the liquid jet, the initially almost monodisperse droplets coalesce during their flight, which leads to a considerable broadening of the pebble size distribution [72]. A typical example of a pebble size distribution is shown in Figure 2.6. The coalescence of nine droplets of initial size (600 µm) is necessary to form a droplet of about 1250 µm diameter. Yet, as a significant amount of even larger pebbles exist, the count of droplets that fuse is surely not limited to nine. As the pebble size distributions show, the coalescence of droplets takes place numerously. Once the surface of a pebble is solid, further coalescence may only occur with still liquid pebbles.

As the droplets are dropped through ambient air and therefore cooled, all droplets will solidify to pebbles eventually. However, this method of cooling has proven to be ineffective, i.e. flight distances of about 14 meters are necessary to cool the pebbles to temperatures at which they do not sinter to each other in the collecting receptacle [23]. Such large flight distances are unachievable in the current facility, which can only provide a flight path of 4 meters. Therefore liquid nitrogen is sprayed from multiple nozzles, surrounding the flight path of the droplets for about 1.5 meters, upon them ⑤. As this cooling method sets in right after the droplets leave the furnace, the coalescence of the droplets is quickly reduced, compared to quenching the pebbles after falling for some time through ambient air. The spray cooling also reduces the thermal shock for the pebbles compared with quenching and is thus the preferable method. Once the pebbles passed the cooling section, they are cool enough to be collected in a receptacle ⑥ at the bottom of the fabrication facility.

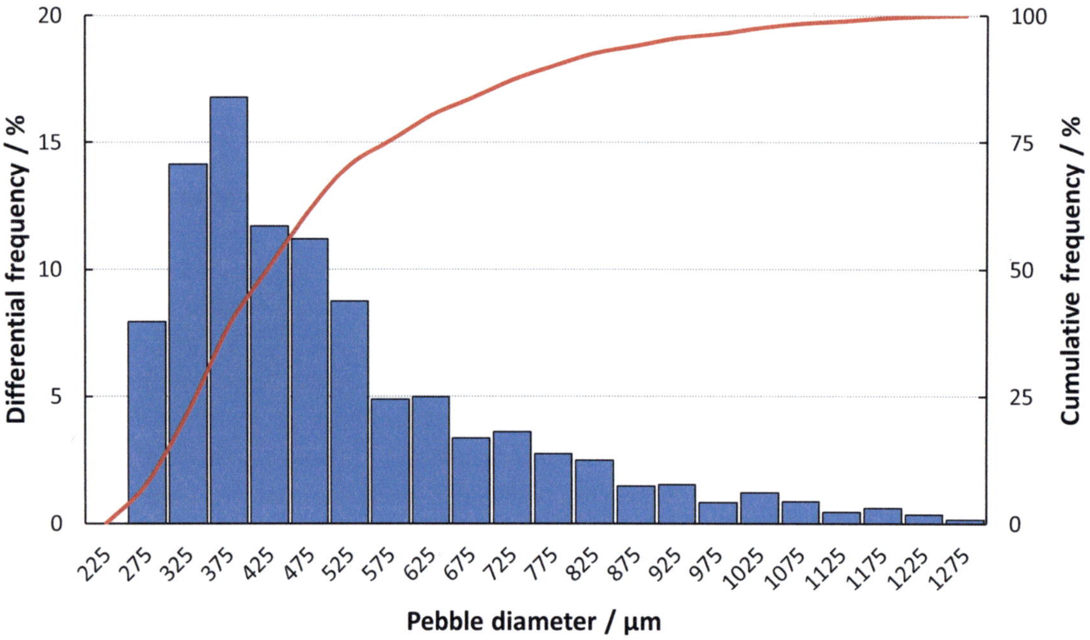

Figure 2.6: Typical pebble size distribution of a batch of KALOS pebbles [23].

2.2.3 Lithium metatitanate

Similarly to lithium orthosilicate, the balanced properties of lithium metatitanate, Li_2TiO_3 or LMT, make it an appealing compound for use as tritium breeder. It was therefore chosen by Japan, India and South Korea for their blanket designs. Within Europe, lithium metatitanate was also developed in this regard, but the development of single phase lithium metatitanate pebbles was stopped in 2005. Nevertheless, lithium metatitanate remains a part of the ceramic breeder development in Europe as second phase of advanced lithium orthosilicate based breeder pebbles (see section 2.2.2.2).

2.2.3.1 Physical and chemical properties of lithium metatitanate

All relevant physical and chemical properties of lithium metatitanate are presented in this part. These data are necessary basic information for almost all characterization that is detailed in the following chapters. Lithium metatitanate crystallizes in two modifications. The low-temperature β-modification is a monoclinic structure, as Kataoka et al. report [78], with the lattice constants detailed in Table 2.3. The structure can be regarded as composed of two principally different layers. A total of four of these layers are stacked in [001]-direction with equal spacing. One of these layers primarily consists of condensed hexagonal rings of edge-linked TiO_6 octahedra.

Table 2.3: Basic crystallographic parameters of the reported modifications of lithium metatitanate. (*: reported for a temperature of 1000 °C [79].)

Phase	Equilibrium temperature of existence	Space group	a [Å]	b [Å]	c [Å]	α / γ [°]	β [°]	Ref.
β-Li_2TiO_3	< 1155 °C	C2/c (15)	5.0623	8.7876	9.7533	90	100.212	[78]
γ-Li_2TiO_3	> 1155 °C	Fm-3m (225)	4.2309*	4.2309*	4.2309*	90	90	[79]
γ-Li_2TiO_3	> 1155 °C	Fm-3m (225)	4.150	4.150	4.150	90	90	[80]
γ-Li_2TiO_3	> 1155 °C	Fd-3m (227)	8.285	8.285	8.285	90	90	[81]

Yet, one lithium ion is located in the center of each of the rings (see Figure 2.7). These TiO_6 layers are separated by layers of lithium ions. Both TiO_6 layers, and also the separating lithium layers, are approximately displaced against each other by [0⅙½]. It is noteworthy, that the β-modification shows a considerable homogeneity range in the pseudo-binary phase diagram Li_2O-TiO_2, especially with respect to a surplus of dilithium oxide. The minimum TiO_2 fraction is reported to be 47 mol%, while 51.5 mol% represent the maximum [82].

The β-modification transforms into the cubic γ-modification at 1155 °C. Two crystal structures are proposed for this modification (Fm-3m metastable at room temperature [80] and in equilibrium at 1000 °C [79] and Fd-3m [81]), yet, both modifications are variants of the rock salt structure with the cations in octahedral vacancies of the anion sublattice and vice versa. The lattice parameters of both structures are also given in Table 2.3. In contrast to the rock salt structure, two cations exist (Ti^{4+} and Li^+) which share the same lattice positions in case of the Fm-3m structure or essentially the same positions for the Fd-3m structure. For both proposed structures, the cations are not fully statistically distributed. In accordance with the greater mixing of the cations in the γ-modification, a significantly larger homogeneity range is found for this phase. The limits in the system Li_2O-TiO_2 were determined by Izquierdo and West [83] as 44 mol% TiO_2 and 66 mol% TiO_2. The stoichiometric γ-modification is stable until 1533 °C and melts congruently, like lithium orthosilicate.

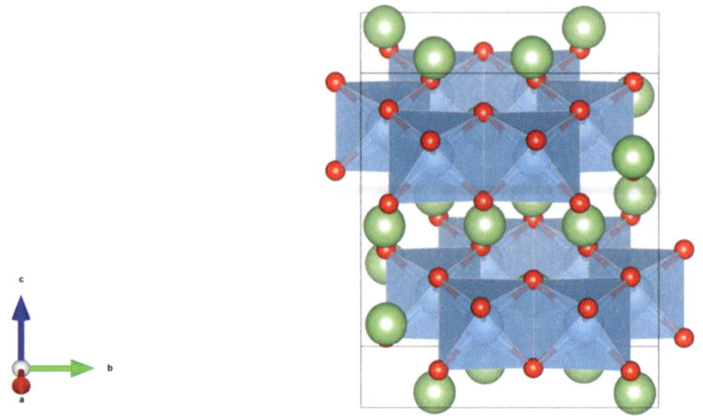

Figure 2.7: Visualization of the crystal structure of β-LMT (Li: green, O: red, Ti: blue) as reported by Kataoka et al. [78].

The mechanical properties of β lithium metatitanate are listed in Table 2.4. Compared to lithium orthosilicate (see Table 2.2 in section 2.2.2.1), lithium metatitanate shows a more than double as high Young's modulus while the Poisson's ratio is relatively similar. If the reported rupture strength is interpreted as tensile failure strength, lithium orthosilicate shows higher strength for comparable microstructure parameters. In contrast to that, the ultimate bending strength is reported to be about 30 % higher for lithium metatitanate as for lithium orthosilicate given the same micro structure parameters.

Table 2.4: Fundamental mechanical properties of lithium metatitanate.

Property	Value	Validity range	Ref.
Young's modulus / GPa	$E = 266.8 \times (1-p)(1-1.2p)^2$	$1 \leq d_g \leq 2$ µm $0.1 \leq p \leq 0.3$ $T = 300\ K$	[87]
Poisson's ratio	$\nu = 0.3(1-p)$	$1 \leq d_g \leq 2$ µm $0.1 \leq p \leq 0.3$ $T = 300\ K$	[87]
Rupture strength / MPa	$\sigma_r = 170 \times (1-2.27p)$	$1 \leq d_g \leq 2$ µm $0.1 \leq p \leq 0.3$ $T = 300\ K$	[87]
Ultimate bending strength / MPa	$\sigma_{ub} = 60$	$p = 0.15$ $d_g = 10$ µm $T = 300\ K$	[87]

Of course, these properties are also highly dependent on the size distribution of defects within the material thus these differences are easily explainable. Chemically, lithium metatitanate is relatively inert to water. In moist air, an uptake of much less than 1 % mass was observed after 3 months [84]. Also the diffusion of lithium into water upon immersion is only about 4 % after 5 months at room temperature [85]. The amount, however, increases in boiling water to about 20 % in 3 months. However, if lithium metatitanate has to be dissolved, strong acids like aqua regia and fluoric acid are needed [86,87] or 5M H_2O_2 [88].

2.2.3.2 Fabrication methods of lithium metatitanate pebbles

The fabrication of lithium metatitanate based pebbles is presented in this section. In contrast to the EU, no fabrication process is considered as reference process. Mainly two processes exist which are used to fabricate lithium metatitanate based pebbles in Japan, namely the 'sol-gel method' and the 'emulsion method'. Either process may be used for the fabrication of the breeder pebbles for the Japanese ITER TBM or a future DEMO blanket.

Sol-gel method

At the Japanese Atomic Energy Agency, JAEA, whose fusion technology branch is now operating under the name of National Institutes for Quantum and Radiological Science and Technology, QST, the fabrication of lithium metatitanate based breeder pebbles was advanced for decades. The so-called sol-gel method (SG) was first applied as a promising fabrication process for the breeder pebbles of the Japanese ITER blanket in 1998 by Tsuchiya et al [89]. Later this method was used by Hoshino and Oikawa [90] to fabricate lithium excessive lithium metatitanate pebbles.

The synthesis of lithium excessive lithium metatitanate powder, as starting material, is also described by Hoshino et al. [90–92]. As raw materials, lithium hydroxide monohydrate, $LiOH \cdot H_2O$, and metatitanic acid, H_2TiO_3, are chosen. These substances are blended with a molar ratio of 2.2 in a polyethylene bottle. The bottle with the substances is kept rotating for 48 hours at room temperature. During the rotating, the reaction of

$$2LiOH \cdot H_2O + H_2TiO_3 \overset{\Delta E}{\rightarrow} Li_2TiO_3 + 4H_2O \tag{2.2}$$

takes place. Because of the high tendency towards polymerization of H_2TiO_3 and Li_2TiO_3 [93], a gel is formed by this reaction. After the gelation reaction is completed, the gel is subsequently dried in air at 500 °C for 5 hours after which pure lithium excessive lithium metatitanate is generated. ICP-AES results indicate that the molar ratio of 2.2 is preserved after the drying. Prior to the actual shaping of the pebbles, a fine grinding step of the powder is necessary. In Tsuchiya et al.'s work [94], an average particle size of 0.62 µm is chosen whereas no information in this respect is given by Hoshino and Oikawa or Hoshino [90,92], but it is presumed that a similar particle size is achieved. Potentially an even smaller particle size is obtained when jet grinding is used [92]. After the grinding, the powder is mixed with polyvinyl-alcohol (PVA) as binder and water to form a slurry. The slurry is subsequently dropped through a nozzle into an acetone bath, where spherical droplets are formed. As the acetone removes the water from the droplets, they begin to gel.

After aging the gel spheres for 1 hour at a temperature between -20 °C and -30 °C, the gelling reaches an optimal degree so that the gel spheres can be removed from the acetone bath. Prior to sintering, a calcination step has to be performed at 700 °C for 6 hours in air, to remove the organic binder. Then the pebbles are sintered at 1100 °C in argon [90] or vacuum [92] for 1 hour. As Hoshino [92] states, a two-step sintering program (1 hour at 800 °C in vacuum and 5 hours at 1080 °C in argon) leads to a significantly higher pebble density. By these thermal treatments, some of the excessive lithium is lost and slightly lower lithium to titanium ratio in the order of 2.12 can be achieved for the pebbles. As anticipated, the so-fabricated pebbles consist of just one single phase [90,92].

Emulsion method

The emulsion method (EM) can be regarded as an advancement of the sol-gel method and was also developed at QST [95]. The principle of the emulsion technique consists of the temperature-triggered gelation of slurry in spherical form, while the slurry contains all raw materials that are to be reacted and/or sintered to dense pebbles in a second step. The production of pebbles without an initial synthesis of the anticipated phase but during the calcination and sintering is termed 'direct pebble fabrication'. The slurry of the emulsion method is water-based and thus all raw materials must be virtually insoluble in water to not deteriorate the gelation behavior of the binder. Therefore for instance, pure lithium orthosilicate or lithium hydroxide cannot be used as raw materials. As binder agarose is used. It gels strongly at room temperature while it is easily dissolvable in water at temperatures above 50 °C. The slurry for the fabrication is prepared by a ball milling step in which all raw materials are blended and milled to a reasonably small particle size. When lithium excessive lithium metatitanate pebbles are to be fabricated, the direct pebble fabrication is not used. Instead a suitable powder is synthesized in the same way as for the sol-gel method and sintered while the raw materials undergo no chemical reaction.

The fabrication devices are setup in a closed, heated Plexiglas box to prevent premature gelling of the slurry. From the slurry, the pebbles are shaped by cutting the slurry with a cross-flow of silicone oil into segments in a T-shaped junction (Figure 2.8).

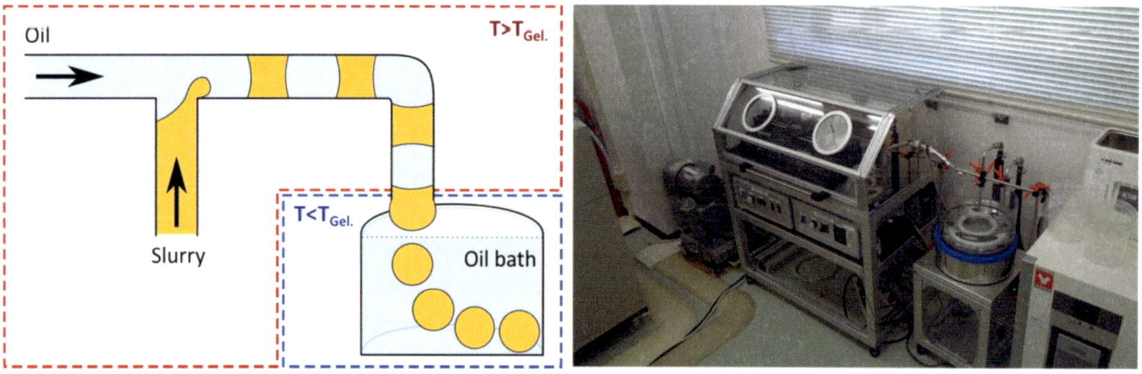

Figure 2.8: Schematic of the wet-chemical pebble fabrication process (left) and the experimental setup at QST, Rokkasho, Japan (right) after Hoshino [95] and as already published [96].

The dispensing speeds of the silicone oil and the slurry determine the later pebble size. In general these parameters are optimized to generate 1000 µm pebbles after sintering. The segmented stream of slurry and silicone oil leaves the Plexiglas box through a feedthrough and is led into a cooled, rotating bath of silicone oil by a heated tubing. The cooling bath temperature is kept at 10 °C by a cooling water flow which surrounds the silicone oil vessel. Once the slurry segments are released from the tubing they form droplets of several millimeters size. These droplets solidify quickly as the gelation of the binder sets in upon cooling in the bath. The rotation of the bath is necessary to prevent the droplets from coalescing by separating them by an adequate margin until the advance of the gelation inhibits coalescence.

Once the gelation is finished, the gelled pebbles and the silicone oil are separated. To fully remove the silicone oil as well as to extract water from the pebbles, a washing procedure is applied. The pebbles are agitated in ethanol to extract water and in a second step isoparaffin is added to the ethanol-pebble mixture to remove the silicone oil. Afterwards, the liquids are drained from the pebbles and the washing is repeated to a total 3 times. In a final step, the pebbles are agitated in pure isoparaffin. The pebbles are subsequently spread out on an absorbent cloth for drying and afterwards put in a circulating air oven at 55 °C to complete the drying. For the sintering of the dried pebbles, various conditions are reported. Generally a sintering temperature of 1100 °C is used in combination with different sintering atmospheres [91,95,97,98]. Mostly the complete removal of carbon from the pebbles is aimed for during sintering, as also traces of lithium carbonate are expected to promote grain growth at elevated temperatures [97]. Needless to say, a reasonably high pebble density and a low grain size are also aimed for.

2.2.4 Lithium aluminate

Lithium aluminate, $LiAlO_2$ or LAO, is used as an alternative to lithium metatitanate as strengthening phase in lithium orthosilicate pebbles, which, as Dienst and Zimmermann [99] reported, shows a significantly higher strength than lithium orthosilicate. Yet, because of the unfavorable activation characteristics of aluminum, lithium aluminate is not considered as a breeder material anymore.

The equilibrium phase of $LiAlO_2$ at room temperature is the α-phase. It is stable until about 800 °C and transforms into the γ-phase at higher temperatures. At 1785 °C lithium aluminate melts congruently [100] which is identical to LOS and LMT. It is reported, that the α-γ phase transformation is virtually irreversible and usually mechanical pressure is needed to promote it [101]. This is in agreement with the observation that the molecular volume significantly increases during the transformation (~30 %). The basic crystallographic data of both phases is given in Table 2.5.

The mechanical properties of γ-lithium aluminate are given in Table 2.6. As the theoretical density ρ_{th}, derived from the crystallographic data, amounts to 2.61 g/cm^3, the Young's modulus of lithium aluminate is significantly higher than that of LOS and LMT. The determined compressive strengths for LOS and LAO samples given in the literature are similar, but with the LAO samples showing a noticeably lower strength.

Table 2.5: Basic crystallographic parameters of the reported modifications of lithium aluminate.

Phase	Equilibrium temperature of existence	Space group	a [Å]	b [Å]	c [Å]	α [°]	β [°]	γ [°]	Ref.
α-$LiAlO_2$	< 800 °C	R-3m (166)	2.8003	2.8003	14.216	90	90	120	[102]
γ- $LiAlO_2$	> 800 °C	$P4_12_12$ (92)	5.1687	5.1687	6.2679	90	90	90	[103]

Table 2.6: Fundamental mechanical properties of polycrystalline γ-lithium aluminate.

Property	Value	Validity range	Ref.
Young's modulus / GPa	$E = 144 \times \rho_{th} \times (1-p)(0.7-p)^2$	$0.1 \leq p \leq 0.5$	[101]
Poisson's ratio	$\nu = 0.22$	$d_g = 0.45$ µm $p = 0.27$	[104]
Compressive strength / GPa	$\sigma_c = \frac{e^{-10p}}{d_g^{0.5}} ln \frac{2200}{T}$	$0.5 \leq d_g \leq 42$ µm $0.05 \leq p \leq 0.3$ $298 \leq T \leq 1373\ K$	[101]

Lithium aluminate is not inert against ambient atmosphere as Denuziere et al. report [101], which is similar for lithium orthosilicate. It was found that there is a small but visible weight gain when the material is exposed to ambient atmosphere for several days. Also a slow decomposition into lithium hydroxide and alumina was observed when lithium aluminate is in contact with water. At room temperature, this decomposition reaction is reportedly slow, however, when boiling water is used, the reaction rate is considerably increased.

2.2.5 Relevant phase equilibria

As the phase equilibria determine any material's behavior, it is important to detail these equilibria for the three lithium compounds that are the anticipated constituents of solid breeder pebbles. Kleykamp et al. [105] published the most comprehensive description of the quasi-binary Li_2O-SiO_2 system (see Figure 2.9). A theoretical analysis of this system by Konar et al. also supports Kleykamp's phase diagram [55].

In total five stable ternary compounds were reported for this system:

- Li_8SiO_6 lithium oxosilicate
- Li_4SiO_4 lithium orthosilicate
- $Li_6Si_2O_7$ lithium orthodisilicate
- Li_2SiO_3 lithium metasilicate
- $Li_2Si_2O_5$ lithium disilicate

In contrast to all other compounds, lithium orthodisilicate is thermodynamically stable within a small temperature window between 1020 °C and 1030 °C. Yet, it is often found as metastable phase instead of lithium metasilicate after a sufficiently fast cooling. None of the stable phases shows a considerable homogeneity range. In the relevant range for tritium breeder compounds, i.e. in the Li_4SiO_4-Li_2SiO_3 region, three non-variant reactions occur. At 1030 °C lithium orthodisilicate forms peritectically from a liquid phase and lithium metasilicate and eutectoidally decomposes into lithium orthosilicate and lithium metasilicate at 1020 °C. Additionally, a eutectic reaction at 1025 °C is reported, at which a liquid phase consisting of about 38 mol% of SiO_2 reacts to lithium orthosilicate and lithium orthodisilicate.

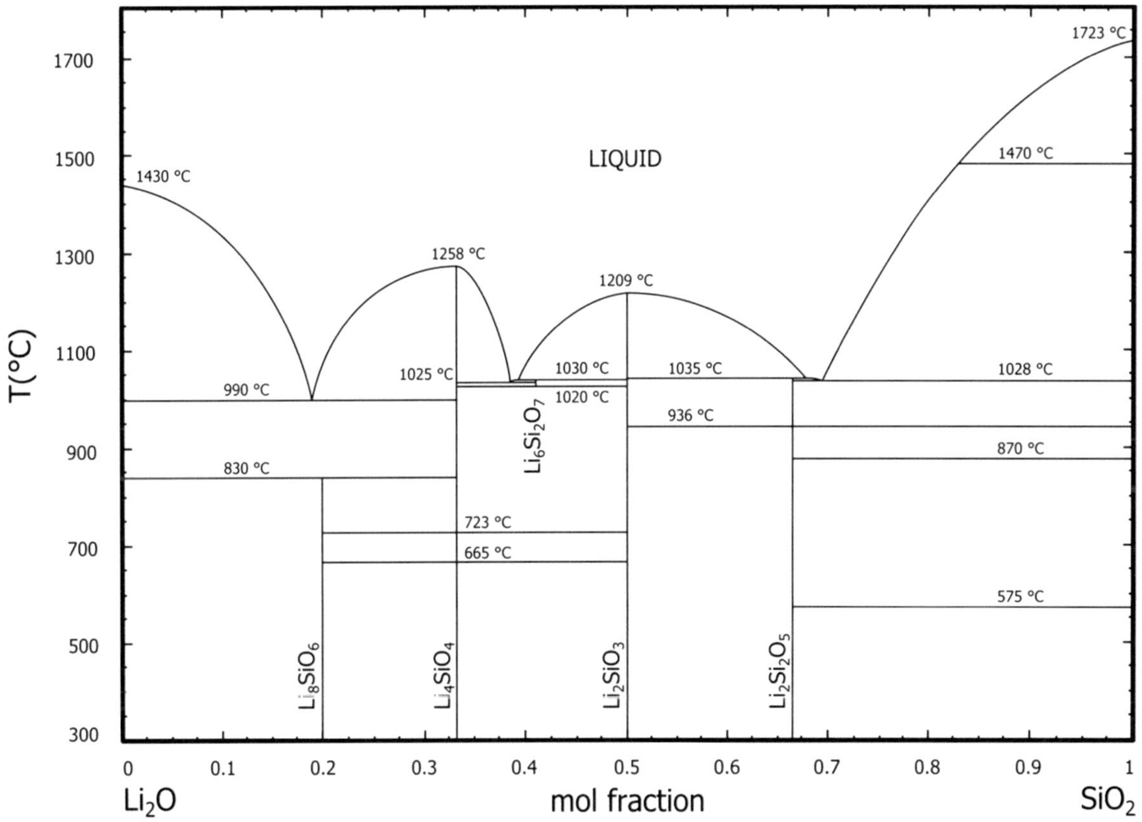

Figure 2.9: Phase diagram of the quasi-binary system Li_2O-SiO_2 reproduced from Kleykamp et al. [105].

For the quasi-binary system Li_2O-TiO_2, Mergos and Dervos [106] presented a phase diagram which combines the data of several other authors (see Figure 2.10). According to this phase diagram, four stable ternary compounds exist:

- Li_4TiO_4 lithium orthotitanate
- Li_2TiO_3 lithium metatitanate
- $Li_4Ti_5O_{12}$ spinell-type lithium titanate
- $Li_2Ti_3O_7$ ramsdellite-type lithium titanate

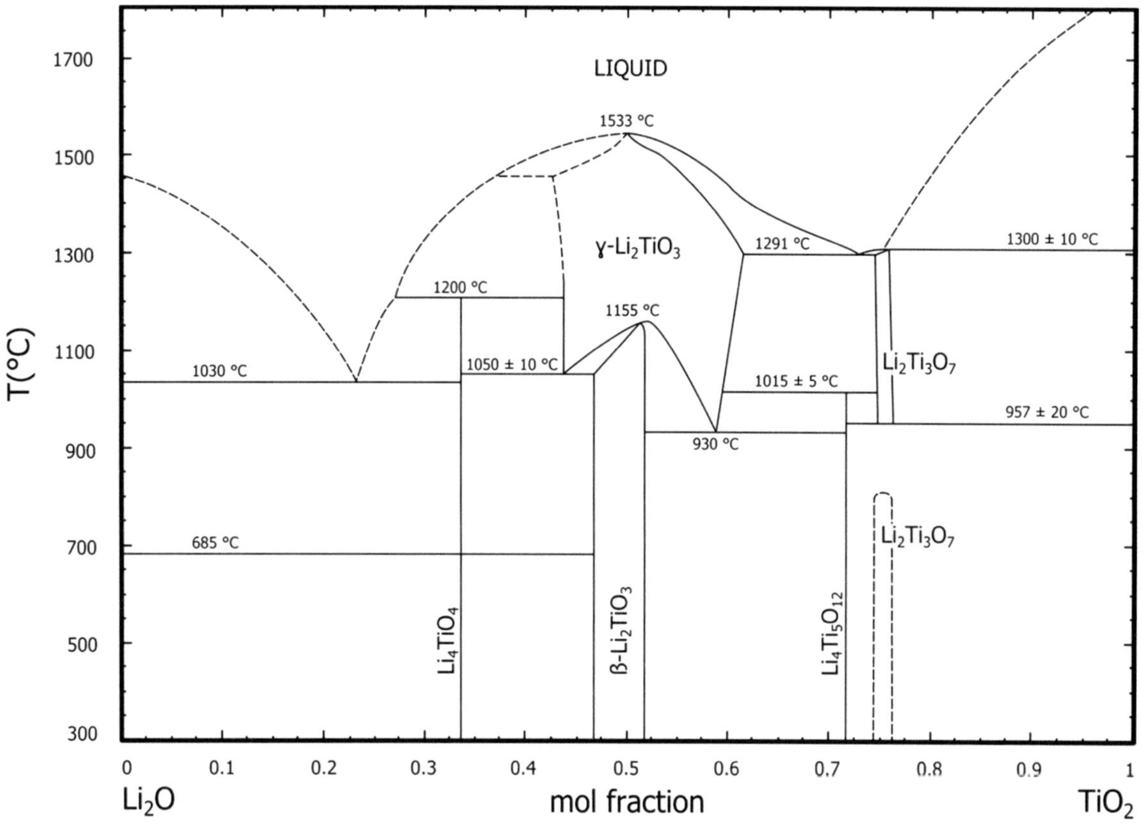

Figure 2.10: Phase diagram of the quasi-binary system Li_2O-TiO_2 reproduced from Mergos and Dervos [106].

While lithium orthotitanate and spinell-type lithium titanate show no visible homogeneity range, lithium metatitanate (see also section 2.2.3.1) and ramsdellite-type lithium titanate show a significant homogeneity range. Within the range of interest for breeder materials, just two eutectoid reactions appear when the γ-phase of lithium metatitanate decomposes into lithium orthotitanate and β-LMT or into β-LMT and spinell-type lithium titanate. It is, however, also necessary to mention that the valence of the titanium ions can be changed with relative ease from +4 to +3 [82,106]. Kleykamp [82] therefore proposed a tentative isothermal section of the ternary system Li-Ti-O at 900 °C (see Figure A.1 in Appendix A).

Principally, for the lithium-rich compounds in the system Li_2O-TiO_2, a significantly low oxygen potential should lead to the formation of $LiTiO_2$. The crystallographic parameters are virtually identical to those of γ-LMT, as shown in Table 2.7, which makes identification by X-ray diffraction rather difficult.

Table 2.7: Basic crystallographic parameters of $LiTiO_2$ with the parameters of γ-Li_2TiO_3 for comparison.

Phase	Equilibrium temperature of existence	Space group	a [Å]	b [Å]	c [Å]	α, β, γ [°]	Ref.
$LiTiO_2$	900 °C, Oxygen deficit	Fd-3m (227)	8.3756	8.3756	8.3756	90	[107]
γ-Li_2TiO_3	> 1155 °C	Fd-3m (227)	8.285	8.285	8.285	90	[81]

As biphasic pebbles consisting of LOS and LMT are prepared for or within this work (see sections 2.2.2.2 and 3.2.2) the phase equilibria in the system Li_2O-SiO_2-TiO_2 are also relevant. The most recent work on this topic was published by Hanaor et al. [108] (see Figure 2.11).

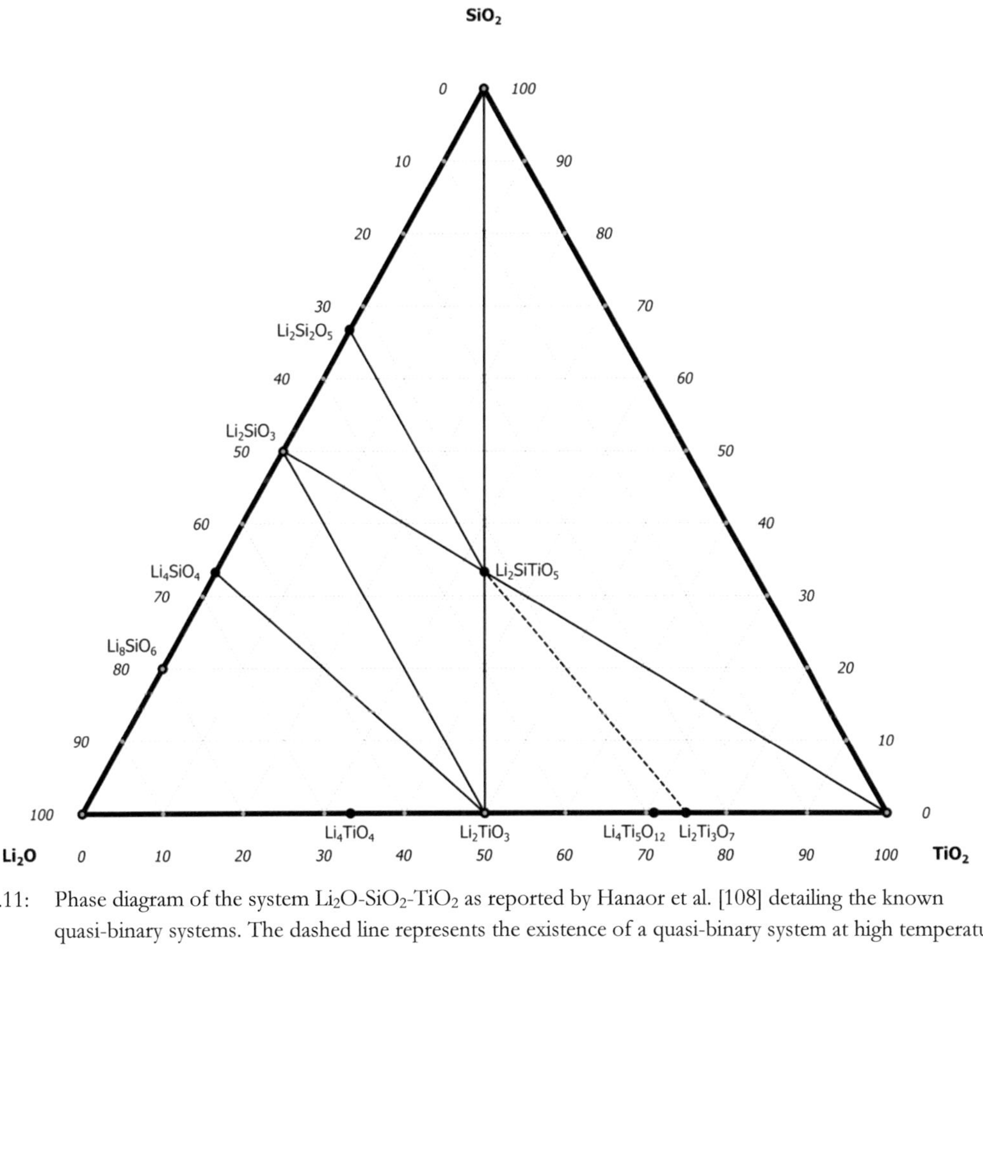

Figure 2.11: Phase diagram of the system Li_2O-SiO_2-TiO_2 as reported by Hanaor et al. [108] detailing the known quasi-binary systems. The dashed line represents the existence of a quasi-binary system at high temperature.

Earlier research of this system mainly focused on the lithium-poor region. In this part, the only quaternary phase, Li_2TiSiO_5, that is reported for this system is found. Between this phase and each of the phases Li_2SiO_3, $Li_2Si_2O_5$, Li_2TiO_3 as well as $Li_2Ti_3O_7$ quasi-binary systems exist. Apart from that, also tie-lines between Li_2SiO_3 and Li_2TiO_3 as well as Li_4SiO_4 and Li_2TiO_3 were shown to exist. A tie-line between Li_4SiO_4 and Li_4TiO_4 or a solid-solution of both seems likely, yet currently no experimental evidence exists.

The phase diagram of the system Li_2O-Al_2O_3 as reported by Kulkarni et al. [100] shows just three stable ternary phases (see Figure 2.12):

- Li_5AlO_4
- $LiAlO_2$
- $LiAl_5O_8$

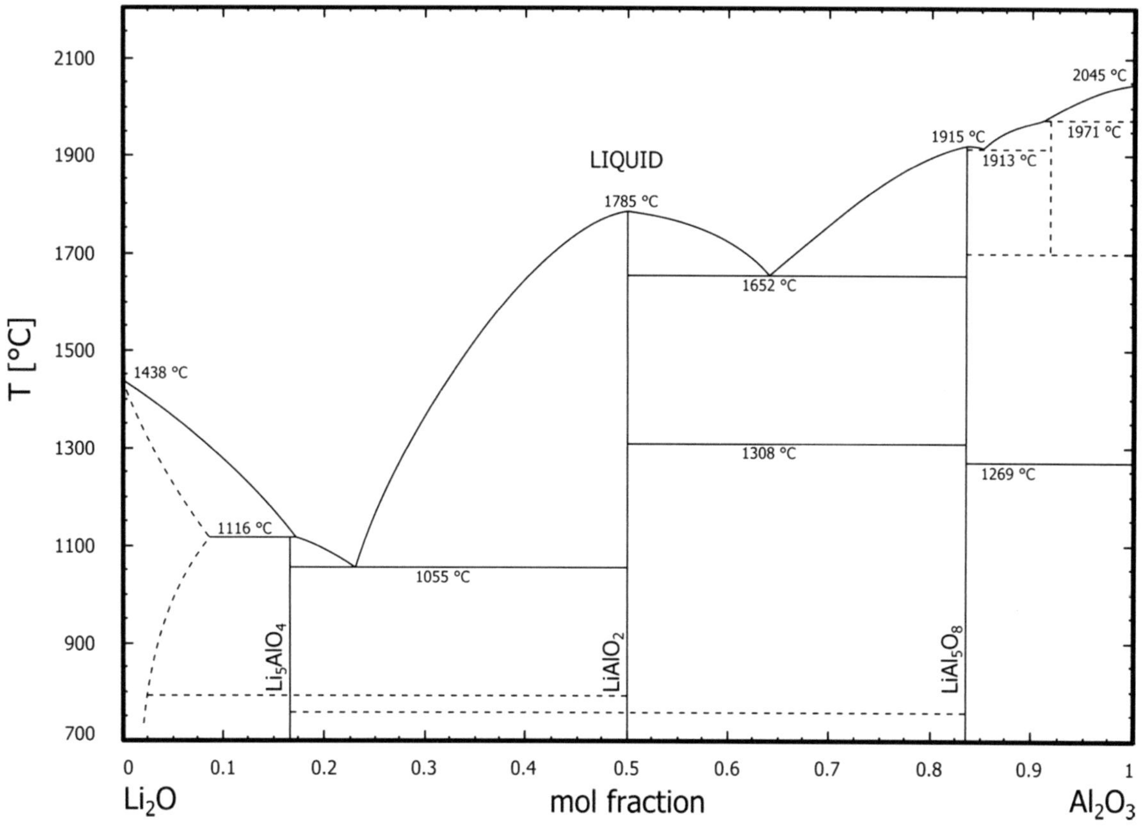

Figure 2.12: Phase diagram of the system Li_2O-Al_2O_3 reproduced from Kulkarni et al. [100].

None of these three phases shows a visible homogeneity range and non-variant reactions near the relevant composition of $LiAlO_2$ do not exists, except for the eutectic decomposition of liquid into Li_5AlO_4 and $LiAlO_2$ as well as $LiAlO_2$ and $LiAl_5O_8$. These eutectic reactions lead to a considerable reduction of the solidus temperature. This is most prominent in case of a lithia excess, which leads to a reduction to 1055 °C. In case of an alumina excess the reduction of the solidus temperature is less prominent with 1652 °C. A comprehensive study of the system Li_2O-Al_2O_3-SiO_2 does not exist. Similar to the case of the system Li_2O-SiO_2-TiO_2, data for the lithium-rich region of the phase diagram is especially scarce.

2.3 Mechanical stresses within spheres

For the evaluation of the quality of ceramic breeder pebbles, the mechanical strength is an important criterion. As it is on one hand easy to load individual pebbles until their failing force is reached in diametral compression and on the other hand difficult to produce comparable test specimen for standardized mechanical test, especially for pebbles that are produced by a melt-based process, the stress fields within a pebble under uniaxial compression have to be detailed. The implications of these stress fields on the failure of the pebbles are also presented.

2.3.1 Stress fields within spheres under uniaxial compression

The stress fields in an elastic spherical sample which is in contact with another elastic half-space were first described by Hertz [109] and Huber [110]. This analysis is still valid and frequently cited in textbooks [111,112]. Hertz simplified the mathematical problem in his analysis by assuming that the contact area where both bodies meet each other is small compared to the dimension of both bodies as well as the relative radii of curvature. The theory also implicitly assumes that the surfaces are frictionless and that therefore only normal stresses are transmitted between both bodies. In Figure 2.13, the geometry and coordinate system of the problem is detailed, with R representing the pebble radius and F the compressive force. With the reduced Young's modulus E^* for two different materials 1 and 2 (with Young's moduli E_1 and E_2 as well as Poisson's ratios ν_1 and ν_2) that are in contact with each other

$$\frac{1}{E^*} = \frac{1-\nu_1^2}{E_1} + \frac{1-\nu_2^2}{E_2} \tag{2.3}$$

and the radius of the elastically levelled contact area

$$a = \sqrt[3]{\frac{3FR}{4E^*}} \tag{2.4}$$

the maximum contact pressure

$$p_0 = \frac{3}{2} p_m = \frac{3F}{2\pi a^2} = \sqrt[3]{\frac{6FE^{*2}}{\pi^3 R^2}} \tag{2.5}$$

is defined as well as the mean contact pressure p_m.

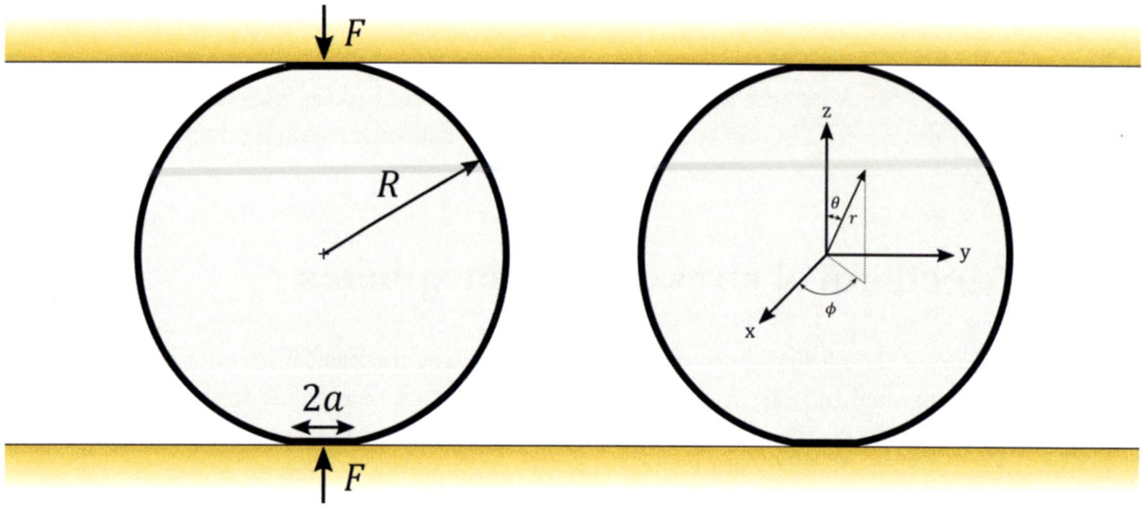

Figure 2.13: Geometry and spherical coordinate system of the problem of an elastic spherical sample which is in contact with an elastic half-space.

The stresses at the surface of the sphere within the contact area are predominantly compressive. Only at the very edge of the flattened contact area at about the radius a, tensile radial stresses are generated with a maximum of

$$\sigma_{contact,max} = p_0 \frac{1-2\nu}{3}. \tag{2.6}$$

They are the highest tensile stress Hertz' theory predicts in general. Therefore a ring-shaped crack is likely to form at the edge of the contact area, once the tensile strength of the material is exceeded. Lawn [113] states that the developed ring crack extends into a cone shaped crack if the load is further increased. The opening angle of the cone crack can be approximated by the angle between the trajectory surface of the σ_2-σ_3 principal stresses and the surface of the contact area. From a geometric point of view, the tip of such an imaginary cone lies outside the material where the crack is actually formed. Furthermore, Lawn et al. [114] show that cone cracks only develop when microplasticity is insignificant in the tested material. This is usually true for fine grained materials.

Yet, the Hertz theory and the derived stress field within spheres under compression focus on the contact area, where the sphere is compressed against another body. In addition to that, catastrophic failure of a Hertzian cone crack is not expected as Verrall [115] as well as Shipway and Hutchings [116] point out. To get a full picture of the stresses within uniaxially compressed spheres, all generated tensile stresses must be looked at. A number of studies were performed on this topic. The first mathematical assessment was performed by Hiramatsu and Oka [117] which was improved by Wijk [118] afterwards. Chau et al. [119] further improved the assessment of Hiramatsu and Oka by incorporating Hertzian contact stresses into it. Yet, this improvement did not lead to a principal change of the stress profiles. Zhao et al. [120] also used the solution of Chau et al. [119] and verified these analytical results by finite element simulations.

From these results, it is also evident that the maximum tensile and maximum shear stresses along the axis of compression are slightly overestimated by the Hertz theory. Furthermore Zhao et al. [120] emphasize that the Hertz theory only applies for the area near the mechanical contact and not for the whole pebble volume. Shipway and Hutchings [116] also calculated the stress fields within spheres under compression by a different approach and found a quantitatively similar but mathematically simpler result as originally published by Hiramatsu and Oka. Yet, both results involve infinite summations of Legendre polynomials and therefore have to be solved numerically. In both cases, the maximum tensile stress is found perpendicular to the axis of loading at about 80 % of the radius from the center of the sphere. As Luscher et al. [121] formulate in their review of both mathematical descriptions as well as other models, the maximum tensile stress, $\sigma_{\theta,\max}$, is reasonably determined by

$$\sigma_{\theta,max} = 1.4\frac{F}{2\pi R^2}. \tag{2.7}$$

The load dependence of both maximum tensile stresses (equations (2.6) and (2.7)) is detailed in Figure 2.14. It can be seen, that at a certain load which depends on the material properties, both stresses are equal, while otherwise one stress significantly exceeds the other one.

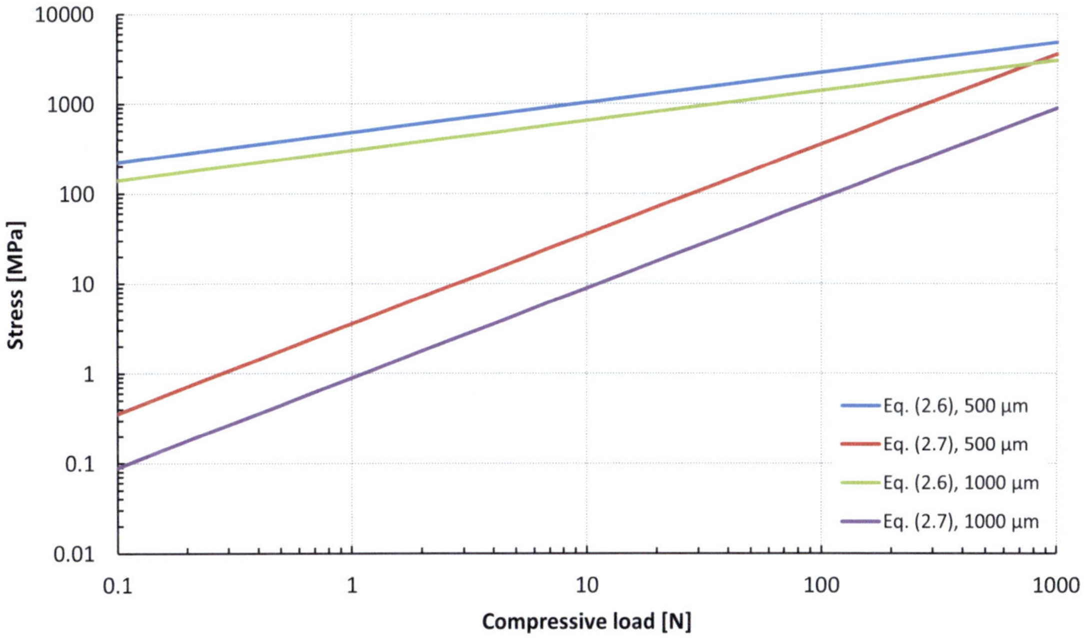

Figure 2.14: Load dependence of the maximum tensile stresses given by the equation (2.6) and (2.7) for pebble sizes of 500 µm and 1000 µm (E^*: 87.8 GPa, ν: 0.25).

Hiramatsu and Oka, who originally stated that a factor of 1.4 is a reasonable approximation for the maximum tensile stress, also note that the Poisson ratio doesn't influence the overall result greatly.

Also Luscher et al. [121] argue that the factor of 1.4 in equation (2.7) seems empirically sound when the radius of the contact area is not exceeding 20 % of the sphere radius. The work of Chau et al. [119], however, seems to lead to a higher factor than 1.4 which is in agreement with the results of Wijk [118]. Shipway and Hutchings [116] also expect larger tensile stresses on the axis of loading when the contact area is small, i.e. the contact radius is less than 15 % of the sphere radius. Additionally Chau et al. [119] as well as Shipway and Hutchings [116] show that there is notable influence of the Poisson ratio on the magnitude of stress unless the radius of the contact area is larger than 30 % of the sphere radius. Yet, in any case a higher Poisson's ratio results in a lower maximum tensile stress. Shipway and Hutchings [116] also note, that the volume in tension significantly depends on the ratio of contact radius to sphere radius. An increased ratio leads to a decreased volume in tension. According to both authors, tensile stresses perpendicular to the axis of loading are also formed on the surface near the equator of the sphere. These stresses do not depend greatly on the flattening of the sphere caps unless the contact radius approaches the sphere radius. These tensile surface stresses are generally lower than the tensile stresses on the axis of compression with a factor of 0.8 instead of 1.4. Yet, for significant flattening, i.e. a ratio of contact radius to sphere radius of 0.6, the surface stresses are larger than the stresses on axis. It has to be noted that Shipway and Hutchings derived their results while keeping the mean contact pressure constant. Because of the high similarity of all results, it can be assumed that all other authors kept the contact pressure constant as well, at least implicitly. It is therefore obvious, that the contact pressure governs the generated stresses.

In addition to the Hertz stresses that may lead to the formation of a cone crack and the tensile stresses on the axis as well as on the equator of spheres under uniaxial compression, also shear stresses within the pebble volume may be discussed. According to Hertz' theory, the maximum shear stress in the system is expected along the axis of symmetry at about half of the radius of the contact area into the volume and amounts to slightly less than a third of p_0. When the stresses within the complete sphere are computed according to Shipway and Hutchings [116], the peak location of the shear stress also approaches the region just below the contact area when the contact area is small. With increasing flattening, the peak location moves towards the center of the sphere. Also, with increasing flattening, the magnitude of the shear stress reduces significantly. These shear stresses are the cause for the formation of microplasticity. According to Lawn [113] the maximum shear stresses from Hertz theory is located just below the contact area and amounts to slightly less than 50 % of p_0 (calculated for a Poisson's ratio of 0.22). Microplasticity does not occur, when the formation of shear stresses is effectively suppressed, i.e. according to the model of Lawn et al. [122] that the friction coefficient for interfacial sliding is low. The nature of the interfacial sliding may be different for different ceramics. However, in many cases the sliding occurs along grain boundaries. If considerable shear stresses establish, shear faults beneath the contact area with a 45° inclination to the axis of compression may form. At the tips of these shear faults cracks are initiated when a certain threshold pressure is exceeded. These cracks may eventually coalesce at high loads as Cai et al. [123] point out. The presence of effects that weaken the grain boundary adhesion of course favors the formation of the shear faults and therefore the formation of microplasticity. Mostly residual stresses, e.g. stresses from thermal expansion mismatch or anisotropy are ascribed to have this weakening effect [123]. But not only the strength of the grain boundaries is important for the formation of microplasticity, also the grain size plays a significant role as already suggested above. The effect of the grain size was tested on several different ceramics and in all cases a rather sharp transition between the formation of a negligible amount of microplasticity and a considerable amount as a function of the grain size was observed. Mostly this transition occurs within a doubling of the grain size [124].

With regard to the application of the ceramic breeder pebbles in a pebble bed, the presence of two diametral mechanical contacts without other mechanical contacts of significance is a rare occurrence. Zhao et al. [120] detail that multiple mechanical contacts significantly change the stresses within a spherical pebble but that the number of contacts may actually be irrelevant. In consequence this means that the strength which is determined by uniaxial compression of pebbles is a relevant criterion for the failing of the pebbles within a compressed pebble bed.

2.3.2 Thermal stresses in spheres

The melt-based fabrication of lithium orthosilicate based pebbles as detailed in section 2.2.2.2 may lead to significant stresses during the solidification and cooling of the pebbles. These stresses may result in the formation of cracks as well as residual stresses in the fabricated pebbles. In turns out, that thermal stresses in spheres are rarely considered in literature. Yet, one basic approach, detailed by Timoshenko and Goodier [112] is often used. Naturally, the temperature field within the sphere defines the stresses and thus it has to be defined in advance. This approach is elaborated for a solid sphere, which can be regarded as the equivalent of a fully crystallized droplet, and for a sphere with spherical cavity. The latter case is used as a simple model of a liquid droplet which crystallizes equally at its surface while the inner volume is still liquid. The model is solved assuming that both solid surfaces, the inner as well as the outer, are free of radial stresses. As long as the crystallization of the melt leads to a reduction of the specific volume of the material, this is a reasonable assumption. Kudinov et al. [125] applied this approach to the quenching of molten oxides in water. Timoshenko and Goodier [112] assumed a steady heat flow from the warmer liquid core to the cooler atmosphere at the droplet surface, which seems to be a good approximation as Genceli et al. [126] find an almost constant heat flow in their model. Yet, Kudinov et al. [125] chose a more elaborate numerical calculation of the temperature profile across the solid crust of the droplet but eventually similar thermal stresses are obtained. The solution of Timoshenko and Goodier [112] yields the radial stress

$$\sigma_r = \frac{\alpha E \Delta T}{1-\nu} \frac{ab}{b^3 - a^3} \left(a + b - \frac{1}{r}(b^2 + ab + a^2) + \frac{a^2 b^2}{r^3} \right) \tag{2.8}$$

and the tangential stress

$$\sigma_t = \frac{\alpha E \Delta T}{1-\nu} \frac{ab}{b^3 - a^3} \left(a + b - \frac{1}{2r}(b^2 + ab + a^2) - \frac{a^2 b^2}{2r^3} \right) \tag{2.9}$$

with the inner radius of the hollow sphere a and its outer radius b. The resultant thermal stresses for a given thickness of the solidified outer layer are exemplarily shown in Figure 2.15. The radial stresses are exclusively compressive with a maximum within the solid crust, whereas the tangential stresses change from tensile to compressive. The maximum values for the tangential stresses are found at the outer and inner surface of crust.

However, these equations do not consider the decreasing specific volume during the crystallization of the melt and the accompanying stresses. Due to this effect, a considerable compressive radial stress will superpose the thermal stresses in the crust that are illustrated in Figure 2.15, while the liquid core of the crystallizing droplet will experience hydrostatic tensile stresses. These stresses counteract the thermal stresses in a hollow sphere that are given by equation (2.11). Evidently, equation (2.11) is only correct if the liquid core does not experience any radial stress, i.e. when both effects compensate each other.

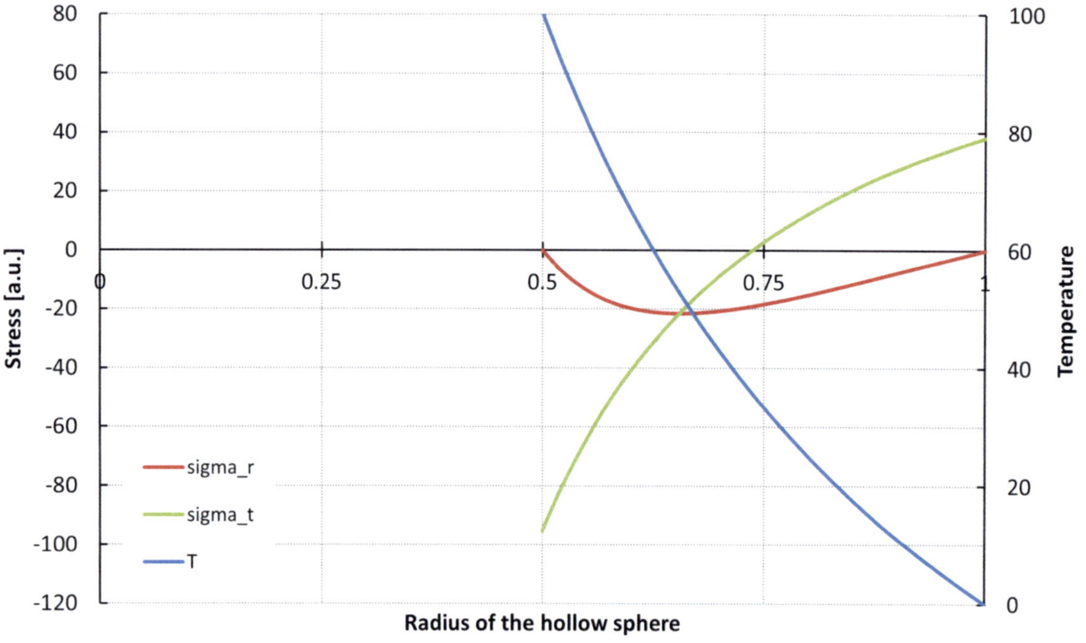

Figure 2.15: Thermal stresses in a hollow sphere according to the equations (2.8) and (2.9) for a temperature difference of 100 K (E: 1 Pa, ν: 0.25, α: 1 K^{-1}).

As Forgac et al. as well as Forgac and Angus [127,128] show, the tensile hydrostatic stress can be relieved by buckling of the crust, cavity formation or plastic deformation of the crust if they are sufficiently higher than the respective compressive thermal stresses. The thermoelastic model of Forgac et al. [128] for the solidification of a sphere correctly considers the radiative, conductive and convective heat transfers as well as the stresses that arise in a spherical body when a moving phase boundary has to be considered.

The model does not provide an analytical solution of the stresses in the sphere during solidification. It predicts, that for a comparably strong cooling of the crust the liquid core may be compressed, but as the solidification of the pebble continues, i.e. the crust grows in thickness, this stress might very well become strongly tensile because the temperature gradients that generate the stresses (also compare with Figure 2.15) become less steep unless the cooling of the surface increases and thus compensates for the growth of the crust.

It also shows that tensile tangential stresses at the surface are increased when the liquid core is compressed. These stresses completely turn compressive once the liquid core experiences mild tensile forces. A strong influence on the development of the stresses originates from the thermal expansion coefficient. If the thermal expansion coefficient of the solid material is large, the generated thermal stresses by the cooling crust will be larger and accordingly the liquid core of the pebble is more likely to experience compressive stresses. Logically, the density ratio of solid and liquid material acts conversely with respect to the generated stresses. Generally, when the liquid core of a crystallizing pebble is compressed, the results of Timoshenko and Goodier [112] (see equations (2.8) and (2.9) as well as Figure 2.15) apply qualitatively at least. However, when the liquid core experiences tensile stress, the tangential stress at the surface of the crust can be either tensile or compressive depending on the parameters of the system, with the maximum tensile stress at the surface, which is similar to the results of the hollow sphere. For the radial stresses the results are equally depending on the system parameters. The radial stresses can be tensile throughout the crust thickness, with the maximum stress at the liquid-solid interface or be compressive and tensile with a maximum of compressive stresses within the volume of the crust.

For the case of a fully crystallized sphere, Grünberg [129] described the thermal stresses that are generated during sudden external heating of a sphere, using the same formalism as Timoshenko and Goodier [112]. Similarly to the case of the hollow sphere, the radial stresses at the surface of the sphere are required to be zero as a boundary condition. In his analysis, Grünberg [129] correctly considers the time dependent heat conduction. Yet, he assumes a fully homogenous and dense single phase material. His analysis stays valid even if cooling is considered rather than heating, which simply leads to a change of sign. Grünberg's solution of the problem leads, at any given time, to stress distributions of the shape shown in Figure 2.16.

While the radial stresses at the surface of the sphere are zero, compressive radial stresses are generated by the thermal expansion of the warmer outer zone of the sphere. These stresses increase until the maximum value is reached at the center of the sphere. The tangential stresses at the center are also compressive and of the same magnitude, which is calculated to

$$\sigma_r(r=0) = \sigma_t(r=0) = 0.771 \frac{\alpha E}{2(1-\nu)} \Delta T. \tag{2.10}$$

From the center of the sphere to the surface, the tangential stresses decrease continuously until they reach a minimum at the surface, which is tensile. This maximum tensile tangential stress is given by

$$\sigma_t(r=R) = \frac{\alpha E}{(1-\nu)} \Delta T \tag{2.11}$$

which is exactly double the maximum tangential stress that equation (2.9) predicts for a given temperature difference as the limit of equation (2.9) is given by

$$\lim_{r=b,a\to b} \frac{\alpha E \Delta T}{1-\nu} \frac{ab}{b^3-a^3}\left(a+b-\frac{1}{2r}(b^2+ab+a^2)-\frac{a^2b^2}{2r^3}\right)=\frac{\alpha E \Delta T}{1-\nu}\cdot\frac{1}{2} \quad (2.12)$$

for an infinitesimally thin crust.

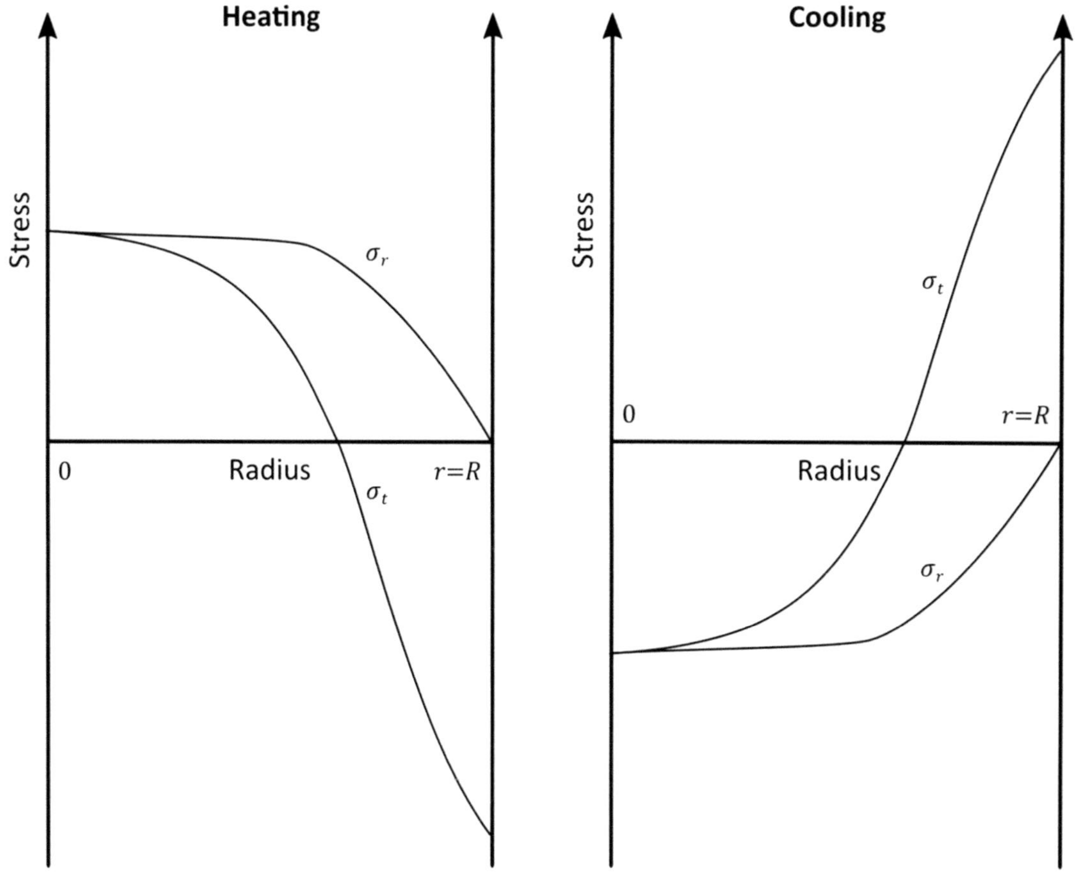

Figure 2.16: Thermal stresses in a heated or cooled dense sphere. Reproduced from Grünberg [129].

2.4 Weibull failure analysis

Weibull analysis is the standard method in industry and in science for statistically assessing, in the broadest sense, life data. For engineering systems, life data unambiguously translates as reliability data. Naturally, there is no restriction on the type of hazard to the life of the system, which, in principal, enables the application of the Weibull analysis to any system. To the great benefits of the Weibull analysis also count its capability of reliably modeling systems with very few samples, which greatly serves cost efficiency. Yet, as any statistical analysis, also the Weibull analysis implicitly assumes that no two samples are exactly alike. The Weibull method was steadily expanded to cope with a number of practical issues, such as interval failures or deficient data in general. While all these, to a greater or lesser extent mathematically demanding, modifications to the original Weibull method are available, the Weibull analysis still provides an easily understandable graphical plot as a result.

The Weibull analysis' especially great success in modelling a wide range of engineering systems derives from its implicit assumption that a system fails according to the 'weakest-link-in-the-chain'-principle [130]. That is, it assumes that all of the system's present failure modes stay inactive and failure only occurs by the first failure mode to be activated. Evidently, this is true for many systems. It is particularly true for the failure of ceramic material parts, whose behavior is usually well described by the Griffith theory of crack propagation [131], which assumes that for a given external stress the dimension of the largest flaw a determines the failure stress

$$\sigma_c = \sqrt{\frac{2E\gamma}{\pi(1-\nu^2)a}} \tag{2.13}$$

which is given for plane strain, with γ being the specific surface energy to generate a unit area of new crack surface. As it is usually more convenient to use the stress intensity factor K, equation (2.13) can be rearranged to obtain this value. In his work, Griffith assumed an infinitely large plate. Therefore, the term Y has to be introduced to normalize the crack size to a characteristic dimension of the tested specimen. Eventually the stress intensity factor

$$K = \sigma Y \sqrt{\pi a} \tag{2.14}$$

is defined as a function of the crack length and the stress that acts on the crack [132].

In Appendix A it is detailed, that the Weibull distribution was not specifically derived for brittle failure, but in a very general sense. Therefore, the maximum elastic energy that is stored within a pebble during mechanical compression at the time of failure can also be used as random variable as Zhao et al. [133] show for lithium orthosilicate pebbles that were fabricated by the EU reference process (see section 2.2.2.2). Yet, this approach requires a reasonable method for estimating the elastic energy for every tested pebble. The resulting Weibull distribution, however, is well suited for numerical methods, such as the discrete element method, with which the mechanical behavior of a pebble bed can be simulated as Annabattula et al. [134] and Zhao et al. [135] show.

2.4.1 The Weibull distribution and the Weibull plot

Waloddi Weibull derived his signature distribution from a series of basic considerations as well as mathematical convenience and simplicity [130]. The details of his reasoning are briefly outlined in Appendix A as well as some basic mathematical properties of the distribution function.

He formulates the distribution function

$$F(x) = 1 - e^{-\frac{(x-x_u)^m}{x_0}} \tag{2.15}$$

with the three independent parameters x_0, x_u and m. Yet, it can be shown mathematically, that the strength distribution of a material which fails according to the condition of equation (2.14) and for which the distribution of (at least) the larger flaws can be approximated by

$$f(a) \propto \frac{1}{a^r} \tag{2.16}$$

is indeed a Weibull distribution with the parameter x_u equal to zero [132]. The latter assumption is usually a sensible approximation as stated by Danzer et al. [136]. In its most relevant form for addressing mechanical strength of ceramics, the Weibull distribution is often given as

$$F(\sigma_c) = 1 - e^{-\left(\frac{\sigma_c}{\sigma_0}\right)^m} \tag{2.17}$$

or as

$$F(\sigma_c) = 1 - e^{-\frac{V}{V_0}\left(\frac{\sigma_c}{\sigma_v}\right)^m} \tag{2.18}$$

which compensates for different sample volumes [132,136]. The derivation of equation (2.18) is also given in Appendix A. The parameter m in these equations is referred to as Weibull modulus in analogy to material characteristics like the Young's modulus, despite the fact that it is no material constant. This analogy mainly stems from an ostensible linear relationship between the mechanical load on a sample and its failure response that is established in a Weibull plot.

The second constitutive parameter of the Weibull distribution is the so-called characteristic strength σ_0 or σ_v. Yet, the characteristic strength σ_v is normalized for the volume of the sample and, as a result, depends only on the material properties. This normalization is not always easily possible and thus σ_0 is the commonly used parameter. Generally a failure mode within Weibull analysis is defined by its Weibull distribution and consequently by the two Weibull parameters. As Munz and Fett [132] illustrate the Weibull modulus is directly defined by the shape of the flaw size distribution as

$$m = 2(r-1) \tag{2.19}$$

and accordingly the characteristic strength follows as

$$\sigma_0 = \frac{K_{Ic}}{Y\sqrt{a_0}(zV)^{\frac{1}{m}}} \tag{2.20}$$

$$\sigma_v = \frac{K_{Ic}}{Y\sqrt{a_0}(zV_0)^{\frac{1}{m}}} \tag{2.21}$$

with z being the flaw density, a_0 being a lower limit of the flaw size distribution which is mathematically necessary for normalization and K_{Ic} being the critical stress intensity factor for mode I loading, i.e. tension normal to the crack plane.

For each data point, i.e. observed failure, which is to be assessed in a Weibull analysis, a failure probability must be estimated. For a Bernoulli process of n samples, which consists of independent and congeneric experiments with two possible results, which are in this case 'intact' or 'failed', the binomial distribution returns the probability of a selected quantity of successes a at a given quantity of trials for which the individual probability for the results 'intact' or 'failed' is known. However, the quantity of successful failures and nonfailures is already known from the experiments, in contrast to the individual probability of failure. The beta distribution, as it is the conjugate prior to the binomial distribution, returns this value. Yet, to assess the experimental data in a Weibull plot, the distribution of the underlying random variable, i.e. the failure probability (σ_c), has to be determined. Inverting the beta distribution into the cumulative beta prime distribution

$$F(x; a, b) = \int_0^{\frac{x}{1+x}} t^{a-1}(1-t)^{b-1}dt \tag{2.22}$$

(with b=n-a+1) leads to these values, given that the individual probability of failure is defined. As the use of median ranks is considered to be most accurate for this case, a value of 50 % is chosen [137]. By ranking the data, the quantity of failures a and nonfailures n-a at a given value of σ_c and therefore the respective failure probability can be deduced directly. To generate the Weibull plot, the Weibull equation (2.17) has to be linearized by minor rearrangement to reach

$$\ln\left(\ln\left(\frac{1}{1-F(\sigma_c)}\right)\right) = m\ln(\sigma_c) - m\ln(\sigma_0). \tag{2.23}$$

By choosing the double logarithm of the inverse reliability as scale on the abscissa and the logarithm of the specimen age as scale on the ordinate, a set of samples following equation (2.17) will plot along a straight line with m as slope. The characteristic strength σ_0 can be read from the y-axis intercept. From this linearization stems the expression of σ_0, as the right side of the linearized Weibull equation turns to zero at this very level of mechanical stress. Therefore, at σ_0, independently of m, 63.2 % of the samples fail which in return justifies the denomination of σ_0 as characteristic strength.

Any Weibull analysis uses the described Weibull plot for illustration. One effect of the presented linearization is the magnification of the region of low failure probability. In real world experiments, however, there are numerous cases in which an initial, low strength failure mode disappears, for instance originating from surface defects, for a second failure mode, which could be volume defects [132]. In such cases the set of samples actually consists of a minimum of two sub-populations of samples where either sub-population has a significantly higher probability to cause the failure in its operating regime with respect to the other sub-population. The resultant Weibull plot for a sample set of this kind will show at least one corner or dogleg and is not modelled well with one Weibull line. It depends on how the failure modes appear in the sample set, which compound distribution would be appropriate [137,138]. However, using several independent Weibull distributions to model such data may be considered to be approximately correct [137].

2.4.2 Deriving the Weibull parameters

For an established Weibull plot, with the data scattering around a straight line, the best fitting Weibull parameters with respect to the data must be determined. This determination is usually performed by algorithms which are implemented into software in order to ensure high repeatability. There is a number of fitting methods, however, the Maximum Likelihood Estimation (MLE) is the generally preferred method by statisticians and materials scientists [132,137].

2.4.2.1 Maximum Likelihood Estimation

In contrast to the fitting of a straight line to data points by a least-squares method and thus deriving the Weibull parameters, the MLE approach directly estimates the Weibull parameters m and σ_0 independently from the Weibull plot. It does so by finding the values of m and σ_0 which maximize the likelihood of obtaining the actual values of the Weibull parameters considering the given data. The basic approach is detailed in its entirety by Abernethy [137], ReliaSoft [139] as well as Munz and Fett [132]. The likelihood function $\mathcal{L}$ for a data set of n failed samples with i failures at $\sigma_{c,}$ is given by

$$\mathcal{L} = \prod_{i=1}^{n} f(\sigma_{c,i}; m, \sigma_0) \tag{2.24}$$

for the Weibull probability density function

$$f(\sigma_c) = \frac{m}{\sigma_0}\left(\frac{\sigma_c}{\sigma_0}\right)^{m-1} e^{-\left(\frac{\sigma_c}{\sigma_0}\right)^m} \tag{2.25}$$

which is the derivative of equation (2.17). Obviously the likelihood is a function of m and σ_0. It is also obvious from equation (2.24) that all failures are equally weighed by the MLE approach, in contrast to a least-squares fit, for which the effect of magnification of the region of low failure probability leads to an unequal weighing as Abernethy [137] points out. In addition it is also evident that the MLE approach is not connected to the Weibull plot and the failure probability (2.22) in any way. Yet, with the derived Weibull parameters, a Weibull plot can be drawn without problems.

The likelihood function is then a function of m and σ_0, its graph is a three dimensional surface with a single peak (see Figure 2.17). To retrieve the coordinates of the top of this maximum, the logarithm of equation (2.24) is differentiated with respect to m and σ_0 and both expressions are equated to zero

$$0 = \frac{\sum_{i=1}^{n} \sigma_{c,i}^{\hat{m}} \ln \sigma_{c,i}}{\sum_{i=1}^{n} \sigma_{c,i}^{\hat{m}}} - \frac{1}{n}\sum_{i=1}^{n} \ln \sigma_{c,i} - \frac{1}{\hat{m}} \tag{2.26}$$

$$\widehat{\sigma_0} = \left(\frac{\sum_{i=1}^{n} \sigma_{c,i}^{\widehat{m}}}{n}\right)^{\frac{1}{\widehat{m}}}. \tag{2.27}$$

In both equations the symbol '^' indicates respective value that was estimated by its maximum likelihood. The value of m can be determined by a number of iterations in a first step and the result is used to calculate σ_0 according to equation (2.27). Yet, with most numerical minimization algorithms a simultaneous variation of both Weibull parameters is possible and thus the preferred choice.

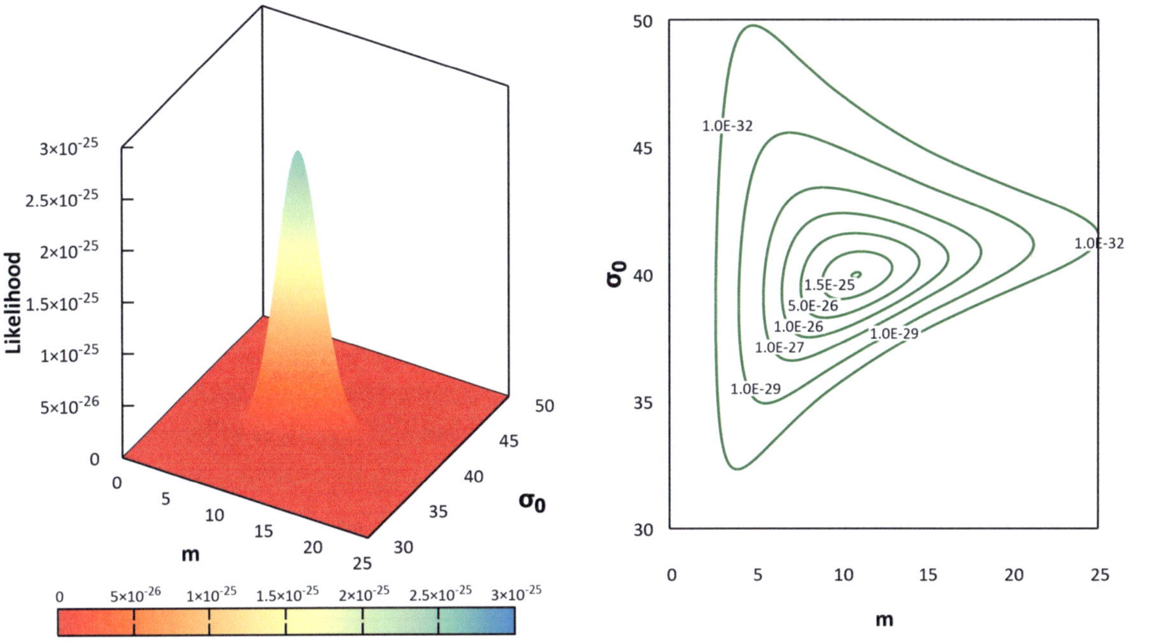

Figure 2.17: Plot (left) and iso-value contours (right) of the likelihood function of the Weibull probability density function for model data (m_{model}: 10, $\sigma_{0,\text{model}}$: 40, n: 20).

Yet, generally the result of an MLE approach is significantly biased for small sample sizes (see Figure 2.17 where the maximum likelihood is not found for the model parameters). In fact an infinite data set is necessary for an exact estimation. Usually more than 100 samples are necessary to prevent the bias from being too large. The bias is especially prominent for the obtained m value, which is estimated too optimistically. In order to reduce this bias for the MLE of parameters of a normal or lognormal distribution a reduced bias adjustment (RBA) factor was introduced in the field of statistical quality control by mathematical reasoning. Similar to these factors, RBA factors U for Weibull distributions are available, which were empirically determined [132,137,139]. The simple multiplication with the RBA factor yields an unbiased estimate

$$\widehat{m_U} = U\widehat{m} \tag{2.28}$$

of the Weibull modulus (indicated by the subscript U). The RBA factors generally depend on the number of failures or observations n, as is shown by

$$U_{Hirose} = \frac{1}{1.0115 + \frac{1.278}{n} + \frac{2.001}{n^2} + \frac{20.35}{n^3} - \frac{46.98}{n^4}} \tag{2.29}$$

for a Weibull analysis in absence of suspensions [140]. For unbiasing the second Weibull parameter, a pragmatic approach can be used, i.e. an additional maximization of the likelihood function with σ_0 as single variable, while keeping the unbiased Weibull modulus constant.

As in some cases it is necessary to remove a number of data points from the analysis, for instance, because the failure mode is different or they are experimentally flawed, the likelihood function of equation (2.24) has to be adapted. The same is true if for some reason the failure of a sample occurred within a stress interval rather than at an exact stress. In principle any suspension or censoring of data points can be categorized into left and right censoring, which stems from the location on the x-axis of the Weibull plot. Any interval failure is mathematically identically treated to censoring low strength data points without a gap between the data points. For both cases, the interval then starts at the minimum uncensored value (subscript L), which will be zero for the latter case, and extends to the maximum uncensored data point next to the censored item (subscript H). Any other censoring can be categorized as being right censored. Here, also the minimum, non-censored value next to the censored data point is used in the calculation (subscript S). The complete likelihood function with R non-censored failures, M right censored items and P left censored items then reads as follows [139]:

$$\mathcal{L} = \prod_{i=1}^{R} f(\sigma_{c,i}; m, \sigma_0) \cdot \prod_{j=1}^{M} \{1 - F(\sigma_{c,S,j}; m, \sigma_0)\} \cdot \prod_{l=1}^{P} \{F(\sigma_{c,H,l}; m, \sigma_0) - F(\sigma_{c,L,l}; m, \sigma_0)\}. \tag{2.30}$$

If the dataset is right censored, the RBA factor may also be adapted to accommodate for this type of censoring with

$$U_{Ross} = \frac{1}{1 + \frac{1.37}{r - 1.92}\sqrt{\frac{n}{R}}} \tag{2.31}$$

according to Ross [141].

2.4.2.2 Likelihood ratio confidence boundaries

Once a Weibull model is selected, the statistical precision of the model's estimates must also be addressed, i.e. measuring the statistical confidence. The confidence represents the frequency of the true value of a given element lying within a previously defined interval, which is thus called confidence interval. However, the statistical confidence is insensitive to systematic errors, therefore these either have to be known or be negligible. In most cases, two-sided confidence boundaries are employed, however in some cases it might be sufficient to only calculate an upper or lower confidence boundary.

Commonly the confidence interval is set as wide as that the confidence δ for the true value lying within it equals 90 %.

There are numerous ways of estimating the confidence interval for a given level of confidence. One widely used assessment of the confidence bounds is detailed in the following for the MLE approach detailed in the previous section (see section 2.4.2.1). For this approach, the likelihood function can directly be used for the estimation of the confidence interval [137,139]. Therefore slices through the likelihood function at a given value of likelihood are used, which can also be referred to as contours similar to elevation contours in a geographic map. The value of likelihood for a given parameter θ, at which the likelihood function is sliced for two-sided confidence boundaries as defined by

$$-2\ln\left(\frac{\mathcal{L}(\theta)}{\mathcal{L}(\hat{\theta})}\right) = \chi^2_{\delta;k} \tag{2.32}$$

is the point at which its ratio to the maximum likelihood equals the value of the chi-squared distribution at δ with k degrees of freedom [139]. As only the value of likelihood function is estimated, k is set to 1 for all cases. These slices are elliptically shaped as shown in Figure 2.17. The confidence interval of each of the two variables is then taken, graphically speaking, by taking the lowest overall extension of the curve in one direction as lower boundary of the given variable and the maximum extension in this direction as upper boundary. As the resultant slice of the likelihood function is usually not symmetrical to either of the axes, the so determined boundaries for a given variable do not share the same value of the second variable. For plotting likelihood ratio boundaries in a Weibull plot, the resultant failure stresses have to be estimated using equation (2.23).

A closed mathematical formulation for this estimation does not exist. Therefore, this problem is solved iteratively. For a given failure probability, the pair of m and σ_0 that lies on the contour and produces the minimum or maximum failure stress is taken for the calculation of the x-value of the lower or upper confidence boundary, respectively. For this method, an additional unbiasing step is not necessary, if the dataset is censored in some way, as long as the complete likelihood function (2.30) is used.

When two datasets are to be compared, the confidence boundaries can help to estimate whether the datasets are significantly different or not. If a gap between the confidence intervals of the datasets exists, the two datasets are clearly distinguishable with the respective level of confidence. Yet, the converse is not necessarily true. As the confidence boundary is fully determined by the contour of likelihood function, the contour itself can equally be used for this purpose. Therefore, if two likelihood contours do not overlap, the failure behavior of the two sample sets is statistically dissimilar with the given level of confidence. Such plots, which show the intersection of the likelihood surface at a given level, are called contour plots and are mostly a lot clearer from a visual point of view compared with a complete Weibull plot, especially when more than two datasets are to be compared.

3 Experimental

In order to address the questions that were raised in the introduction, a number of experiments were performed. These experiments as well as necessary complementary work are detailed in this chapter. At first, the available and additionally prepared advanced breeder pebble samples within this work are introduced. All of these samples were fabricated by the processes which were described in the previous chapter. Also the synthesis of pure powders as raw materials or for dedicated analysis is reported. For the investigation of the long-term development of the advanced breeder pebbles under relevant conditions, a dedicated annealing experiment was performed. The experimental setup and annealing parameters are presented as well as the data acquisition system and data processing of the obtained data.

As the mechanical strength of breeder pebbles is an issue of high importance, the performed determination of the crushing strength is reported in full. This comprises an estimation of the mechanical properties of the biphasic pebbles as well as the procedural method for the Weibull analysis, in which the concepts that were shown in the previous chapter are implemented. To gain insight into the phase composition of the supplied and fabricated pebbles as well as the evolution of the phase composition during the annealing of the pebbles, a thorough X-ray diffraction analysis of the samples was performed. The used structure models that are refined in Rietveld analyses are composed of several established models which are also illustrated in this chapter. Temperature varied X-ray diffraction analysis was performed to determine the thermal expansion behavior of the pure phases of the pebbles.

Furthermore, the analysis of the pebble porosity, morphology as well as microstructure is also vital to understand the resultant pebble properties. Therefore, the applied methods to investigate these three characteristics are also introduced.

3.1 Summarization of the performed characterization

The investigated pebble samples and the performed characterization techniques are summarized in Table 3.1 and Table 3.2, to get a quick overview of the attained results.

Table 3.1: Summary of the experimental characterization of the samples that were fabricated by the KALOS process, which nominally contain 20 mol% and 30 mol% lithium metatitanate, as well as the lithium excessive lithium metatitanate samples that were fabricated by the emulsion method (EM). (*: 500 μm pebbles only.)

Sample batch	Sample characteristics (annealing duration)	Analyzed property	**Elemental analysis**			**Phase analysis**				**Morphology**		**Microstructure**		**Pebble density**			**Mechanical strength**	**Fractography**
		Specification	Main constituents	Main constituents	Impurities	Qualitative	Quantitative	Amorphous content	Thermal expansion	Surface	Pebble size distribution	Cross-section	Grain size	Inner density	Apparent density	Density		
		Analyzing technique	XRF	ICP-OES	ICP-OES	XRD	XRD	XRD	HT-XRD	SEM	OM	SEM	SEM	He-Pycnometry	Mercury porosimetry	OM	Uniaxial compression	SEM
KALOS 20 mol% LMT	As-received		✓	✓	✓	✓	✓	✓		✓		✓		✓	✓	✓	✓	
	4 days		✓	✓	✓	✓	✓	✓		✓		✓		✓	✓	✓	✓	
	32 days		✓	✓	✓	✓	✓	✓		✓		✓		✓	✓	✓	✓	
	64 days		✓	✓	✓	✓	✓	✓		✓		✓		✓	✓	✓	✓	
	128 days		✓	✓	✓	✓	✓	✓		✓		✓		✓	✓	✓	✓	✓
KALOS 30 mol% LMT	As-received		✓	✓	✓	✓	✓	✓	✓	✓		✓		✓	✓	✓	✓	✓
	4 days		✓	✓	✓	✓	✓	✓		✓		✓		✓	✓	✓	✓	
	32 days		✓	✓	✓	✓	✓	✓		✓		✓		✓	✓	✓	✓	✓*
	64 days		✓	✓	✓	✓	✓	✓		✓		✓		✓	✓	✓	✓	
	128 days		✓	✓	✓	✓	✓	✓		✓		✓		✓	✓	✓	✓	
Li-excessive LMT EM	As-received		✓	✓	✓	✓	✓	✓		✓		✓	✓	✓	✓	✓	✓	✓
	4 days		✓	✓	✓	✓	✓	✓		✓		✓	✓	✓	✓	✓	✓	
	32 days		✓	✓	✓	✓	✓	✓		✓		✓	✓	✓	✓	✓	✓	
	64 days		✓	✓	✓	✓	✓	✓		✓		✓	✓	✓	✓	✓	✓	✓
	128 days		✓	✓	✓	✓	✓	✓		✓		✓	✓	✓	✓	✓	✓	

Table 3.2: Summary of the experimental characterization of the lithium excessive lithium metatitanate samples that were fabricated by the sol-gel method (SG), as well as all biphasic samples (nominal content given in the table) that were fabricated by the emulsion method (EM) and the nominally pure LOS and LMT powders.

	Analyzed property	**Elemental analysis**			**Phase analysis**				**Morphology**		**Microstructure**		**Pebble density**			**Mechanical strength**	**Fractography**
	Specification	Main constituents	Main constituents	Impurities	Qualitative	Quantitative	Amorphous content	Thermal expansion	Surface	Pebble size distribution	Cross-section	Grain size	Inner density	Apparent density	Density		
Sample batch	Sample characteristics (annealing duration or nominal composition) / Analyzing technique	XRF	ICP-OES	ICP-OES	XRD	XRD	XRD	HT-XRD	SEM	OM	SEM	SEM	He-Pycnometry	Mercury porosimetry	OM	Uniaxial compression	SEM
Li-excessive LMT SG	As-received	✓	✓	✓	✓	✓	✓		✓		✓	✓	✓	✓	✓	✓	✓
	4 days	✓	✓	✓	✓	✓	✓		✓		✓	✓	✓	✓	✓	✓	
	32 days	✓	✓	✓	✓	✓	✓		✓		✓	✓	✓	✓	✓	✓	
	64 days	✓	✓	✓	✓	✓	✓		✓		✓	✓	✓	✓	✓	✓	
	128 days	✓	✓	✓	✓	✓	✓		✓		✓	✓	✓	✓	✓	✓	✓
LOS/LMT EM	10 mol% LMT	✓	✓		✓	✓			✓	✓	✓		✓		✓	✓	
	20 mol% LMT				✓	✓			✓	✓	✓		✓		✓	✓	
	30 mol% LMT				✓	✓			✓	✓	✓		✓		✓	✓	
	40 mol% LMT				✓	✓			✓	✓	✓		✓		✓	✓	
	50 mol% LMT	✓	✓		✓	✓			✓	✓	✓		✓		✓	✓	✓
	60 mol% LMT				✓	✓			✓	✓	✓		✓		✓	✓	
	70 mol% LMT				✓	✓			✓	✓	✓		✓		✓	✓	✓
	80 mol% LMT				✓	✓			✓	✓	✓		✓		✓	✓	
	90 mol% LMT	✓	✓		✓	✓			✓	✓	✓		✓		✓	✓	✓
LOS/LAO EM	10 mol% LAO	✓	✓		✓	✓			✓	✓	✓		✓		✓	✓	
	20 mol% LAO				✓	✓			✓	✓	✓		✓		✓	✓	
	30 mol% LAO	✓	✓		✓	✓			✓	✓	✓		✓		✓	✓	
LOS powder	100 mol% LOS				✓	✓		✓									
LMT powder	100 mol% LMT				✓	✓		✓									

3.2 Advanced breeder pebble samples

In the following, the available and newly fabricated breeder pebble samples are introduced which were characterized in this work. The pebbles were either produced by the KALOS process or the emulsion method as well as the sol-gel method. Standard grades of these pebbles (see section 3.2.1) were available for extensive characterization, which includes the analysis of the changes of their properties during simulated operation (see section 3.3).

Since the fabrication of the pebbles by the KALOS process involves the formation of a melt, compared to the classical powder based fabrication of the other processes, disparate microstructures of the pebbles result. It is therefore sensible to compare equal compositions of LOS-LMT mixtures. Furthermore, the powder based fabrication processes are not limited with respect to the composition as the KALOS process in its current state is. On account of this, the fabrication of LOS-LMT mixtures beyond the KALOS limit will give insight in the possibilities to reinforce LOS pebbles by LMT as a second phase or in other words how much increasing the lithium density by adding LOS might lower the mechanical strength of LMT pebbles.

As biphasic lithium orthosilicate based ceramic breeder pebbles are currently only considered with either lithium metasilicate or lithium metatitanate as second phase (see section 2.2.2.2), there might also be a potential to reinforce the pebbles by another mechanically strong phase. In the past lithium aluminate ($LiAlO_2$, LAO) was investigated as breeder ceramic and it showed remarkable strength compared to other breeder ceramics as detailed in section 2.2.3.1. Therefore, lithium aluminate is added as an alternative to lithium metatitanate as second phase, using the high versatility of the emulsion method.

3.2.1 Available pebble samples

The melt-based KALOS process is described in section 2.2.2.2. In total three batches of nominally 20 mol% lithium metatitanate and 30 mol% lithium metatitanate containing lithium orthosilicate pebbles were supplied for the experiments that were carried out within this work. It was necessary to produce a second batch of 20 mol% lithium metatitanate containing lithium orthosilicate pebbles to obtain the necessary amount of pebbles. Both batches (A & B, see Table 3.3) were mixed before the samples were used in any experiment. All relevant fabrication parameters for the produced batches of KALOS pebbles are given in Table 3.3. Since the KALOS process is an experimental process, the properties of these particular samples may deviate to a certain degree from batches that are produced later as the process is continuously being optimized [75].

From QST, two grades of lithium excessive lithium metatitanate pebbles were available. One of the grades was produced by the emulsion method while the other grade was produced by the sol-gel method (see section 2.2.3.2). In Table 3.4, the fabrication parameters of these pebbles are given as well as the pebble size, which varied just slightly. The composition was determined by QST using ICP-AES.

Table 3.3: Fabrication parameters of the different KALOS pebble samples.

Sample name	Nominal composition (bal. Li_4SiO_4)	Processing temperature (crucible)	Nozzle diameter	Processing gauge pressure
KA-20-LMT (batch A)	20 mol% Li_2TiO_3	1310 °C	300 µm	400 mbar
KA-20-LMT (batch B)	20 mol% Li_2TiO_3	1360 °C	300 µm	400 mbar
KA-30-LMT	30 mol% Li_2TiO_3	1450 °C	300 µm	400 mbar

Table 3.4: Fabrication parameters of the two lithium excessive lithium metatitanate grades supplied by QST.

Sample name	Composition	Sintering temperature	Sintering atmosphere	Sintering duration	Approximate pebble size
EM-LMT	$Li_{2.16}TiO_{3.08}$	1100 °C	He/1 % H_2	2 h	1250 µm
SG-LMT	$Li_{2.11}TiO_{3.055}$	800 °C 1080 °C	Vacuum Argon	1 h 5 h	1000 µm

3.2.2 Wet-chemical fabrication of mixed Li_4SiO_4/Li_2TiO_3 and $Li_4SiO_4/LiAlO_2$ pebbles

In addition to the supplied advanced breeder pebbles, the direct emulsion method (see section 2.2.3.2) was adapted and used to fabricate pebbles with a mixture of lithium orthosilicate and lithium metatitanate as already published in [96]. Also pebbles with additions of lithium aluminate were prepared in this way. Table 3.5 summarizes all grades of samples that were prepared as well as the amount of sintered pebbles that were obtained. These samples were fabricated during a research visit of QST, Rokkasho, Japan. Because of the limited time of this visit, the fabricated amount of samples is rather modest.

The raw materials which are introduced into the slurry are commercial lithium metatitanate, lithium carbonate, aluminum oxide and previously synthesized lithium metasilicate (see section 3.2.3). In contrast to the work of Hoshino [95] in which SiO_2 is used as raw material, lithium metasilicate is used instead as in this way less carbon dioxide is generated during the sintering and the thus generated porosity is minimal. Upon sintering, it is expected that the reactions

$$Li_2CO_3 + Li_2SiO_3 \xrightarrow{\Delta E} Li_4SiO_4 + CO_2 \tag{3.1}$$

$$Li_2CO_3 + Al_2O_3 \xrightarrow{\Delta E} 2LiAlO_2 + CO_2 \tag{3.2}$$

form the anticipated ternary lithium oxides. As binder agarose is used as it gels strongly at room temperature while it is easily dissolvable in water at temperatures above 50 °C. A solution of 1.48 % agarose in distilled water provides enough strength upon gelling to handle the produced gel spheres while premature gelation during the shaping process is no issue. From this solution the slurry for the fabrication is prepared by a ball milling step. 1200 g of zirconia balls of 5 mm size and 17.91 g of the raw material powders, which equals a solid content of 15 % in the final slurry, are put into a 500 ml polyethylene bottle together with the agarose solution.

Table 3.5: Fabrication parameters of the lithium orthosilicate based pebble samples that were produced by the adapted emulsion method.

Sample name	Nominal composition (bal. Li_4SiO_4)	Sintered amount of pebbles	Sintering temperature	Sintering atmosphere	Sintering duration
EM-10-LMT	10 mol% Li_2TiO_3	0.415 g	1000 °C	Air	5 h
EM-20-LMT	20 mol% Li_2TiO_3	0.639 g	1000 °C	Air	5 h
EM-30-LMT	30 mol% Li_2TiO_3	0.620 g	1000 °C	Air	5 h
EM-40-LMT	40 mol% Li_2TiO_3	0.765 g	1000 °C	Air	5 h
EM-50-LMT	50 mol% Li_2TiO_3	0.628 g	1000 °C	Air	5 h
EM-60-LMT	60 mol% Li_2TiO_3	0.740 g	1000 °C	Air	5 h
EM-70-LMT	70 mol% Li_2TiO_3	0.653 g	1000 °C	Air	5 h
EM-80-LMT	80 mol% Li_2TiO_3	0.724 g	1000 °C	Air	5 h
EM-90-LMT	90 mol% Li_2TiO_3	0.935 g	1000 °C	Air	5 h
EM-10-LAO	10 mol% $LiAlO_2$	0.637 g	1000 °C	Air	5 h
EM-20-LAO	20 mol% $LiAlO_2$	0.489 g	1000 °C	Air	5 h
EM-30-LAO	30 mol% $LiAlO_2$	0.582 g	1000 °C	Air	5 h

Yet, to compensate for lithium losses throughout the fabrication and sintering, the concentration of lithium carbonate was increased by a factor of 1.75 on its stoichiometric ratio, i.e. from 1:1 to 1.75:1 for Li_2CO_3:Li_2SiO_3 as well as for Li_2CO_3:Al_2O_3. The milling is then carried out for 20 hours in a heated environment with a rotating speed of 90 rpm. The ball milling step reduces the particle dimensions of the powders to a suitable size and additionally ensures a homogenous mixture of the various raw materials. After the ball milling the resultant slurry is drained from the bottle through a mesh and transferred to the fabrication setup.

All fabrication devices are heated to 60 °C to prevent premature gelling of the slurry. Independent of the slurry composition, the flow speed of the slurry is set to 80 ml/h and the cutting silicone oil is dispended with a speed of 200 ml/h into the T-junction by automated syringes. With the dispensing speeds, the segment or droplet size can be regulated. For pure lithium metatitanate pebbles, these parameters lead to 1000 µm pebbles after sintering. Since the sintering behavior of the chosen compositions was not known, the dispensing speeds were kept constant for all of them. The so-generated slurry segments are then gelled at 10 °C, washed with ethanol and isoparaffin and dried as described before (see section 2.2.3.2).

The dried pebbles are sintered in alumina boats at 1000 °C for 5 hours independent of the pebble composition, as the sintering behavior of each individual composition was unknown at that time. To reach the sintering temperature, the samples are heated at a rate of 100 K per hour. After sintering, the pebbles cool to room temperature naturally. During sintering the agarose binder is combusted and leaves the pebbles in gaseous form. It is also expected that the excessive lithium carbonate evaporates or reacts with the crucible material.

3.2.3 Synthesis of Li_4SiO_4, Li_2SiO_3 and Li_2TiO_3 powders

For the wet-chemical fabrication of breeder pebbles as well as for the characterization of the various substances, pure powders had to be prepared as the quality of the commercially available powders was insufficient for the purposes of this work.

3.2.3.1 Lithium metasilicate

Lithium metasilicate is prepared by a solid state synthesis. As raw materials, commercially available silicon dioxide powder and lithium carbonate are chosen, because lithium carbonate can withstand higher temperatures without melting compared to lithium hydroxide. To cope with the lithium losses during the synthesis, the stoichiometric ratio of lithium carbonate to silicon dioxide is increased by a factor of 1.65 from 1:1 to 1.65:1. For the synthesis, the powders are dispersed in ethanol and blended by a covered automated alumina mortar. During the blending, the particle size of the two powders might also be reduced, which is favorable for the subsequent synthesis as the reactivity of the powders is increased. The blending is carried out for 20 hours in the automated mortar until the mixture turns into a viscous mush and a significant amount of the ethanol is evaporated. The so-blended powders are then calcined in an alumina crucible at 1000 °C for 3 hours in air, with a heating rate of 100 K per hour. Upon calcining, the two powders react according to

$$Li_2CO_3 + SiO_2 \xrightarrow{\Delta E} Li_2SiO_3 + CO_2 \qquad (3.3)$$

to form lithium metasilicate. After the calcining, the porous body is ground to a fine powder in an alumina mortar.

3.2.3.2 Lithium orthosilicate & lithium metatitanate

Similarly to lithium metasilicate, lithium orthosilicate and lithium metatitanate can be synthesized with relative ease via solid state reactions. Lithium hydroxide monohydrate and silicon dioxide for lithium orthosilicate or titanium dioxide for lithium metatitanate are chosen as starting powders to form the anticipated phases according to

$$4LiOH \cdot H_2O + SiO_2 \overset{\Delta E}{\rightarrow} Li_4SiO_4 + 6H_2O \quad (3.4)$$

$$2LiOH \cdot H_2O + TiO_2 \overset{\Delta E}{\rightarrow} Li_2TiO_3 + 3H_2O. \quad (3.5)$$

Again, to compensate for possible lithium losses during the synthesis, the molar ratio of lithium hydroxide monohydrate to silicon dioxide is increased by a factor of 1.65, from 4:1 to 6.6:1 for the synthesis of lithium orthosilicate and from 2:1 to 3.3:1 for lithium metatitanate. Mixtures of the starting powders, according to these ratios are weighted in to a total of 15 g each and are dispersed in ethanol. The blending is carried out in a planetary ball mill in closed alumina beakers with 30 g of 10 mm alumina balls for 15 hours. During this time, the powders are not only blended, but the particle size is significantly reduced, which increases the reactivity of the powders upon calcination. Subsequent to the blending, the alumina balls are removed. As the closed alumina beakers do not allow for a significant evaporation of the added ethanol, compared to the synthesis in an automated mortar, the beakers are left open for 24 hours to remove most of the ethanol from the powder mixtures after the blending.

The so-prepared powder mixtures are pre-reacted in alumina crucibles at 430 °C for 3 hours, well below the melting point of lithium hydroxide at 450-471 °C as specified by its ICSC (International Chemical Safety Card). After cooling to room temperature, the final calcination is carried out at 800 °C for another 3 hours. The temperatures are ramped with 5 K/min upon heating and 10 K/min during cooling. After the calcination, the porous body is ground to a powder in an agate mortar.

3.3 Long-term annealing in relevant atmosphere

For the evaluation of the long-term stability of multiple samples of ceramic breeder pebbles, a dedicated experimental setup was assembled, which is able to continuously operate for at least five months with a constant gas purging of the samples and to allow for the acquisition of crucial process data in high frequency. These experiments are referred to as long-term annealing (LTA) experiments in the following.

3.3.1 Annealing set-up and data acquisition

The annealing was performed in a modified rotary furnace with three separate alumina tubes in which a controlled atmosphere is established during the experiments (see Figure 3.1). These tubes are connected to influent and effluent gas tubing by water-cooled flanges. The temperature of each alumina tube is monitored by an individual thermocouple attached to the outside of the tubes. The temperature-sensing tips are placed in the middle of the furnace where the hottest region is expected. On the influent and effluent gas tubing of each alumina tube a digital mass flow controller and a digital absolute pressure controller is installed, respectively. Between these controllers a defined pressurized atmosphere is established, independent of the environmental atmospheric pressure. Within this section of the tubing, four optical oxygen sensors are installed, as the oxygen measurements are sensitive to pressure changes. One of the sensors is placed on one of the influent lines and three are placed on the effluent side of each alumina tube.

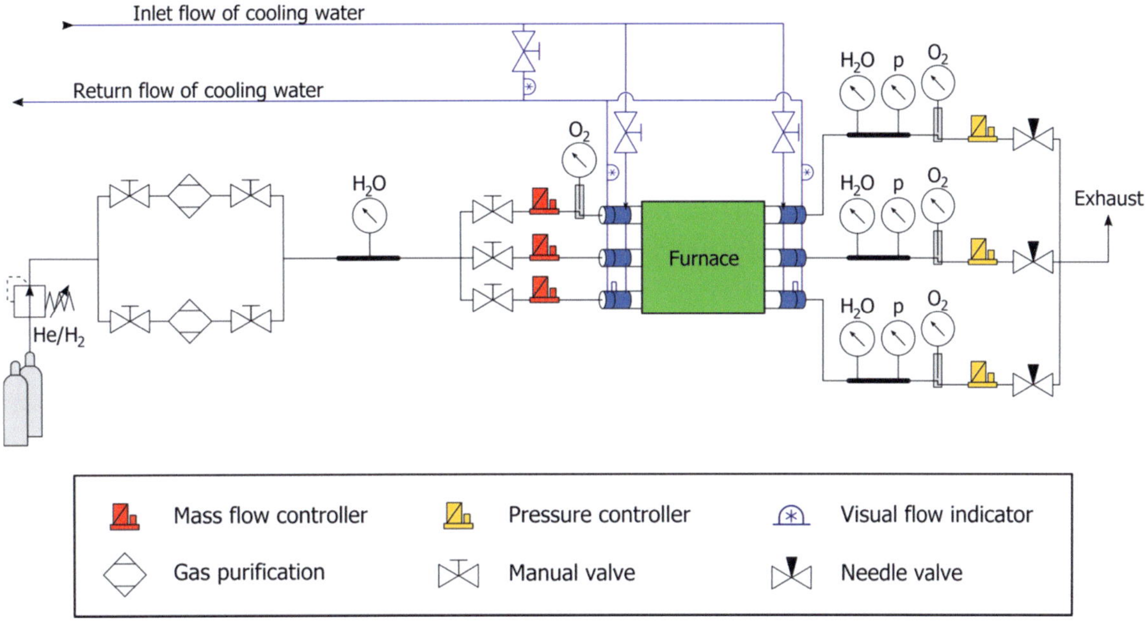

Figure 3.1: Schematic of the annealing setup. The annealing furnace is depicted in green.

To also monitor the dew point and the relative humidity of the purge gas, four alumina moisture probes are installed. The moisture sensors, as well as the oxygen sensors, monitor the temperature of the purge gas stream. Yet, only the probes that monitor the effluent gas streams are placed in the pressurized section for construction-conditioned reasons.

After the pressure controllers, a needle valve on each line is installed to dampen pressure oscillations, as light gases are otherwise difficult to regulate by the feedback control of the controllers. The influent helium-0.1 %-hydrogen gas mixture, simulating the purge gas flow in an ITER TBM, is supplied by premixed pressurized gas bottles. Potentially carried water is removed from the gas stream by molecular sieve beds. All sensors and controllers are connected to a PC-based data acquisition and control system running LabView 2013 software which was written specifically for this setup. The software is effectively subdivided into six blocks which are consecutively executed. A schematic illustration of the program sequence is given in Figure 3.2.

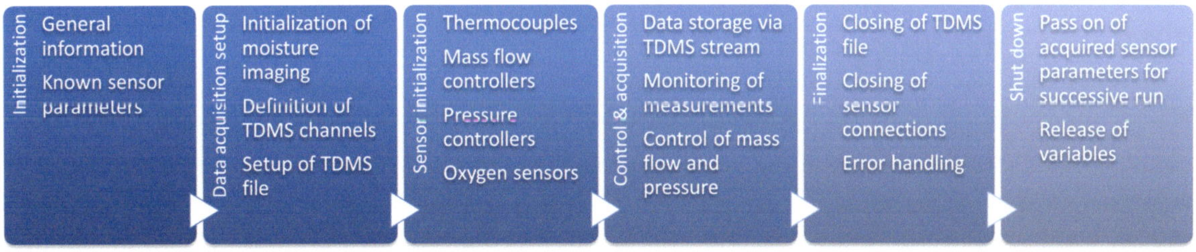

Figure 3.2: Schematic measurement software sequence.

The main part of the software comprehends the actual controlling and monitoring of the sensors and controllers as well as the storage of the data on the hard drive of the PC. The communications with each of the sensors/controllers as well as the data storage are decoupled from each other to prevent interferences between the subroutines and to query each sensor/controller according to its measurement speed. According to the respective sensor or controller update rate, all attached equipment is queried at least twice each minute by the respective data acquisition software subroutine. The data is stored in binary, hierarchical TDMS (Technical Data Management Streaming) files every ten seconds. To effectively handle the acquired amount of data, the data is processed after finishing the actual experiment to produce a reduced and resampled dataset of process values, in addition to the raw data. Yet, as this procedure involves the usage of considerable amounts of memory and calculation time for large data sets, it is not realized within the data acquisition and control software but as separate software, also written in LabView 2013. By this step, a common axis of time, which is identical for each measured quantity, is established and the data points are condensed to a 5-minute increment. In a first step, the data is filtered through a symmetric median-filter employing a rank of 7 to remove spike artifacts. The filtered output is then interpolated with respect to the new common axis of time, which starts with the first recorded measurement, using cubic hermite splines.

The signals of the moisture probes are processed further to take the measured temperature and pressure into account. From the measured dew point t_d the resultant vapor pressure e can be converted using the following equation [142]:

$$t_d = \left(243.12 \times ln\left(\frac{e}{611.2}\right)\right) \Big/ \left(17.62 - ln\left(\frac{e}{611.2}\right)\right). \tag{3.6}$$

By applying the approximate, integrated form of the *Clausius-Clapeyron equation*

$$ln\left(\frac{p_2}{p_1}\right) = \frac{\Delta H_{m,v}}{R} \times \left(\frac{1}{T_1} - \frac{1}{T_2}\right) \tag{3.7}$$

the vapor pressure of water in each alumina tube can be converted to a constant temperature of 25 °C (with $\Delta H_{m,v}$=43990 J/mol [143]). The conversion of the vapor pressure into volume parts per million by using the actual pressure within the system leads to a pressure independent value of the humidity. As the absolute atmospheric pressure at the moisture probe on the inlet is not known, a constant value of 1000 mbar is chosen for the calculations.

3.3.2 Annealing parameters and methodology

During the long-term annealing experiments, the pebbles are annealed at 900 °C, which is the maximum expected temperature within the breeder zone of the Japanese ITER TBM and just slightly lower than the maximum expected temperature of 920 °C within the breeder zone of the European Helium Cooled Pebble Bed ITER TBM (see also section 2.1) [51]. During the operation of the furnace, the alumina tubes are constantly purged with reference purge gas. The volume flow of the gas is adjusted individually for each alumina tube to 1200 ml/h.

The absolute pressure within each alumina tube is adjusted to 1200 mbar, which is slightly above the constantly changing environmental atmospheric pressure to prevent the intrusion of ambient atmosphere gases into the system. The annealing furnace is held at 300 °C during the introduction of the samples. To purge the introduced oxygen and water from the system, the samples are held at 300 °C within the purge gas stream for 24 hours. This procedure is also applied after each sampling step when the system is opened in the same way. For sampling, the annealing at 900 °C is interrupted and the temperature is lowered to 300 °C to avoid thermal shocks for the samples and for safely removing the samples from the furnace. Of course, the gas tight set-up must be opened for the sampling. Therefore the pressure within the alumina tubes is slowly reduced to ambient pressure within 120 minutes by shutting down the influent gas stream and opening the pressure controller valves, before opening the individual lines. The actual pressure reduction rate largely depends on the flow rate through the needle valves at the exhaust of the system. After extracting a sample from the furnace, the pebbles are immediately put in a desiccator to avoid reactions with the ambient atmosphere. The extracted pebbles are afterwards split into several representative sub-samples according to the different characterization techniques.

Only the periods at 900 °C are counted as annealing time. The heating and cooling rate prior to and after sampling is set to 5 K/min. However the cooling of the furnace, especially at lower temperatures, is significantly slower than the aspired cooling rate as the kiln is not actively cooled. The actual mean cooling rate is about 0.43 K/min. Prior to the insertion in the alumina tubes, the samples are dried in a vacuum furnace at 300 °C and 100 Pa for 1 hour. The pebble samples are held in special platinum-rhodium alloy boats and alumina boats (see Figure 3.3).

Figure 3.3: Photographs of the platinum boats (left) and the alumina boats (right).

The platinum boats are designed to minimize the usage of space while allowing for direct purging of the pebbles. Therefore, a platinum-iridium net forms the front- and backside of the platinum boats. For each sampling and sample type, a dedicated platinum or alumina boat is prepared with 15 grams of pebbles. As stated in section 3.2.1, two types of KALOS pebbles with discriminative compositions, namely 20 mol% lithium metatitanate (KA-20-LMT) and 30 mol% lithium metatitanate (KA-30-LMT), available for the LTA experiments, as well as two types of lithium excessive lithium metatitanate samples namely emulsion method derived pebbles with a lithium to titanium ratio of 2.16 (EM-LMT) and sol-gel derived pebbles with a lithium to titanium ratio of 2.11 (SG-LMT).

One part of the annealing campaign focuses on the annealing in helium-0.1 %-hydrogen purge gas atmosphere for a maximum of 128 days, with intermediate sampling steps after 4 days, 32 days and 64 days. As the mixture of helium with 0.1 % hydrogen is the designated purge gas for all solid breeder TBMs in ITER, it is carried out with the ambition to anneal the pebbles in conditions as relevant as possible with regard to the designated application. In order to evaluate the performance of all four pebble types in the same environment, a total of 16 samples are annealed in two separate tubes (see Table 3.6). There is a difference of orders of magnitude between the expected concentration of generated tritium and the hydrogen addition of 0.1 % in the purge gas [40,144] and therefore the largest fraction of the purge gas does not contain the bred tritium although all possible hydrogen isotopologues are expected to form.

Table 3.6: Annealing duration, sample placement and annealing atmosphere during the LTA experiments. The third tube of the furnace was not used.

	Samples	Annealing atmosphere	Annealing duration	Number of samples	Boat type
Tube 1	KA-20-LMT & KA-30-LMT	0.1 % H_2 in He	4 to 128 days	8	Platinum
Tube 2	EM-LMT & SG-LMT	0.1 % H_2 in He	4 to 128 days	8	Alumina

3.4 Mechanical testing of individual pebbles

Uniaxial crush load measurements, i.e. the mechanical loading of a pebble until it breaks, are a standard characterization technique for all pebble samples, since the breaking of the pebbles during their application in a fusion blanket is a major concern. The complex state of mechanical load on the pebbles within a pebble bed is not reproducible by such experiments. However, these experiments can be used as a basis for pebble bed simulations and they are valuable indicators for the pebble quality. The crush load of single pebbles is determined with a universal testing machine, UTS 10T, equipped with a 100 N load cell with a class of accuracy of 0.1. According to DIN EN 60051-1 this equals a possible error of ±0.1 N for all determined forces. For the measurements, the pebbles are individually placed between two sapphire plates by using vacuum tweezers. Subsequently the two sapphire plates are approached by a traverse displacement rate of 0.2 mm/min until a compressive preload of 0.5 N is reached. Obviously, no crush loads below 0.5 N can therefore be observed.

As the pebbles are too small to precisely measure their strain during mechanical loading with the available experimental setup, the determination of material properties is not possible. It is also very challenging to precisely track the strain of the pebbles during loading with the available experimental setup or to reliably calibrate the distance between the sapphire plates. Therefore the obtained data of traverse displacement were disregarded for any evaluation. It was stated in section 2.2.2.1 that lithium orthosilicate may take up a considerable amount of moisture or react with water in general. Therefore the mechanical behavior of the pebbles might change. In order to prevent a deterioration of the measurements, the pebbles are carefully dried in a vacuum furnace at 10 Pa and 150 °C, to remove any moisture from the pebbles prior to the testing. The drying furnace is heated at a rate of 0.8 K/min and cooled to room temperature at a rate of 5 K/min.

Of course the determined crush load is significantly dependent on the pebble size. Therefore the pebbles are screened to sizes of 500 µm and 1000 µm, except for all pebbles that were produced by the emulsion method. For precisely screening the pebbles, a representative subsample is sieved with meshes of the respective width in a first step. For the mechanical testing, only the pebbles that are caught between the threads of the meshes are used, which are gently removed from the mesh in a second step. The batches of pebbles produced by either the emulsion method or the sol-gel method (see sections 2.2.3.2 and 3.2.2) do not feature pebbles of 500 µm diameter as the pebble size distribution is very narrow compared to melt-based fabrication. Therefore no screening of the pebbles was performed before the crush load testing. These pebbles are approximately of 1000 µm size, or 1250 µm in the case of the emulsion method pebbles. For the pebble diameter of the multi-phase pebbles produced by emulsion method (see section 3.2.2) varies considerably and no common pebble size exists for all samples. Therefore, the mean pebble diameter is determined by optical means (see section 3.6) and no sieving is carried out for these pebbles.

The actual measurement of the crush load is started upon reaching the preload. The mechanical loading of the pebble is continued with a traverse displacement rate of 0.2 mm/min until a typical load drop for a yield point is detected by the measurement software (see Figure 3.4 and Figure 3.5). Yet, the traverse displacement rate is not indicative of the strain of the pebble as the measurement setup is also strained significantly and the experimental error of a calibration series is considerably larger than the expected displacements. The measured load at the so-determined yield point is constituted as the crush load of the specific pebble. Yet, this value does not necessarily constitute the point of ultimate pebble failure but it does so in most instances. In cases in which the pebble does not fail completely when a considerable load drop is observed, the so-determined load may be referred to as 'apparent crush load' for an 'apparent failure' (see Figure 3.4 and Figure 3.5). In order to reach the point of complete failure, the load is increased further to reach the maximum load that the pebble withstands. However, the two determined loads of failure may lie too close to each other to consider them as significantly different. Therefore, assuming a Hertzian contact, a 5 % difference in the mean contact pressure that results from both loads is defined as necessary difference to consider both failure loads as significantly different. From experience this point is not as well defined as it may appear, because pebbles can also fail 'quasi-continuously' into powder after the initial load drop and ultimately reach arbitrarily high values for this reason. Such measurements are routinely aborted as soon as a 'quasi-continuous' failure is obvious (see Figure 3.4 and Figure 3.5). Furthermore, any sort of load drop must be associated with a significant change of the pebble, which can lead to the establishing of non-Hertzian contacts that may require a significantly different theoretical treatment.

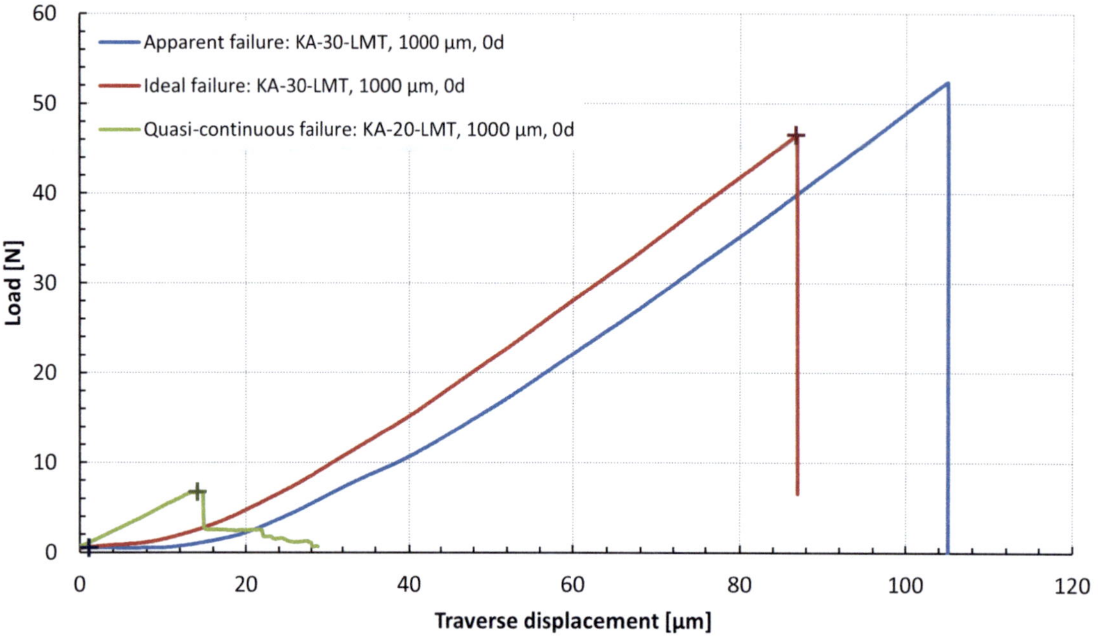

Figure 3.4: Examples of three experimentally obtained load-displacement curves for illustrating the three observed failure types. On each curve, a '+' indicates the actually determined failure load.

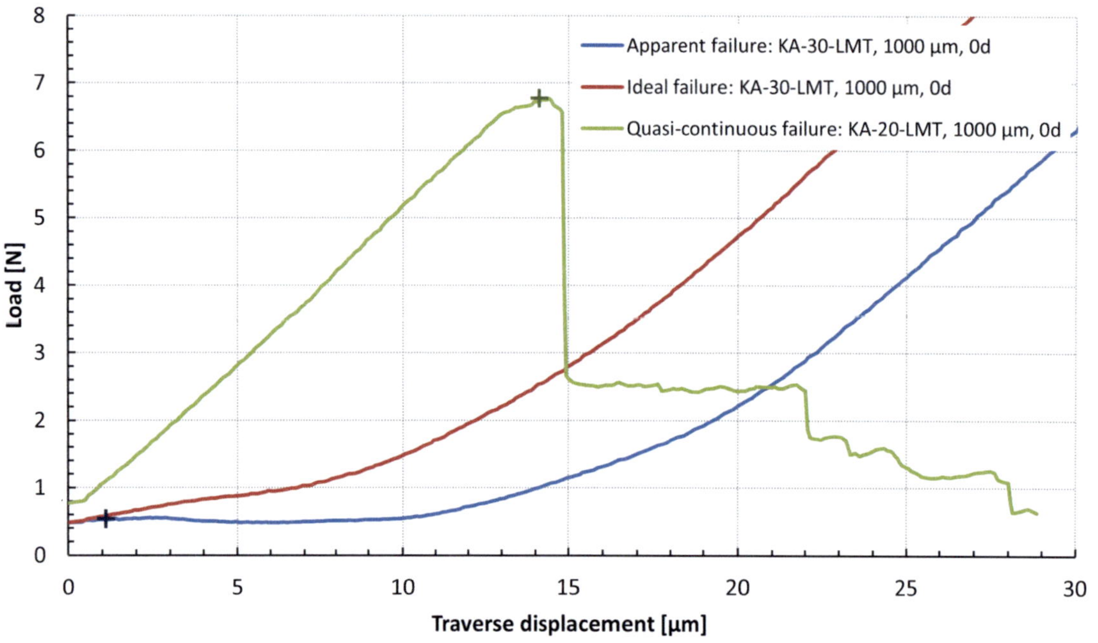

Figure 3.5: Examples of three experimentally obtained load-displacement curves for illustrating the three observed failure types, with special emphasis on the low load region. On each curve, a '+' indicates the actually determined failure load.

Consequently these values are not used by default, but are generally available. A rough discrimination of the case of 'quasi-continuous' failure from abrupt brittle failure can be made by evaluating whether the traverse displacement is unusually high. Yet, it is difficult to unambiguously discriminate between both types of non-ideal failure on the basis of the measurement data.

A total of at least 40 pebbles are analyzed in this way for each pebble size of each sample. The exact number of the tested pebbles is given in Table B.1 in Appendix B. To increase the insight into the pebble strength distribution and the way the mechanical performance of the pebbles changes after treating them in different environments, a Weibull analysis is performed for every sample of pebbles. To carry out such an analysis, the concepts described in section 2.4 are applied. The statistical programming language R (version 3.2.3, x64) [145] together with the 'Abrem'-packages [146] is used to automate the Weibull analysis for all samples. The 'Abrem'-packages are developed on the basis of 'The New Weibull Handbook' by Robert B. Abernethy [137]. Since these packages do not cover all concepts completely, the packages were significantly enhanced (see Appendix B). However, the packages serve as a well-established basis of R-code.

As explained in section 2.4, the 2-parameter Weibull function (2.17) is usually fully appropriate to model a given set of data, thus only this function is used. Additionally, consistency in the analysis is ensured if only one model is chosen. Abernethy [137] recommends using median rank regression, X on Y, for samples sizes up to 100 failures, since it is the most accurate method for such sample sizes. Yet, he also states that a reduced bias MLE is a legitimate alternative and as the complete likelihood function easily deals with censoring and suspensions, the MLE approach is chosen. Consequently, upper and lower likelihood ratio boundaries for a confidence interval of 90 % are calculated for every performed parameter estimation. To graphically evaluate whether two or more of the so-fitted datasets are statistically significantly different, contour plots are calculated in addition for the respective datasets.

As a direct comparison of the crush load values of the pebbles may be misleading because of their different size, the crush load values of all samples are converted into the mean Hertzian contact pressure of a perfect sphere according to equation (2.5). By this the mechanical strength of pebbles of different size can be compared, as all stresses within a sphere in diametral compression depend on it (see section 2.3.1). Yet, for calculating the contact pressure, the elastic constants of the various materials have to be known. These constants are only known for the pure materials as listed in Table 2.2, Table 2.4 and Table 2.6. Therefore, the constants of the two-phase materials are estimated by calculating the Hashin–Shtrikman [147] upper and lower boundaries for the bulk and the shear modulus of multiphase materials. These boundaries are narrower than the boundaries of the Voigt and Reuss model which consider the extreme cases of iso-strain and iso-stress for all phases, respectively [148].

With Hashin–Shtrikman boundaries, the upper and lower estimates of the Young's modulus and the Poisson's ratio can be calculated and the mean value of both boundaries represents the final estimate of the material, even though the boundaries on the Poisson's ratio may be relatively lavish for some cases as Zimmerman points out [149]. Hashin and Shtrikman assumed a homogenous and isotropic material in their model, which is also assumed during the conversion between the elastic constants. Yet, as the porosity is too high to disregard it for the samples, also the porosity has been incorporated into the Hashin-Shtrikman model so that effectively a three phase material is considered for the LOS/LMT and LOS/LAO compositions, like in the work of Wang and Pan [148]. Because the expected contact area is small compared to the pebble size and because there is some ambiguity about the absolute value of the open porosity of the samples (see section 3.6), only the determined closed porosity of a sample is considered for the performed estimation. The mechanical properties detailed in Table 2.2, Table 2.4 and Table 2.6 are given as a function of porosity. Yet, as the porosity is introduced as a third phase, it is assumed that the correlations given in these tables are also valid for zero porosity.

The elastic properties of the pores are in principle easily estimated with a Poisson's ratio of zero and a Young's modulus of almost zero. Yet, in cases where the bulk modulus and/or the shear modulus of a phase take very low values, the lower boundary of the Hashin-Shtrikman estimate approaches the Reuss boundary which is effectively zero in these cases and thus unrealistically low. Yan and Han [150] propose the use of a weighted average instead of the arithmetic mean for such cases, which requires at least one reliable (experimental) value for the determination of the weighting factor. Yet, for the porosity dependence of the Young's modulus a polynomic relation of third degree is proposed for both compounds, LOS and LMT. Therefore it is reasonable to assume that the porosity dependence of a composite material of both compounds can equally be described by a function of this kind. This requires a significantly higher lower boundary which is achieved by choosing a Young's modulus of (at least) 4 GPa for the pore phase. The resultant mechanical properties are detailed in Appendix B.

3.5 X-ray diffraction analysis

X-ray diffraction (XRD) analysis is carried out with a Bruker D8 Advance and copper Kα radiation emitted by a 1600 W sealed tube in parallel beam focus configuration, with the exception of the samples that were fabricated at QST (see section 3.2.2). For these samples, a Rigaku UltimaIV device is used which is equipped with a standard 0-D point detector. All measurements are performed in Bragg-Brentano geometry on powdered samples. A detailed list of operation parameters for the D8 Advance is given in Table 3.7. As specified, the instrument is equipped with a 1-D detector, which greatly increases the effective acquisition time of diffracted radiation, because it covers a segment of the angular range at once. However, in contrast to other detectors, a nickel filter has to be applied to subdue the copper Kβ fraction from the emitted X-rays, which reduces the overall intensity of the radiation to a certain extent.

Table 3.7: Chosen goniometer parameters for characterizing the samples.

Goniometer parameters	Bruker D8 Advance
Radiation	Cu Kα, $\lambda \approx 1.54$ Å
Filters	Ni-filter to suppress Cu Kβ radiation
X-ray tube operating voltage	40 kV
X-ray tube operating current	40 mA
Filament length	12 mm
Detector	LynxEye 1-D position sensitive detector
Angular coverage	2.94° (2θ)
Number of strip channels used	192
Detector upper / lower discriminator voltage	0.250 V / 0.110 V
Goniometer radii (tube / detector)	280 mm / 280 mm
Sample length in equatorial plane	25 mm (routine & quant.), 12 mm (temp. varied)
Receiving slit length	16 mm
Fixed divergence slit angle	0.3° (routine & temp. varied), 0.6° (quant.)
Primary Soller slit angle	2.5°
Secondary Soller slit angle	2.5°
Sample oscillation speed	Off (routine & temp. varied), 1080°/min (quant.)
Sample oscillation amplitude	180°
Counting time per step	0.5 s (routine & temp. varied), 5 s (quant.)
Measurement geometry	Bragg-Brentano

3.5.1 Room temperature measurements

Room temperature XRD measurements are routinely carried out for every pebble sample and for all synthesized powders. For such measurements, the pebbles or synthesized powders are ground by hand in an agate mortar to a fine powder before the analysis. The powder is then, depending on the available sample mass, either pressed into a grooved sample holder or spread out on a silicon single crystal after dispersing the powder in Canada balsam which is dissolved in xylene. All measurements were performed between 10° and 80° (2-theta) and at a constant counting time of 0.5 seconds for an increment of 0.02° in 2-theta. The room temperature measurements of the long-term annealing samples are taken between 2-theta angles of 10° and 140° by default and at an increment of 0.015° in 2-theta. To simultaneously achieve a reasonable total measurement time and a constantly low signal-to-noise ratio also for high 2-theta values, the counting time per increment is continuously increased from 0.1 seconds to 2.9 seconds in 6 steps.

For the quantitative analysis of selected samples, the measurements are performed between 14° and 100° (2-theta) as this range spans the most prominent and relevant reflexes of Li_4SiO_4 and Li_2TiO_3. The chosen sample rotation speed and counting time per step lead to an illumination of the complete powder sample within two 2-theta increments, which represents a reasonable compromise between good particle statistics and overall measurement time. To maximize the quality of the long-term annealing samples, about 1 gram of each sample is powdered by a swing mill (Retsch MM400) for 10 minutes at a frequency of 20 Hz. The milling is performed in 35 ml yttria-stabilized zirconia grinding beakers with a single 20 mm grinding ball of the same material, while the material is immersed in about 1 gram of anhydrous toluene. Through this milling procedure, the samples are fragmented into, relatively equally sized particles within a range of about 1 µm to 10 µm diameter (see Figure B.1 in Appendix B). The fragments together with the toluene form paste-like slurry, which is easily and densely spread into a grooved acryl sample holder. Before the XRD measurement is started, the toluene is evaporated at room temperature within about 10 minutes.

3.5.2 Temperature varied measurements

For temperature varied measurement, the powder was similarly ground by hand as for routine analyzes and pressed into a grooved platinum-rhodium sample holder of 12 mm length. To account for the smaller sample holder diameter, the temperature varied measurements were performed, as detailed in Table 3.7, with the divergence slit angle set to 0.3°. Sample spinning/oscillating was not used, because of the relatively fragile experimental setup. The temperature varied measurements are carried out with an Anton Paar HTK 1200N high-temperature oven chamber mounted on the Bruker D8. Although it is possible to change the atmosphere in the chamber, the measurements are carried out in ambient air. Prior to the measurement, a height calibration is carried out to ensure that the sample is well lightened by the X-ray beam and that during the heating of the whole setup, the thermal expansion of the setup does not drive the sample out of the X-ray beam. For the height or Z-calibration, the sample position perpendicular to the equatorial plane is adjusted.

The calibration is performed with a 2-theta angle of 0°, so that a direct X-ray beam hits the detector without diffraction at the sample. The sample is then driven through the beam while acquiring the counting rate. At the height at which half of the maximum counting rate is detected, the sample is shielding half of the X-ray beam from the detector. This height is chosen as initial Z position and is not deliberately adjusted throughout the measurements. However, as thermal expansion of the alumina support of the sample and the platinum-rhodium sample holder as well as the possible densification of the powder sample inevitably lead to a change of the Z position at increased temperatures, these changes have to be accounted for by a correction of the acquired data in a later step.

All measurements are taken from 10° to 80° (2-theta) with an increment of 0.02° and a counting time of 0.5 seconds per increment. During such a measurement, the temperature is kept constant. The temperature varied measurements are started and finished with the acquisition of a diffractogram (also referred to as 'diffraction histogram' with reasonable justification [151]) at room temperature. Between these measurements, 37 diffractograms are acquired from 100 °C to 1000 °C and back to 100 °C in

50 K increments. These measurements are performed with the Li_4SiO_4 and Li_2TiO_3 powders synthesized by a solid state method as described in section 3.2.3.

For the as-received KA-30-LMT samples, temperature varied XRD measurements were performed identical to those of the synthesized Li_4SiO_4 and Li_2TiO_3 powders regarding the measurement parameters and the sample preparation. Yet, in contrast to the measurement parameters given in Table 3.7, the counting time was doubled to 1 s per step. The measurements were performed between 750 °C and 800 °C by steps of 25 K and thereafter from 800 °C to 900 °C by steps of 10 K in ambient atmosphere. Also room temperature measurements were performed before and after finishing the temperature varied measurements.

3.5.3 Analysis of the acquired XRD data

The acquired XRD data are analyzed with two software tools. To qualitatively evaluate the present phases and to get an overview over the acquired data, the Bruker DIFFRAC.EVA v4.1 software in conjunction with the ICDD-PDF-2 Release 2010 database is used. For quantitative Rietveld analysis of the measurements, TOPAS v.5 is applied. The crystal structures which are used in the Rietveld analysis are taken from the ICSD database (data release 2015.2). All pebble samples annealed in the LTA experiments and the respective untreated samples are analyzed for their phase composition by Rietveld analysis. The lithium orthosilicate and lithium metatitanate powders synthesized by the solid state method (see section 3.2.3) are primarily analyzed for the changes of their lattice parameters as a function of temperature, however, also the phase content is analyzed in addition.

3.5.3.1 Models employed in Rietveld analysis

For a successful Rietveld analysis, an appropriate model has to be employed to be able to explain the obtained experimental data. Such a model can be built from an assembly of a number of models which focus on a specific detail. The models that are used in the Rietveld analysis will be presented in brief in the following.

Fundamental parameters approach

An in-depth Rietveld-analysis of selected XRD data is generally carried out by employing a fundamental parameters approach (FPA). The fundamental parameters approach utilizes physically based models to generate the measured line profile. Therefore three functions, describing the geometric instrument profile, the wavelength profile and the specimen (broadening) function, are convoluted together as described by Cheary, Coelho and Cline [152]. Each of these functions is again a convolution of several other functions, each reflecting a specific effect, for instance the finite X-ray source width or the flat specimen error.

To obtain the geometric instrument profile, the experimental setup of the Bruker D8 as given in Table 3.7 is fully modelled by the respective functions (see Appendix B). Therefore a so-called full axial model is used which takes the Soller slits and the sample geometry into account as well as the filament dimension and the receiving slit length of the detector. Also the fact that a one dimensional detector is used for the measurements which covers an angular range of 2.94° is taken into account. In addition to that, the sample may be displaced from the exact center of the goniometer. Such possible sample displacements in z-direction are considered and corrected in the refinements. Usually the sample displacement amounts to less than ±0.5 mm. As it is the most accurate available mathematical description, Berger's [153] Cu Kα wavelength profile, which consists of five components, is used in every refinement. Since there are no monochromators or mirrors used in the experimental setup, the X-rays can be considered to be non-polarized.

The specimen function consists of several contributions, which are only implemented when it is justified to do so. Yet, in every performed refinement, the background of the measurement is fitted with a Chebyshev polynomial with a sufficient number of coefficients to satisfyingly describe the background's shape. For each of the identified phases the lattice parameters and thus the peak positions are refined, as well as the scale parameter. As a basic principle, the peak shape is modelled by Gaussian and Lorentzian contributions from crystallite size broadening as well as Gaussian and Lorentzian contributions from intrinsic strain broadening also called microstrain broadening.

Preferred orientation

In addition to these refinements, it turned out to be necessary to consider the preferred orientation (PO) of crystallites for pure LMT samples. The simplest case of a preferred orientation is an ordering of the crystallites in one crystallographic direction. The effect on the diffractogram is described by Dollase [154] who utilizes the March function [155]. Only the March parameter r is refined in this approach with the preferred crystallographic plane being a previously known or guessed input parameter. Refining this parameter at the same time as the previously mentioned parameters does not lead to significant interferences. According to Zolotoyabko [156] the March parameter can be translated into the degree of preferred orientation

$$\eta = 100\% \times \sqrt{\frac{(1-r)^3}{1-r}}. \tag{3.8}$$

Modelling of stacking faults

β-LMT apparently develops stacking faults, which is obvious by comparison with stacking faults that develop in monoclinic Li_2MnO_3 [157]. Therefore a proper refining model has to take this effect into account. TOPAS v.5 can deal with stacking faults by a supercell approach, which extends the unit cell normal to the stacking fault planes by stacking a sufficient number of unit cells on top of each other. In all but the initial unit cell predefined layers may be displaced according to a stacking vector which is refined in the process. As stacking faults do not follow the symmetry elements of the space group of the initial structure, the supercell must not include these symmetry elements. Therefore the space group of the supercell is set to P1, the least restricted space group. As a result, the initial structure has to be translated into the space group P1 in advance. This is done conveniently by applying the

TRANSTRU application of the Bilbao Crystallographic Server [158–160]. As shown in section 2.2.3.1 the crystal structure of β-LMT consists simplistically of two TiO_6 layers and separating lithium layers. The separating lithium ion layers are not displaced during the refinement, while both TiO_6-layers are freely and independently displaced in [100] and [010]-direction. The implementation of this model into TOPAS macro code is shown in Appendix B.

The refinement is carried out by applying a simulated annealing algorithm, which is an iterative technique as described by Coelho [161]. During simulated annealing, the refined parameters are randomized after convergence. Subsequently the refinement is performed again with these altered parameters. The best fitting set of parameters, however, is stored. This process is carried out numerous times, until a global minimum in the difference between experiment and model is achieved. Without applying simulated annealing in this case, a satisfying adaption of the model to the experimental data cannot be achieved, since multiple local minima exist with the supercell approach. From the refined stacking vector, the stacking fault probability can be determined. For doing so, the number of actual stacking faults within the sequence of stacked unit cells is related to the number of displaced planes.

Modelling of anisotropic peak broadening

Lithium orthosilicate usually displays strong anisotropic peak broadening. There are numerous ways of modeling this issue. However there are hardly any physically based models available. One of the most common ways is to apply spherical harmonics on the Gaussian and the Lorentzian component of a pseudo-Voigt function. Yet, this approach bears no physical considerations and is therefore not used. Instead a phenomenological approach of Stephens [162] is used to model the anisotropic peak broadening of LOS. The model of Stephens was derived for the simplest case which is strain broadening. It is assumed that each grain is individually strained. For symmetry restrictions of the crystal systems, a maximum of 15 independent anisotropic strain parameters exist for a triclinic system which reduces to 9 for a monoclinic system. These parameters are introduced into a Gaussian as well as a Lorentzian function and an additional parameter is introduced which allows for an interpolation between both functions to generate a pseudo-Voigt function with the anisotropic contribution.

During the refinement, the anisotropic strain parameters as well as the interpolation parameter are refined. If the Stephens model is used in a Rietveld refinement, the usual models for isotropic grain size and strain do not provide an additional degree of freedom in the model as their contribution on the peak shape is refined to insignificance. Therefore, they can be removed from the model. As for the refinement with stacking faults, a simulated annealing of the anisotropic strain parameters is necessary to achieve a satisfying result. Even though this model was originally derived for anisotropic strain broadening, it can model anisotropic peak broadening in general satisfyingly well. In the present case, compared to a purely non-physical modelling of the anisotropic peak broadening, the model of Stephens achieves similar or better results.

3.5.3.2 Quantitative phase analysis

Generally, a Rietveld refinement which contains more than one crystalline phase will easily return the weight percentage of each phase with respect to all identified phases, once a reasonably good fit is achieved. No external information is needed for quantification as it is assumed that the content of all phases adds up to 100 %. This method is also referred to as Matrix Flushing Method [163,164]. It is clear that it is not always the case that the sample exclusively consists of the modelled phases. Especially when part of the sample is non-crystalline or when a phase does not yield sharp diffraction peaks.

Determination of non-crystalline fraction

As basically all obtained diffractograms in this study show broad features that are most probably related to an amorphous or seemingly amorphous feature within the sample, it is necessary to quantify its amount. Ideally the use of an internal standard or spiking phase would be the method to use for these cases. However, the many diffraction peaks that are generated by LOS and LMT lead to a significant peak overlap and therefore to reduced precision in the Rietveld analysis. Also microabsorption, which is the effect that the intensity of strongly X-ray absorbing phases is underestimated relative to weakly X-ray absorbing phases, is another issue which complicates the use of an internal standard.

Therefore, the so-called Degree of Crystallinity (DOC) method is used as described by Riello [165] as well as Madsen et al. [166]. The method directly estimates the amorphous content by designating one or more phases as amorphous and relating the area of all crystalline phases $A_{crystalline}$ to the sum of the area of all crystalline and amorphous ($A_{amorphous}$) phases:

$$DOC = \frac{A_{crystalline}}{A_{crystalline} + A_{amorphous}}. \tag{3.9}$$

The weight percentage of the amorphous content W_{amorph} is then given by

$$W_{amorph} = 1 - DOC. \tag{3.10}$$

Consequently, the weight fractions of the crystalline phases that were determined by the Matrix Flushing Method have to be multiplied by the DOC to reach the actual value. For the modelling of the diffractogram, the 'amorphous phases' do not necessarily have to have a defined crystal structure as basis, but an arbitrary one that describes the measured intensities, as only the determined area is used for the evaluation of the DOC. Therefore, the usually broad peaks that represent the amorphous contribution were modelled with crystal structure independent functions. For all samples for which the Degree of Crystallinity was determined, two sets of 5 pseudo-Voigt functions were used. For each set of the functions, the shape of all 5 functions was refined, while the intensity of each of the 5 functions was refined individually.

Yet, there are also contributions to the measurement background which do not originate from the amorphous phases of the sample. These contributions mostly originate from the used measurement setup, for instance the scattering of the X-ray beam in air, and can be assumed to be constant.

To remove these effects from the incorporation into the determination of the amorphous area, a standard substance (annealed Y_2O_3) was measured with the same measurement program as the samples of interest (see Appendix B). As this substance can be considered to be fully crystalline, the observed background in this measurement was assumed to solely originate from the measurement setup. It was modelled with a Chebyshev polynomial of reasonable order and a 1 over x function.

The so-determined background was kept constant for all refinements. However, as the sample preparation is never completely identical, a 2θ-independent shift in the intensity has to be taken into account. Therefore one parameter was introduced and refined which compensates for these shifts. The translation of this procedure into TOPAS code is shown in Appendix B. Furthermore, it has to be noted that titanium considerably absorbs Cu-Kα radiation [167] which is translated into fluorescence radiation. This radiation also contributes to the background signal and reduces the signal-to-noise ratio. The implementation of a 2θ-independent shift also compensates for this effect to a certain degree.

Refinement strategy for the lithium excessive pebble samples

The general refinement strategy for the samples that were annealed up to 128 days is illustrated in Figure 3.6. The analysis is started with the lithium excessive pebbles fabricated with the sol-gel method and the emulsion method. For these samples, the stacking faults in the β-LMT structure are modelled using the supercell approach detailed in the preceding section (3.5.3.1). The lattice parameters of the β-LMT structure, the preferred orientation of (001)-planes (also see section 3.5.3.1) and Gaussian as well as Lorentzian contributions to the peak shape via grain size and microstrain are principally refined. As a matter of course, also the scale parameter of the β-LMT phase is refined.

However, as a simultaneous modelling of the amorphous area would lead to considerable interferences, this step was performed after determining the background of the measurement by an unrestricted, individual Chebyshev polynomial which is kept constant during the refinement of the stacking fault structure. After the structure is determined to a satisfying degree, it is kept constant and the initial background polynomial is replaced by the standard-derived background functions and the shift parameter is refined. Afterwards, the amorphous area is determined as explained above. To reach a global minimum of the refinement, also here a simulated annealing algorithm is used. In a final step, the shift parameter is refined together with the amorphous area.

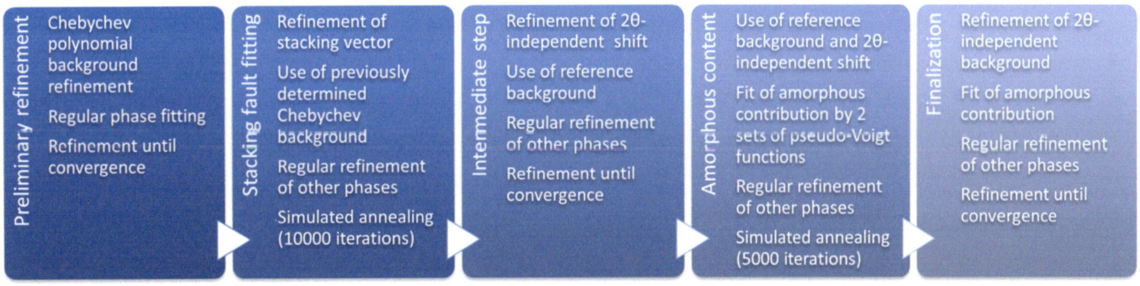

Figure 3.6: Illustration of the refinement strategy for the lithium excessive samples.

Refinement strategy for the LOS containing pebble samples

For the LOS containing samples of this series, the refinement strategy is in principle similar to the approach that is detailed in Figure 3.6. However, as there is significant peak overlap of LOS and β-LMT peaks in the range 20° to 25° 2θ, it is not sensible to refine the stacking fault structure of the β-LMT phase as the refinement cannot separate the effects unambiguously. Therefore, for all samples that show β-LMT in the diffractograms, a previously determined stacking fault structure of β-LMT is used. The structure is taken from the refinement of the emulsion method pebble samples that were annealed in the same way as the respective LOS containing ones produced by the KALOS process. Yet, as the β-LMT grains are not necessarily oriented along a certain direction, the March parameter is always refined. Self-evidently, the scale parameters of LOS and LMT are always refined as well as the lattice parameters. Also the contributions of the grain size and microstrain to the peak shape of the LMT phase peaks are always refined. These contributions are also refined for the LOS phase until the anisotropic peak broadening is addressed with the Stephens model in later refinement steps.

In a first step of the refinement, the constant shift of the predefined background is determined. As the applied milling step (see section 3.5.1) leads to the formation of a small fraction of lithium carbonate (LCO) for the LOS containing samples, which the comparison with hand-ground samples clearly illustrates (see Figure B.3 in Appendix B), also LCO has to be included in the refinements. In the second step solely the lattice parameters of LCO are refined together with its scale parameter, which is always refined afterwards. In another step, the peak shape contributions originating from the grain size are refined in addition. This careful refinement of the lithium carbonate phase is necessary to prevent a modelling of the amorphous or background intensities by this phase as the detectable amount of lithium carbonate is relatively small (usually 2 to 5 wt.%). For the same reason, neither the lattice parameters nor the grain size are refined in the following steps. In these steps, the Stephens model is applied to model the anisotropic peak broadening of the LOS phase using the simulated annealing approach. After a satisfying modelling of the LOS phase is achieved, the parameters of the model are kept constant and the amorphous content is refined as previously described. In a final step, the parameters of the Stephens model, the parameters of the Pseudo-Voigt functions as well as the lattice parameters and grain size parameters of the lithium carbonate are refined simultaneously together with the constant shift of the background to equilibrate the refinement.

Analysis of the biphasic emulsion method pebbles

For the biphasic pebbles that were fabricated by the emulsion method no well described XRD device was available. Therefore, a less physical model was used for the quantitative analysis. To model the peak shape of all identified phases, the modified Thompson-Cox-Hastings pseudo-Voigt function was used [168] instead of the FPA approach. Therefore no refinement of the crystallite size and microstrain was performed. Yet, the same wavelength profile was used as for all other models. Also the measurement background was addressed in the same way.

The analysis was mainly directed on the determination of the lattice parameters of the identified phases. However, as for the other LMT containing pebbles, a significant formation of stacking faults within the β-LMT phase was observed. As the EM-90-LMT sample shows the strongest intensity from this phase, it was chosen as a 'standard' for all EM-X-LMT samples and the stacking faults were modelled just for it. For all other samples, the so-derived model of the β-LMT phase was used and left unchanged. Identical to the KA-X-LMT, EM-LMT and SG-LMT samples also a possible preferred orientation of (001)-planes was considered for the β-LMT phase for all samples. Since the diffracted intensity of the β-LMT phase is significantly stronger than that of the LOS phase, the LOS signal almost vanishes for the EM-90-LMT sample. It was therefore not possible to obtain reasonably accurate lattice parameters for the LOS phase for this sample. It is however assumed that the LOS lattice parameters that are obtained for the EM-80-LMT sample are a reasonable estimate of those of the EM-90-LMT sample. For the LAO containing samples no special models were needed to reproduce the diffractograms satisfyingly.

3.5.3.3 Thermal expansion analysis

As in the previous section, the Rietveld method is used to fit the diffractograms, obtained at various temperatures, with an appropriate model system. In this case, a simulated annealing approach was unnecessary to achieve satisfying fits. The model also employs the fundamental parameters approach as described in section 3.5.3.1, taking into account the measurement conditions and setup, as well as impurity phases in the samples. As mentioned in section 3.5.2 the sample is displaced by the thermal expansion of the alumina support at elevated temperatures. Therefore this issue is addressed during the refinements. The focus of the analysis essentially lies on the determination of the lattice parameters of the phases of interest. For this purpose the LeBail method would also be applicable, however, as pure samples are hard to obtain, overlapping peaks of impurity phases might be an issue which could disturb the fitting quality, thus this method is not chosen. Still, a complete modelling of the phases with possible preferred orientation, stacking faults and anisotropic peak broadening is not performed as this would require immense computational time without affecting the peak positions notably. To describe the thermal expansion of the lattice of the lithium orthosilicate or lithium metatitanate phase, respectively, each independent lattice parameter is treated separately. The exponential equation

$$x = x_{RT} \times \left(1 + \vartheta\left(A_x + B_x\vartheta - C_x\, e^{-D_x\vartheta}\right)\right) \tag{3.11}$$

with four fitting parameters A_x, B_x, C_x and D_x is oftentimes used for describing the temperature dependence of a given lattice parameter x with respect to its room temperature value x_{RT}, with the temperature ϑ given in degree centigrade [169]. Therefore this equation is also used for the two phases Li_4SiO_4 and β-Li_2TiO_3. As noted in section 2.2.2.1, a phase transformation takes place in Li_4SiO_4 between about 665 °C and 725 °C. Therefore, a set of low-temperature (≤ 700 °C) and high-temperature (> 700 °C) fitting functions is used for Li_4SiO_4. The respective linear thermal expansion coefficient

$$\alpha_x = \frac{1}{x}\frac{dx}{d\vartheta} = A_x + 2B_x\vartheta + (C_x D_x\vartheta - C_x)e^{-D_x\vartheta} \tag{3.12}$$

is given as the derivative of the description in equation (3.11) with respect to the temperature normalized by the instantaneous dimension x. For each diffractogram, the furnace temperature, which is measured just below the platinum sample holder, is used to calculate the lattice parameters. As in calibration experiments (see Appendix B) with pure silicon and aluminum oxide whose thermal expansion is well described in literature [169,170], marginal discrepancies between the anticipated and actual sample temperatures were determined, it is legitimate to rely on the thermocouple of the furnace. To improve the quality of the determination of the fitting parameters, all 37 diffractograms of a sample are fitted simultaneously, while directly refining the fitting parameters A, B, C and D instead of the individual lattice parameters. By this measure, an unequal weighing of the results is achieved which effectively depreciates outliers.

Yet, all other parameters, such as the sample height displacement, the Gaussian and Lorentzian fraction of grain size and strain dependent peak broadening, as well as the lattice parameters of impurity phases are refined individually for every diffractogram. Also the background is fitted individually for each diffractogram applying a Chebyshev polynomial of the first kind of 3rd order. As opposed to the main phases, the strain broadening of the peaks of the impurity phases is not considered in order to limit the number of parameters and to reduce the number of correlating parameters. Also the strain broadening of impurity phases rarely contributes significantly to the quality of the fit.

For the analysis of the temperature varied measurements of the KA-30-LMT sample, a similar but simplified approach as presented in section 3.5.3.2 is used. The simplifications concern the removal of the refinement of the stacking faults in lithium metatitanate as well as the amorphous content. Also no impurity phases have to be considered. Furthermore, it is necessary to simultaneously consider both modifications of lithium metatitanate during the refinements. In contrast to the analysis of the synthesized powders just the individual lattice parameters and scale factors are obtained and no fit of the complete thermal expansion is attempted.

3.6 Other characterization techniques and equipment

The density of the samples is determined by helium pycnometry (Porotec Pycnomatic ATC), thus their closed porosity can be deduced. The theoretical density of the samples is calculated from their crystal structure (see section 2.2) and according to their by XRF (X-ray fluorescence spectroscopy) and ICP-OES (inductively coupled plasma optical emission spectroscopy) determined composition. For samples EM-LMT and SG-LMT, the lithium excess is neglected for these calculations. The theoretical density of multiphase phase pebbles, ρ_{mix} is calculated according to

$$\rho_{mix} = \sum_{i=1}^{n} m_i \Big/ \sum_{i=1}^{n} \left(\frac{m_i}{\rho_i}\right) \tag{3.13}$$

from the molar masses of the composing phases, m_i, and their densities, ρ_i. By mercury porosimetry (Thermo Electron Corp. Pascal 440) the open porosity of the pebbles is measured as well as the accessible pore size distribution assuming cylindrical pores. By default, cross sections and the surfaces of the pebbles are investigated by scanning electron microscopy, SEM, (Zeiss SUPRA 55). The cross sections of the 20-LMT and 30-LMT samples are etched by pure water, while it is not necessary to etch the cross sections of the other samples. The cross section samples were prepared by embedding a subsample of whole pebbles in epoxy resin in vacuum with subsequent grinding and polishing steps using water-free ethanol instead of water. Also fragments of selected pebbles, for which the failure load is known, are imaged by SEM (Zeiss SUPRA 55). Since all of these samples are electrical insulators, a thin layer of gold-palladium is sputtered onto the surface of the samples to be able to investigate the samples without disturbing electrical charges. For determining the grain size of the lithium excessive lithium metatitanate pebbles with reasonable accuracy, a selection of cross section images was investigated by the linear intercept method as, for instance, described by Macherauch and Zoch [171]. During the analysis, about 100 to 700 grains were determined for each sample and obvious pores were excluded. The accuracy of the analysis relies significantly on the proper determination of the number of grains that are intercepted by one line and thus the degree of ambiguity is estimated to be low. Together with the high accuracy of the length measurements, an uncertainty of the mean grain size of ±1 % is estimated.

In contrast to the comparably broad pebble size distribution of the pebbles fabricated by the KALOS process (see Figure 2.6), the emulsion method allows for a well-controlled pebble size. Still, the emulsion method is also prone to experimental deviations, which can lead to differences in the obtained dried pebble diameter. For instance coagulations within the gelation bath are possible, but also systematic changes because of the slurry composition are a possibility. Because of the low available sample amount it was not possible to select pebbles of a certain size for the crush load experiments. Therefore, the pebble size distributions of all biphasic pebbles that were fabricated by the emulsion method are determined. By measuring the size distributions of the dried pebbles as well as the sintered pebbles, the shrinkage during sintering can also be determined. For pebble size analysis, a number of optical micrographs of a monolayer of pebbles are taken with a Leica M205C microscope. Of course, such micrographs represent 2-dimensional projections of the pebbles only. The pebble diameter is then obtained by image analysis applying the Leica QWin Suite (version 3.1.0) software. In total the size of about 2000 individual pebbles is measured in this way. Technically, a resolution of 5.5 μm/pixel is achieved for the used magnification and thus an uncertainty of ±11 μm is estimated for the measurement of a single pebble.

To determine a measure for the green density of the dried pebbles (see section 3.2.2) as well as the pebble density of the sintered samples geometrically, the optical size measurements are also applicable. After measuring the size of a reasonable number of pebbles (less than 100) via image analysis, all analyzed pebbles of a given composition were weighted together. In case of the dried pebbles, the obtained mass is corrected for the amount of binder and excessive lithium carbonate as their weight fractions are known. The amount of possibly present residual water cannot be estimated. Yet, as the pebbles are carefully dried it is assumed that the water content is insignificantly low and can thus be neglected. From the diameter of each measured pebble, the volume is estimated by approximating a pebble as perfect sphere. The estimated total volume of the subsample is obtained by summating the ideal sphere volume of all pebbles. However, the so-determined volume is most likely overestimated as any macroscopic deviation from perfect sphericity is not compensated for. As a result, the so-obtained density value is decreased with respect to the true density value by an unknown amount of negative 'imaginary density'. The so-obtained density is the mean value of the analyzed pebbles. For the green pebbles, their density is then, nonetheless, given as a fraction of the density value that is obtained by the measurements, i.e. the sum of the green density and the imaginary density, to theoretical density of the used raw materials (see section 3.2.2).

For the sintered pebbles and the KALOS pebbles, the optical size measurements are used to determine the total porosity of the samples ϕ_{total}, which can be calculated from the theoretical density ρ_{th} and the actual pebble density ρ_{exp} as

$$\phi_{total} = \frac{\rho_{th} - \rho_{exp}}{\rho_{th}}. \tag{3.14}$$

Naturally, the total porosity of a sample is comprised of the open porosity, which is accessible by the environmental atmosphere, and the closed porosity, which is not. Or, referring to equation (3.14), the equivalent mass to the missing volume of an open pore m_{op} or closed pore m_{cp} is implicitly subtracted from the ideal sample mass m_{ideal} of a perfect sphere with volume V_{ideal} that is free of pores. Yet, as for the determination of the green density, any macroscopic deviation from perfect sphericity of the pebbles is not compensated for and thus also the equivalent mass $m_{nonideal}$ for these deviations has to be subtracted and leads to

$$\phi_{total} = \frac{\rho_{th} - \rho_{exp}}{\rho_{th}} = \frac{\rho_{th} - \frac{m_{sample}}{V_{sample}}}{\rho_{th}} = \frac{\rho_{th} - \left(\frac{m_{ideal} - m_{op} - m_{cp} - m_{nonideal}}{V_{ideal}}\right)}{\rho_{th}}. \tag{3.15}$$

Alternatively, equation (3.15) can be written in terms of porosity with ϕ_{op}, ϕ_{cp}, $\phi_{nonideal}$ as partial porosities for open pores, closed pores and deviations from perfect sphericity, respectively, as

$$\phi_{total} = \frac{\frac{m_{ideal}}{V_{ideal}} - \left(\frac{m_{ideal} - m_{op} - m_{cp} - m_{nonideal}}{V_{ideal}}\right)}{\rho_{th}} = \phi_{op} + \phi_{cp} + \phi_{nonideal}. \tag{3.16}$$

Since any significant deviations from ideal spherical shape that cannot be reasonably regarded as open pores, are nevertheless (misleadingly) determined by this method as a form of porosity, this quantity ($\phi_{nonideal}$) will be referred to as 'imaginary porosity'.

When the amount of closed porosity ϕ_{cp} is known from helium pycnometry measurements, it can be subtracted from the optically determined total porosity value to yield the sum of the open porosity ϕ_{op} and the imaginary porosity $\phi_{nonideal}$. For all determined density values, a total measurement error of ±5 % is assumed while neglecting the deviations from ideally spherical shape. This method's convenience is that it is not restricted by the available sample amount in contrast to the mercury porosimetry and that experimental errors are very low, while they occur more easily for mercury porosimetry. For this reason, the optical size measurements were needed to determine the total density of the biphasic pebbles that were fabricated by the emulsion method. In order to compare the determined values by both methods, the optical measurements were performed for all pebble samples.

The chemical composition is determined by XRF and ICP-OES focusing on the elements silicon, aluminum and titanium. Furthermore, the lithium content was determined by ICP-OES as well as the following impurity elements: sodium, aluminum and rhodium, iridium, platinum and gold. All ICP-OES and XRF measurements as well as the corresponding analysis was carried out by the 'Chemical analysis' group at IAM-AWP at KIT. Each sample was measured by XRF at least two times after a fluxing digestion. For the ICP-OES measurements, the samples were prepared by an acid digestion within an ultrasonic bath, before measuring each one at least three times. The oxygen content of the samples is determined indirectly by subtracting the detected masses of lithium, silicon and titanium from the original sample mass and assigning it to oxygen. The detection limits are calculated individually for the KA-20-LMT and KA-30-LMT, the EM-LMT and SG-LMT as well as the biphasic EM samples as the preparation for the measurements is not identical. For the EM/SG-LMT samples, magnesium was also included in the analysis as some of the samples showed a significant concentration of this impurity.

4 Results

In the following, the obtained results of the analyzing methods that are introduced in the experimental part are shown and explained. At first the elemental composition of the samples is presented and subsequently the results of the X-ray diffraction analysis are detailed. These sections are followed by the investigation of the pebble morphology and the pebble porosity. The final section of will show the results of the mechanical analysis of the samples.

4.1 Elemental analysis

In the following, the elemental analysis results are shown with the focus on the main constituents of the samples. From the element concentrations of silicon, titanium and aluminum, the virtual phase content of the main phases Li_4SiO_4, Li_2TiO_3, and $LiAlO_2$ are calculated assuming perfect stoichiometry. The measured lithium concentration is related to the content of silicon, titanium and aluminum as a simple ratio (also called 'lithium ratio'). This ratio is especially useful when non-stoichiometric lithium metatitanate is concerned (see Figure 2.10), i.e. for the EM-LMT and SG-LMT samples. For all other samples, the difference between the measured lithium ratio and the expected ratio, on the basis of the measured silicon, titanium and aluminum concentrations (again assuming perfect stoichiometry), is a good indicator for a shortage or surplus of lithium. For all given values, an individual error of ±1 % is a conservative estimate for the measurements. The standard deviations of the individual measurements are generally far below 1 % and rarely exceed 0.5 %. The complete data of the RFA and ICP measurements are given in Appendix C.

Table 4.1 shows the so-derived virtual phase compositions of the nominally 20 mol% LMT (and thus nominally 80 mol% LOS) containing KALOS samples for different annealing (see also section 3.3.2). The values can be considered to be constant as a function of annealing duration and the determined concentration of LOS is just about 1 mol% higher than anticipated. The lithium concentration of these samples is also slightly higher than the concentrations of silicon and titanium suggest if only both compounds, LOS and LMT, are present in the samples. The surplus seemingly diminishes with increasing annealing duration, but it is still present after 128 days. As most prevalent impurity, the analysis determined aluminum with a relatively constant value for these samples of 262±2 ppm in mean for all five samples. Platinum and rhodium impurities were determined to be significantly less for these samples with 4.5±0.1 ppm and 6.6±0.2 ppm in mean, respectively.

Lithium orthosilicate pebbles, fabricated by the KALOS process, that feature nominally 30 mol% of lithium metatitanate and which were annealed in the same way as stated above, show about 2 mol% less lithium orthosilicate than anticipated (see Table 4.2). In contrast to the nominally 20 mol% LMT in LOS pebbles, the 30 mol% LMT in LOS pebbles show a slight trend of an increase of the lithium metatitanate content with longer annealing durations.

Table 4.1: Resultant virtual phase contents, as-determined by the elemental analysis, of the nominally 20 mol% LMT containing KALOS samples that were annealed for up to 128 days. Additionally, the lithium ratios are given.

KA-20-LMT	0 d	4 d	32 d	64 d	128 d
Virtual Li_4SiO_4 content [mol%]	81.1 ± 0.8	81.7 ± 0.8	81.1 ± 0.8	81.4 ± 0.8	81.2 ± 0.8
Virtual Li_2TiO_3 content [mol%]	19.0 ± 0.2	18.4 ± 0.2	18.91 ± 0.2	18.6 ± 0.2	18.8 ± 0.2
Lithium ratio	3.815	3.796	3.753	3.746	3.719
Lithium ratio difference (measured minus expected)	0.193	0.163	0.131	0.118	0.096

Yet, only the sample that was annealed for 128 days shows a significantly different lithium metatitanate content, with respect to the estimated error of the measurements. The lithium ratio difference shows a slight deficit of lithium for these samples. It is most prominent for the sample that was annealed for the longest time. Yet, in absolute terms, the deficit is reasonably small. Similar to the KA-20-LMT pebbles, aluminum was determined to be by far the most abundant impurity with 225±6 ppm in mean, and platinum and rhodium showed a rather similar amount compared to the LMT-poorer pebbles with 4.3±0.4 ppm and 5.9±0.1 ppm in mean, respectively.

Table 4.2: Resultant virtual phase contents, as-determined by the elemental analysis, of the nominally 30 mol% LMT containing KALOS samples that were annealed for up to 128 days. Additionally, the lithium ratios are given.

KA-30-LMT	0 d	4 d	32 d	64 d	128 d
Virtual Li_4SiO_4 content [mol%]	68.3 ± 0.7	67.8 ± 0.7	67.9 ± 0.7	68.1 ± 0.7	66. 9 ± 0.7
Virtual Li_2TiO_3 content [mol%]	31.7 ± 0.3	32.2 ± 0.3	32.1 ± 0.3	31.9 ± 0.3	33.1 ± 0.3
Lithium ratio	3.323	3.337	3.318	3.309	3.231
Lithium ratio difference (measured minus expected)	-0.043	-0.019	-0.040	-0.052	-0.107

For both lithium excessive samples (EM-LMT and SG-LMT), an initial surplus of lithium is observed from the ICP measurements (see Table 4.3 and Table 4.4). This surplus reduces continuously with prolonged annealing in He/H_2 atmosphere. A linear relationship seems to reasonably represent the observed decreases. For the samples fabricated by the emulsion method, the initial surplus of lithium is considerably lower than for the pebbles produced by the sol-gel method. As the rate of the decrease is similar for both pebble samples, this turns into an overall deficit of lithium for the emulsion method pebbles after 64 days of annealing. For the SG-LMT pebbles, a substantial amount of impurity elements was detected. Magnesium, aluminium and sodium are the three most prevalent impurity elements with an average content of about 500 ppm, 350 ppm and 140 ppm, respectively. The amount of impurities for the EM-LMT samples is considerably lower with a sodium content of about 70 ppm and an aluminium content of about 20 ppm while magnesium was not detectable.

Table 4.3: Resultant lithium ratios, as-determined by the elemental analysis, of the lithium excessive samples, fabricated by the emulsion method, that were annealed for up to 128 days.

EM-LMT	0 d	4 d	32 d	64 d	128 d
Lithium ratio	2.090	2.030	2.014	1.991	1.966
Lithium ratio difference (measured minus expected)	0.090	0.030	0.014	-0.009	-0.034

Table 4.4: Resultant lithium ratios, as-determined by the elemental analysis, of the lithium excessive samples, fabricated by the sol-gel method, that were annealed for up to 128 days.

SG-LMT	0 d	4 d	32 d	64 d	128 d
Lithium ratio	2.208	2.133	2.150	2.102	2.056
Lithium ratio difference (measured minus expected)	0.208	0.133	0.150	0.102	0.056

Aluminium as an impurity was detected in all KALOS as well as all single phase emulsion method and sol-gel method samples. Its mean concentration within 20 mol% LMT and 30 mol% LMT KALOS samples amounts to 262 ppm and 225 ppm respectively. The aluminum concentration is, taking the standard deviation of the measurements into account, constant as a function of annealing time. The same is true for the rhodium and the platinum content found in these samples. With a mean rhodium content of 7 ppm and 6 ppm as well as a mean platinum content of 5 ppm and 4 ppm for the KALOS samples with 20 mol% LMT and 30 mol% LMT respectively, there are also no significant differences between the samples. Both noble metal impurities are very likely introduced during the fabrication due to the contact of the melt with the platinum/rhodium alloy crucible (see section 2.2.2.2). The aluminum impurities are probably introduced by the raw materials as the pebbles were kept in platinum boats (see section 3.3.2). Since the platinum content in the samples is no function of the annealing time, a significant reaction between the pebbles and the boats can be excluded.

The EM-LMT and SG-LMT samples show significantly different impurity concentrations. The aluminum content of EM-LMT is very low with a mean concentration of only 39 ppm. However, the as-received sample shows a significantly increased aluminium concentration. There seems to be no systematic for this value and consequently it is regarded as runaway value. Apart from that, the aluminum value slowly and steadily increases to 28 ppm during the annealing. The sodium concentration of the EM-LMT pebbles is only for the non-annealed and the 4 days sample within detectable limits (39 ppm). The mean concentration amounts to 184 ppm. The SG-LMT samples show significantly higher impurity concentrations. The aluminium and the magnesium content are constant over the duration of annealing with mean values of 345 ppm and 507 ppm respectively. The developing of the sodium concentration resembles that of the EM-LMT samples, falling below the detection limit of 39 ppm after 4 days of annealing from a mean value of 339 ppm.

It is evident, that for the biphasic emulsion method samples, the elemental composition matches the anticipated composition very well, with only small deviations (see Table 4.5). The analysis also shows a slight surplus in lithium, which is highest for the 10 mol% lithium metatitanate containing sample and is continuously reduced with increasing lithium metatitanate content. The overall surplus, however, is reasonably small. Although in principal all available biphasic samples that were fabricated with the emulsion method could have been analyzed, the samples were only tested for a possible systematic, concentration dependent variation of the composition.

Table 4.5: Resultant virtual phase contents, as-determined by the elemental analysis, of various samples that contain either LMT or LAO as second phase beside LOS. Additionally, the lithium ratios are given.

	EM-10-LMT	EM-50-LMT	EM-90-LMT	EM-10-LAO	EM-30-LAO
Virtual Li_4SiO_4 content [mol%]	90.3 ± 0.9	50.5 ± 0.5	9.2 ± 0.1	90.1 ± 0.9	70.4 ± 0.7
Virtual Li_2TiO_3/$LiAlO_2$ content [mol%]	9.8 ± 0.1	49.5 ± 0.5	90.8 ± 1.0	9.9 ± 0.1	29.3 ± 0.3
Lithium ratio	3.879	3.041	2.192	3.550	3.104
Lithium ratio difference (measured minus expected)	0.074	0.030	0.009	-0.153	-0.007

It was therefore considered to be sufficient to analyze the extremities and the center composition. The compositions of the remaining samples can then be interpolated from these results with reasonable accuracy. For the lithium aluminate containing pebbles, the virtual phase composition also matches the nominal composition very well (see again Table 4.5). Yet, there is a noticeable deficit in lithium for the nominally 10 mol% LAO containing sample, while the lithium content for the 30 mol% LAO containing sample is very close to the expectations. For these samples, the analysis of possible impurity elements was not performed. Also for the LAO containing samples, the composition of the not analyzed sample can be obtained by interpolation with reasonable accuracy.

4.2 Phase analysis

This section focuses on the results that were obtained by X-ray diffraction (see section 3.5). At first the qualitative results are presented, followed by the quantitative analysis of the diffractograms that were obtained at room temperature. The results on the thermal expansion of lithium orthosilicate and β-lithium metatitanate are presented in the last subsection.

4.2.1 Qualitative phase analysis

In Figure 4.1 the results of the qualitative phase analysis of the pebble samples that were fabricated by the KALOS process as a function of annealing duration are detailed. It is obvious, that both anticipated phases, lithium orthosilicate and lithium metatitanate, are present. Except for some relatively constant amount of Li_2CO_3, which can unambiguously be attributed to a reaction with ambient atmosphere during milling (see section 3.5.1), no other impurity phases are visible. While all of these samples, except the as-received samples, clearly show the β-modification of lithium metatitanate, the as-received samples predominantly show the rock-salt type γ-modification with a mildly stretched lattice (see Figure 4.1). As the γ-modification represents the high-temperature structure of lithium metatitanate, this phase transformation is expected as a result of the annealing. Other than that, the effect of the annealing is rather subtle as the qualitative phase composition is unchanged.

For the lithium excessive pebble samples, fabricated by either the emulsion method or the sol-gel method, the phase composition is also rather stable throughout the annealing in He/H_2 atmosphere, as shown in Figure 4.2. Yet, hints of lithium orthotitanate, Li_4TiO_4 or LOT, seem to be present in these pebbles. Apart from this observation, no impurity phases are present in the samples. In contrast to the samples that were fabricated by the KALOS method, these samples do not show any phase transformations, as the low temperature β-modification is already established in the as-received state.

The results of the XRD analysis of all biphasic samples that were fabricated by the emulsion method are given in Figure 4.3A. The lower LMT containing samples do not show any sign of impurity phases while both anticipated phases Li_4SiO_4 and β-Li_2TiO_3 are clearly present. As expected, the intensity of the Li_4SiO_4 reflexes increases with decreasing nominal LMT content and vice versa.

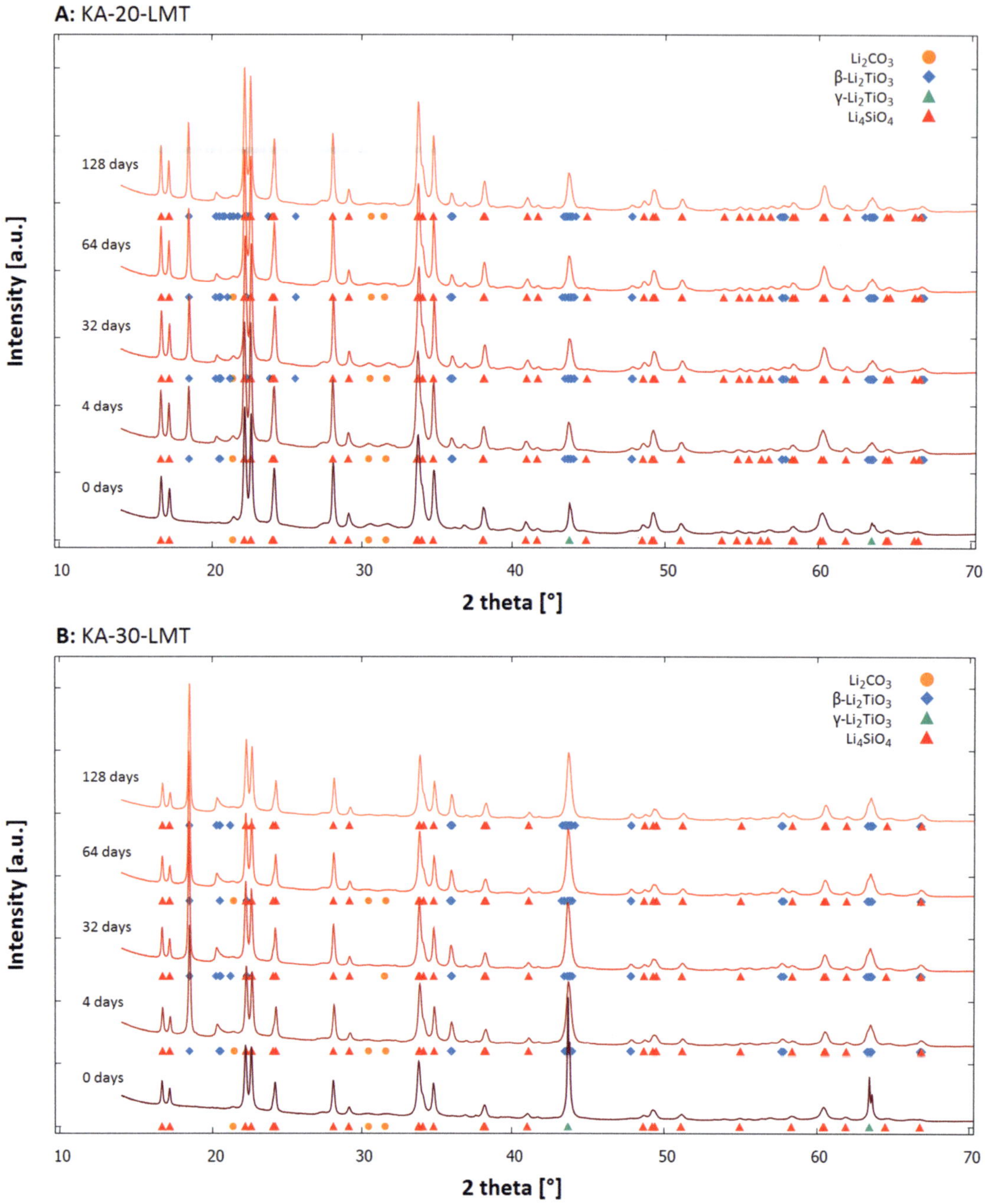

Figure 4.1: XRD analysis of the LOS+LMT pebbles fabricated by the KALOS process with nominally 20 mol% LMT (A) and nominally 30 mol% LMT (B). For each diffractogram the annealing duration of the sample is given in the graphs. Just 15° to 70° 2θ are shown to increase the visual quality of the diffractograms. Reflexes of identified phases are indicated, if they are significantly strong. The peak positions of all identified phases were refined stacking faults in β-Li_2TiO_3 were addressed (see section 3.5.3.2). The equilibrium crystal structures of Li_2CO_3, β-Li_2TiO_3, γ-Li_2TiO_3 and Li_4SiO_4 are detailed by Idemoto et al. [172], Kataoka et al. [78], Morales et al. [80] and Deng et al. [56], respectively.

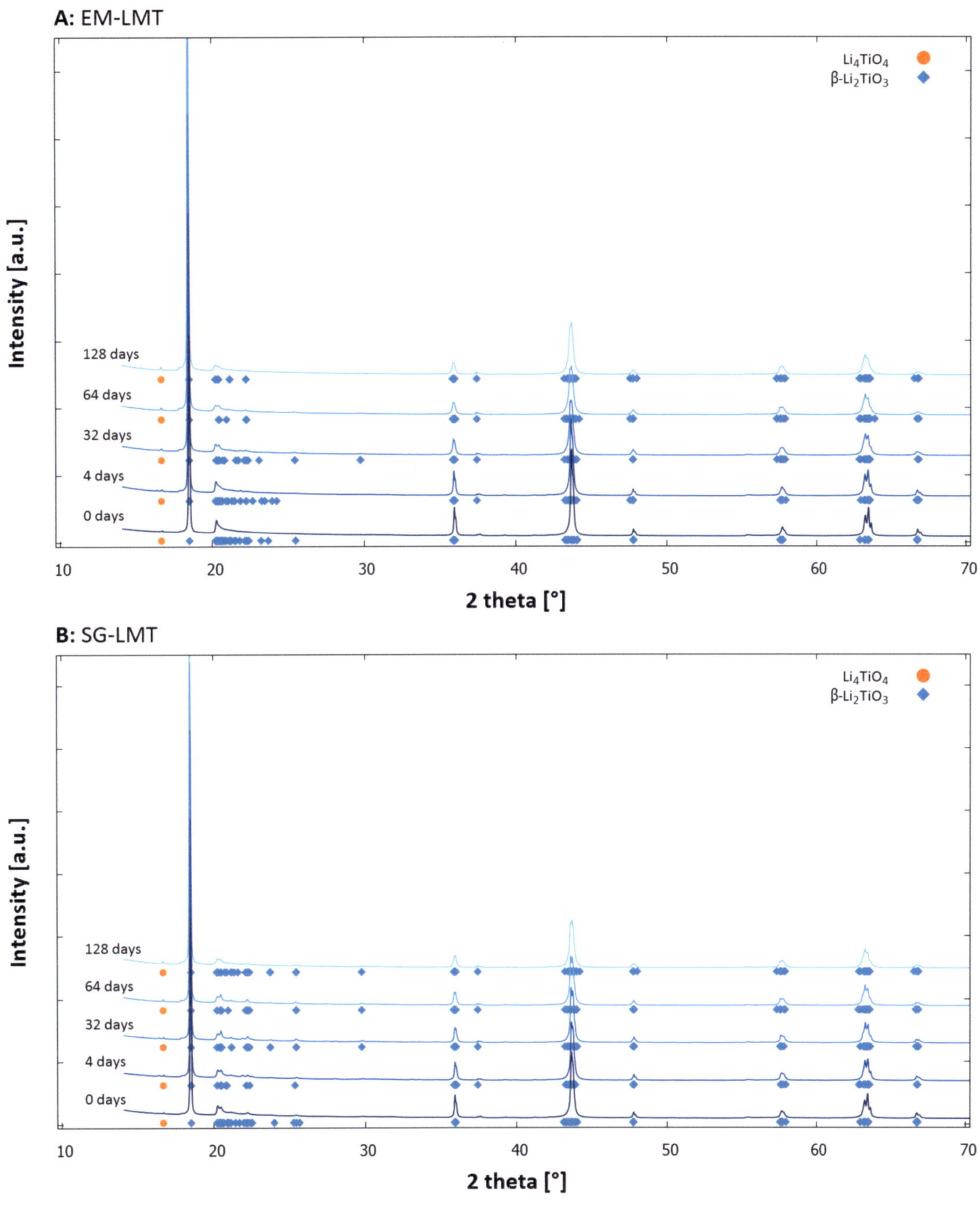

Figure 4.2: XRD analysis of the lithium excessive samples fabricated by the emulsion method (A) and the sol-gel method (B). For each diffractogram the annealing duration of the sample is given in the graphs. Just 15° to 70° 2θ are shown to increase the visual quality of the diffractograms. Reflexes of identified phases are indicated, if they are significantly strong. The peak positions were refined and stacking faults were addressed (see section 3.5.3.2). The equilibrium crystal structure of β-Li2TiO3 is detailed by Kataoka et al. [78].

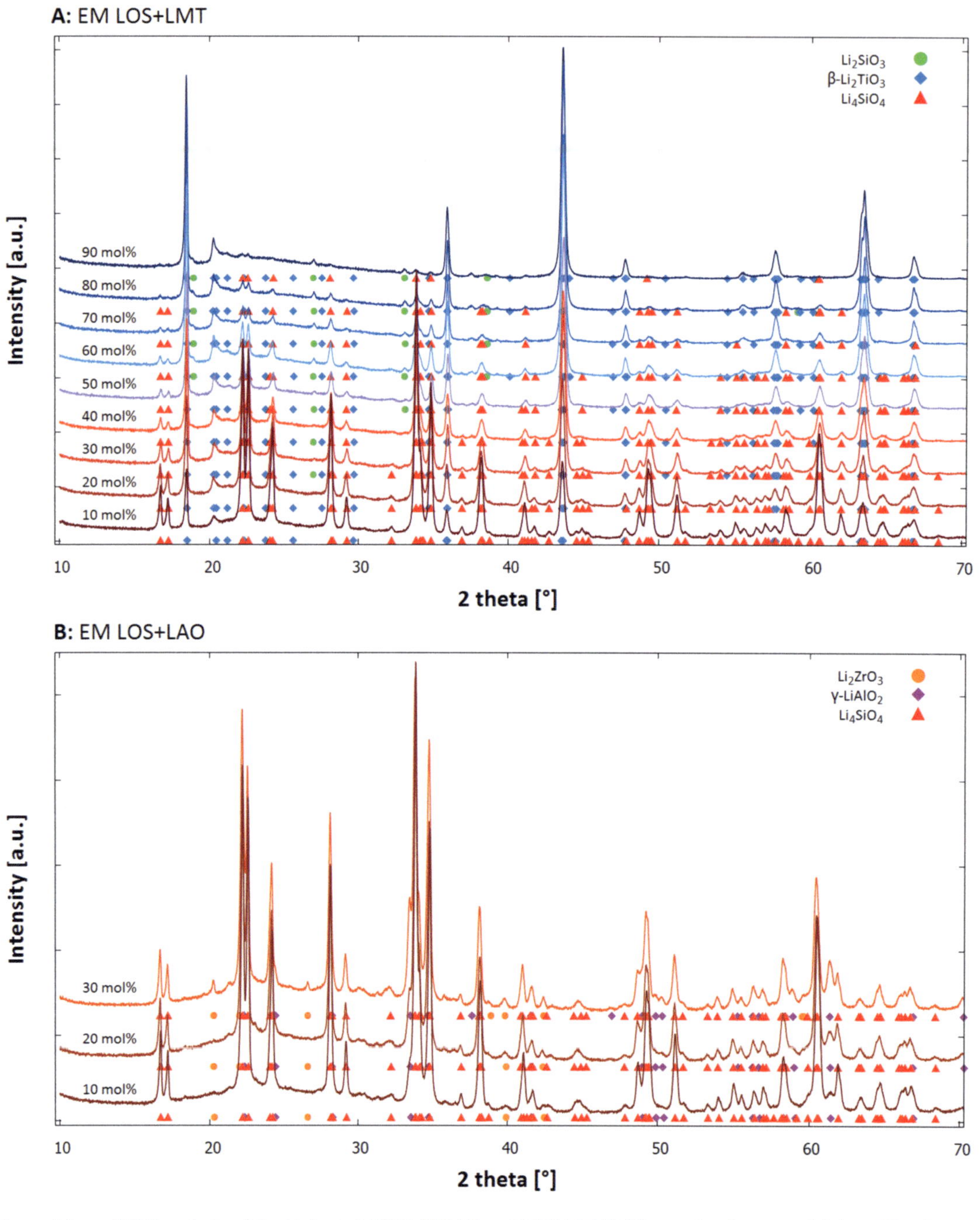

Figure 4.3: XRD analysis of the biphasic LOS+LMT (A) and LOS+LAO (B) samples that were fabricated by the emulsion method as already published in [96]. For each diffractogram the nominal content of LMT or LAO of the sample is given in the graphs. Just 15° to 70° 2θ are shown to increase the visual quality of the diffractograms. Reflexes of identified phases are indicated, if they are significantly strong. The peak positions of all identified phases were refined and stacking faults in β-Li_2TiO_3 were addressed (see section 3.5.3.2). The equilibrium crystal structures of β-Li_2TiO_3, γ-$LiAlO_2$, Li_2ZrO_3 and Li_4SiO_4 are detailed by Kataoka et al. [78], Heiba et al. [174], Marezio [103] and Deng et al. [56], respectively.

Yet, there are slight indications of lithium metasilicate, Li_2SiO_3, for the samples that nominally contain 60 to 90 mol% lithium metatitanate (Völlenkle detailed the crystal structure of Li_2SiO_3 in [173]). However, the amount of lithium metasilicate does not seem to increase as a function of lithium metatitanate concentration. In Figure 4.3B, the XRD analysis of all three lithium aluminate containing samples is shown. Generally the two major phases are lithium orthosilicate and γ-lithium aluminate, as anticipated.

The lithium aluminate phase shows a seemingly stretched lattice as the peak positions for higher diffraction angles are displaced towards lower angles. Yet, the stretching of the lattice does not seem to depend on the added amount of lithium aluminate. It is unclear whether there is a significant solubility of lithium orthosilicate in lithium aluminate, which could explain the stretching of the lattice. Yet, Zhao et al. as well as other authors report that there is considerable solubility of lithium aluminate in lithium orthosilicate [175]. In this case, however, this solubility does not seem to manifest. For the nominally 30 mol% lithium aluminate containing sample, a third phase is visible, which is identified as lithium zirconate, Li_2ZrO_3 (the crystal structure is given by Heiba and El-Sayed [174]). The lithium zirconate phase is supposed to originate from reaction with abraded particles of the ZrO_2 milling balls that are used during the preparation of the slurry (see section 3.2.2). The verification or quantification of zirconium within these samples by ICP-OES was not performed. However, there is clear evidence from EDX spectra for the presence of a small amount of zirconium within these samples (see section 4.3.3). Since all other samples of biphasic emulsion method pebbles, do not show any zirconium impurities, it is reasonable to assume that their appearance is closely related to the alumina additions to the slurry, which are very probably much more abrasive for the milling balls than the other raw materials.

4.2.2 Quantitative phase analysis

The quantitative phase analysis focuses on the main constituents of each sample. Because of the observed strong anisotropic peak broadening, it was not possible to introduce microstructural parameters into the refined model. Unconcerned from the anisotropic peak broadening are the peak positions and it was therefore possible to derive the lattice parameters (a_i, b_i, c_i, β_i of phase i) of each of the phases and accordingly the volume of the unit cell of each phase (V_i of phase i). Also the molar fraction x of each crystalline phase is one result of these analyzes. All determined parameters from the Rietveld analysis of the diffractograms are presented in Appendix C.

The refinement of the degree of crystallinity (DOC) results in a reasonable estimate of the (apparently) amorphous content of the samples. It is assumable that both of the main constituents contribute to the observed amorphous content. Therefore, with equation (3.10), the phase content of a given sample equals

$$W_{cryst.} + W_{amorph.} = DOC + (1 - DOC) = \sum_{i=1}^{n} W_{i,cryst.} + \sum_{i=1}^{n} W_{i,amorph.}. \tag{4.1}$$

The total content, amorphous and crystalline, of every phase i can be estimated from the elemental analysis (see section 4.1) and thus the amorphous content x_{amorph} of each phase i is determinable, if the molar mass of the crystalline and amorphous phases do not differ significantly. As already mentioned in section 3.5.3.2, a significant amount of lithium carbonate was formed on the surface of the lithium orthosilicate containing samples after the milling during the preparation, as no significant lithium carbonate reflexes were found during the qualitative analysis (see section 4.2.1). The determined content of LCO lies between 2.5 wt.% and 5 wt.%. Since lithium carbonate is not found in the lithium excessive LMT samples after the milling, all formed LCO must originate from the reaction

$$Li_4SiO_4 + CO_2 \overset{\Delta E}{\rightarrow} Li_2CO_3 + Li_2SiO_3. \tag{4.2}$$

Therefore, all determined LCO is attributed to LOS and added to the observable amount of LOS. Since it cannot be determined whether the LOS from which the LCO formed would have been determined as crystalline or amorphous, its amount is distributed on both parts proportionately with respect to the determined DOC.

For all samples that show the β-LMT phase, stacking faults in the structure were modelled as explained in section 3.5.3.1. An evaluation of the stacking vector of a layer of the structure with respect to the preceding layer leads to the number of actually faultily stacked layers and as an additional result an estimate of the stacking fault probability. Yet, the powdering step in the swing mill clearly leads to the generation of stacking faults and as a result there is no correlation of the observed stacking fault probability and the annealing time. As a consequence, the development of the stacking fault probability is not presented in detail in the following. Of course, the healing of stacking faults with increasing annealing duration is expectable and the diffractograms of the less finely ground lithium excessive samples (see Figure 4.2) are mildly suggestive of this healing. For the samples that were analyzed after the extensive milling, the stacking fault probability varies between 31 % and 62 %.

The obtained refined parameters of the samples that were fabricated by the melt-based process KA-LOS are given in Figure 4.4 as a function of the annealing duration. The unit cell volume of the LOS phase is obviously increased over the reference value of 1164.31 Å^3 (from the lattice parameters given in Table 2.1). Yet, while the nominally 20 mol% LMT containing samples show a stretched LOS lattice of 1179.58 Å^3 (equates to a tensile strain of 1.31 %) in the as-received state which continuously decreases to 1175.48 Å^3 (equates to a tensile strain of 0.96 %) over the course of annealing, the lattice volume of LOS in the initial state for the nominally 30 mol% LMT containing sample is less. It is determined to be 1171.91 Å^3 (or a tensile strain of 0.65 %) which is reduced during the annealing to 1167.66 Å^3 (or 0.29 %). The β value varies ineffectually at a slightly higher value than the reference of 99.058° for all annealed samples. More precisely, the deviations amount to a maximum of 0.20 % for the as-received KA-20-LMT sample and 0.02 % for the as-received KA-30-LMT sample. As a result, the changes of this parameter have a negligible effect on the overall lattice volume. The major change in the LOS unit cell volume is found after 4 days of annealing for these samples. For either set of samples, the observed increase in lattice volume becomes manifest in all lattice parameters and is similarly reduced as a function of annealing duration. The strain in the lattice parameters for any annealing duration decreases in the order $b > c > a$, although the differences between these parameters are relatively low.

In contrast to the LOS phase, the unit cell volume of the LMT phase is decreased in both sets of samples with respect to the reference value of 71.47 Å^3 (see Table 2.3), although just mildly. It is determined as 70.96 Å^3 and 71.06 Å^3 in the as-received state which amounts to a compressive strain of -0.72 % and -0.58 % for the nominally 20 mol% and 30 mol% LMT containing samples, respectively. After 4 days of annealing and after the phase transformation from the γ- to the β-modification, no significant deviation from the reference unit cell volume of β-LMT is 427.0 Å^3 is detectable (see Figure 4.4). However, for both samples that were annealed for 128 days, an increased a parameter of 5.072 Å (equates to a tensile strain of about 0.2 %) is obtained for both sets of samples. This increase is offset by an almost equally large decrease of b. After 128 days of annealing, the β parameter also changes significantly to about 99.9° which is significantly below the reference value of 100.21°. With the exception of this observation, for all other samples that were fabricated by the KALOS process, the parameter β is just slightly lower than the reference value.

The seemingly amorphous content of the samples is reduced by the annealing. The content of amorphous LOS in the as-received state is higher for the nominally 30 mol% LMT containing sample compared to the 20 mol% LMT containing one. Yet, for the KA-30-LMT samples, the amorphous LOS fraction remains at a relatively constant level (about 11 mol%) after 4 days of annealing, while this constant level (about 12.5 mol%) is not reached before 32 days of annealing for the KA-20-LMT samples. Concerning the LMT phase, the amorphous fraction seems to reach a constant value after 128 days of annealing for both sets of samples. The major reduction is seen after 4 days of annealing. After 64 days of annealing a no considerable reductions of these values are observable. The amorphous content attributed to LMT after 128 days of annealing is about 7.1 mol% for the nominally 20 mol% LMT containing sample whereas it is only about 3.4 mol% for the nominally 30 mol% LMT containing one. For all samples, the estimated standard deviation of each refined parameter is given in Figure 4.4, Figure 4.5 and Figure 4.6. These errors can be relatively small, especially for parameters that contribute significantly to the quality of the fit. However, they should not be mistaken as a measure for the overall uncertainty of these values as it is significantly higher. This is particularly true for the determined amorphous content, but an unambiguous estimation of the overall uncertainty cannot be given.

The lithium excessive samples show relatively few differences in their crystallographic changes as a function of annealing duration (see Figure 4.5). Both samples show a slightly increased unit cell volume with respect to the reference values, which varies slightly but insignificantly. The trends of the three lattice parameters are equal for both samples. A slight increase is found for the a parameter in the as-received state, which is continuously reduced as a function of annealing duration and eventually reaches a value close to the reference value.

In contrast to this development, the b parameter steadily increases from a slightly lower value than the reference reports to a slightly larger value with increased annealing. For the c parameter there is a similar increase found for the samples fabricated by the emulsion method, while it fluctuates close to the reference value for the samples fabricated by the sol-gel method. The angle β increases slightly from a significantly reduced value (EM-LMT: -0.30 %, SG-LMT: -0.14 %) with ongoing annealing to values close to the reference, which are still lower than the literature value.

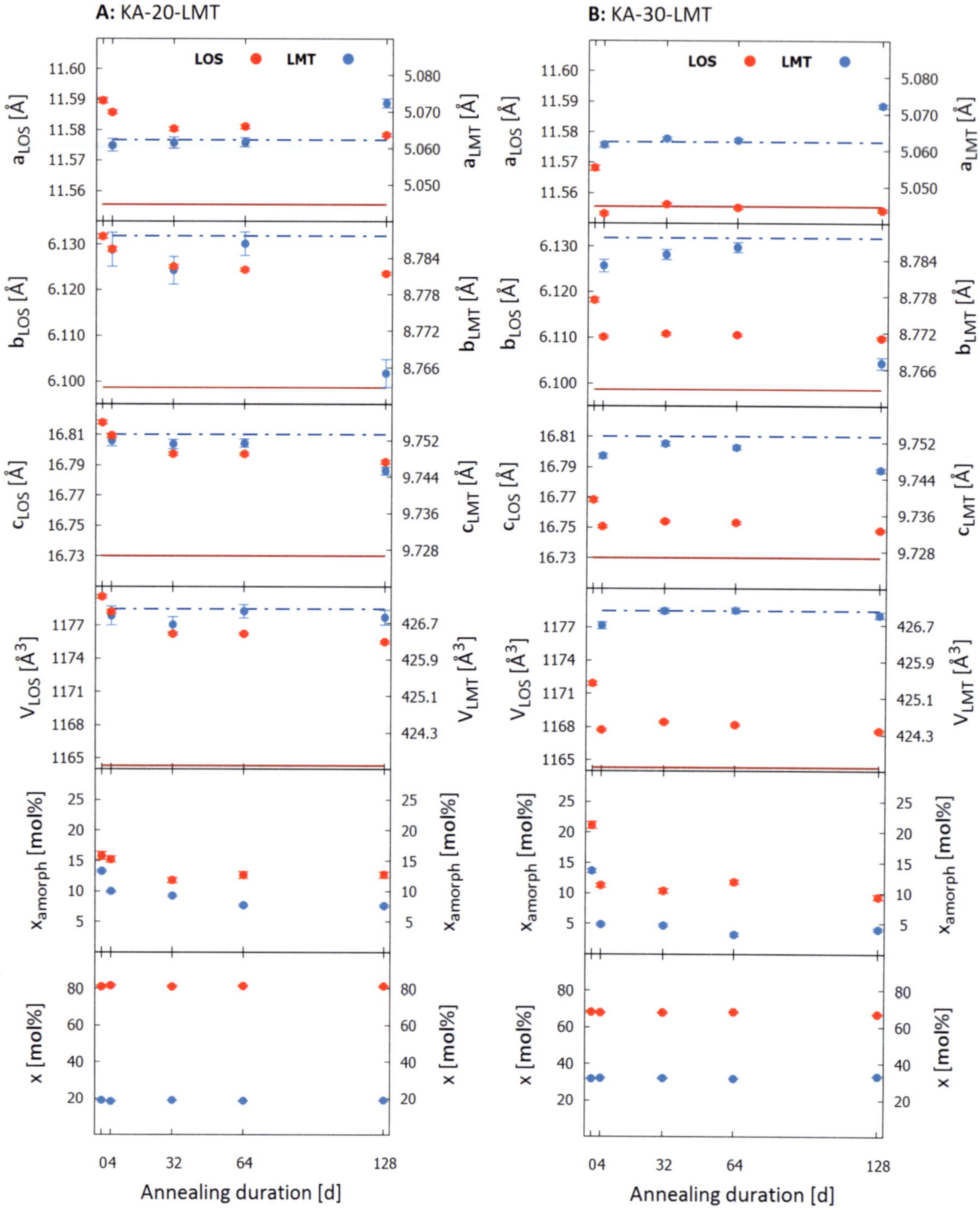

Figure 4.4: Resultant parameters from the refinements of the pebbles that were fabricated by the KALOS process (A: nom. 20 mol% LMT, B: nom. 30 mol% LMT) and were annealed for up to 128 days. The error bars indicate the estimated standard deviation of each refined parameter. The lines illustrate the reference lattice parameters and unit cell volume of LOS (solid lines) and LMT (dash-dot lines) according to Deng et al. [56] and Kataoka et al. [78] ,respectively. The determined lattice parameters and unit cell volume of the LMT phase in the as-received samples are not displayed for the sake of visibility.

The degree of crystallinity of the EM-LMT samples slightly increases after 4 days of annealing, which is displayed by the reduction of the amorphous content in the sample. Then, however, a further increase of the DOC does not seem to occur. On the contrary, the amorphous content seems to increase again after 128 days of annealing, with regard to the value after 64 days of annealing. Yet, as the values after 32 days of annealing and 128 days of annealing are very similar, it is also possible that the determined value after 64 days of annealing represents a deviation from an otherwise constant level to lower values. For the SG-LMT samples, the decrease of the amorphous content after 4 days of annealing is much more prominent than for the EM-LMT sample and continues until 32 days of annealing. Then, similarly to the EM-LMT samples, a relatively constant level of the amorphous fraction is established. However, the same issue with regard to the value after 64 days of annealing is also found for this set of samples. If the values for long annealing durations are interpreted as relatively constant, both constant levels are not too different with about 7.3 mol% and 6.8 mol% for the EM-LMT and SG-LMT samples, respectively.

Both sets of samples, however, show a continuous increase of the observable preferred orientation of (001)-planes in parallel to the powder sample surface with increased annealing duration. In the as-received state, both samples, EM-LMT and SG-LMT show a degree of preferred orientation of about 10 % which sharply increases after 4 days of annealing and reaches a value close to 30 % after 128 days of annealing. A considerable degree of preferred orientation was not observed for the samples that were ground by hand and therefore the observed preferred orientation is most probably an effect of the sample preparation.

In Figure 4.6, the determined crystallographic parameters of the biphasic pebbles that were fabricated by the emulsion method are shown. The unit cell volume of LOS of the LMT containing pebbles does not vary greatly and is, with a mean volume of 1166.12 Å^3 generally about 0.16 % larger than the reference lattice parameters suggest. As a function of nominal lithium metatitanate concentration, there is a slight increase in unit cell volume until 50 mol% LMT observable. For higher LMT concentrations, the unit cell volume stays relatively constant when the estimated standard deviation of the refinement is taken into account. Because the intensity of the LOS phase in the diffractogram of the nominally 90 mol% LMT containing sample is comparably low, the refinement of the lattice parameters and the unit cell volume shows considerable ambiguity. As a result, these values should not be taken into further consideration. The observed slight increase in the unit cell volume is an effect of the expansion in b- and c-direction as the a parameter is virtually constant with about 11.544 Å. Along the c-axis, the lattice of the LOS phase expands seemingly linearly with increasing LMT content until a nominal content of 60 mol%. For higher LMT contents, the uncertainty of the refined values increases and seemingly stays at a constant value, but a continued linear increase is equally justified. In contrast to the simple behavior of the c parameter, the b parameter shows a kind of bell curve behavior with the maximum at nominally 50 mol% LMT with a maximum value of 6.112 Å (that equates to a strain of 0.23 %) compared a minimum value of 6.105 Å (that equates to a strain of 0.10 %). The angle β matches the reference value at low LMT concentrations very well and slightly decreases steadily with increasing LMT additions to 98.97°, i.e. a reduction of -0.09 %. Similarly as for the LOS phase, the unit cell volume of β-LMT does not seem to change too much.

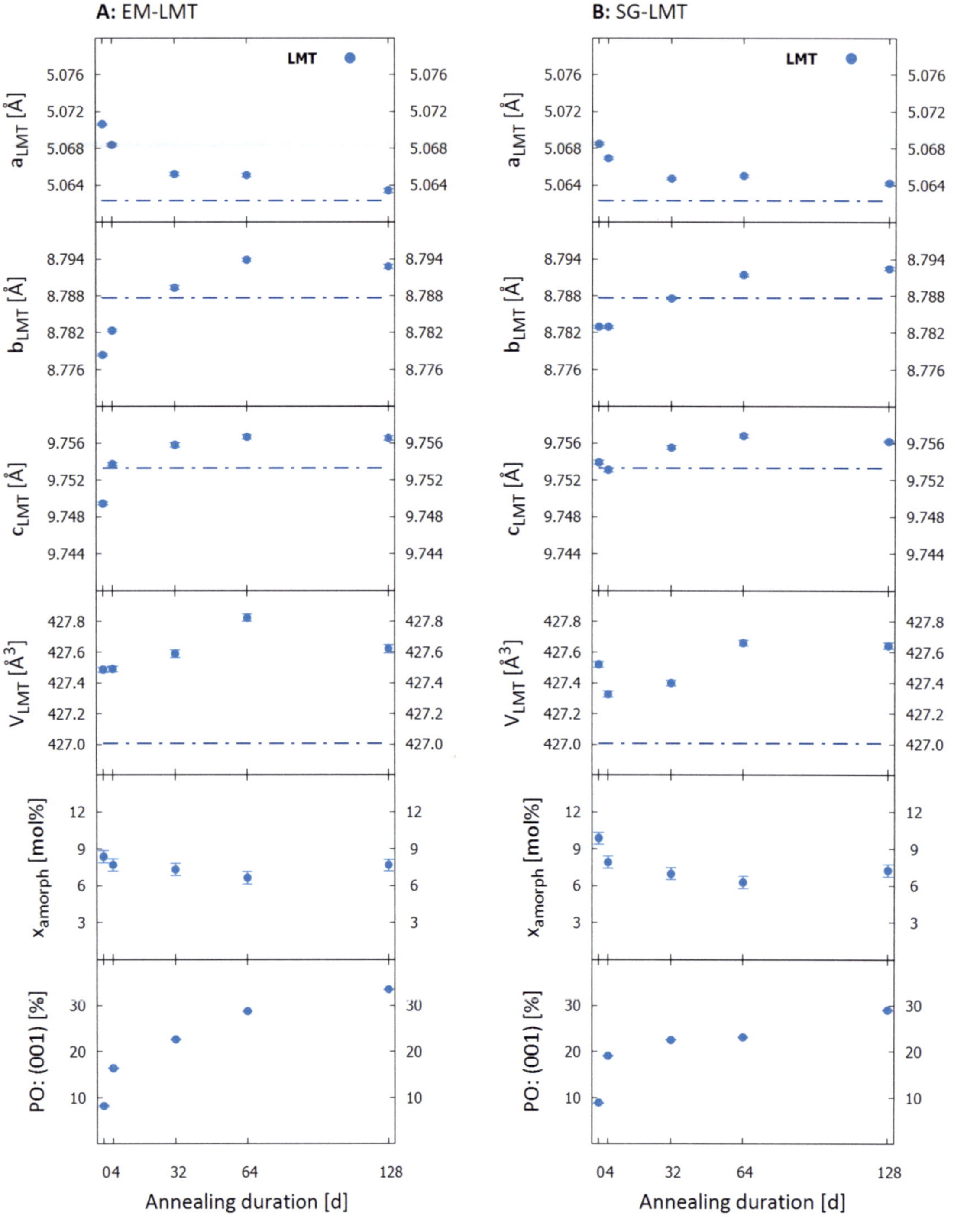

Figure 4.5: Resultant parameters from the refinements of the lithium excessive pebbles that were fabricated by the emulsion method (A) as well as the sol-gel method (B) and were annealed for up to 128 days. The error bars indicate the estimated standard deviation of each refined parameter. The dash-dot lines illustrate the reference lattice parameters and unit cell volume of LMT according to Kataoka et al. [78]. The probability of preferred orientation (PO) is calculated according to equation (3.8).

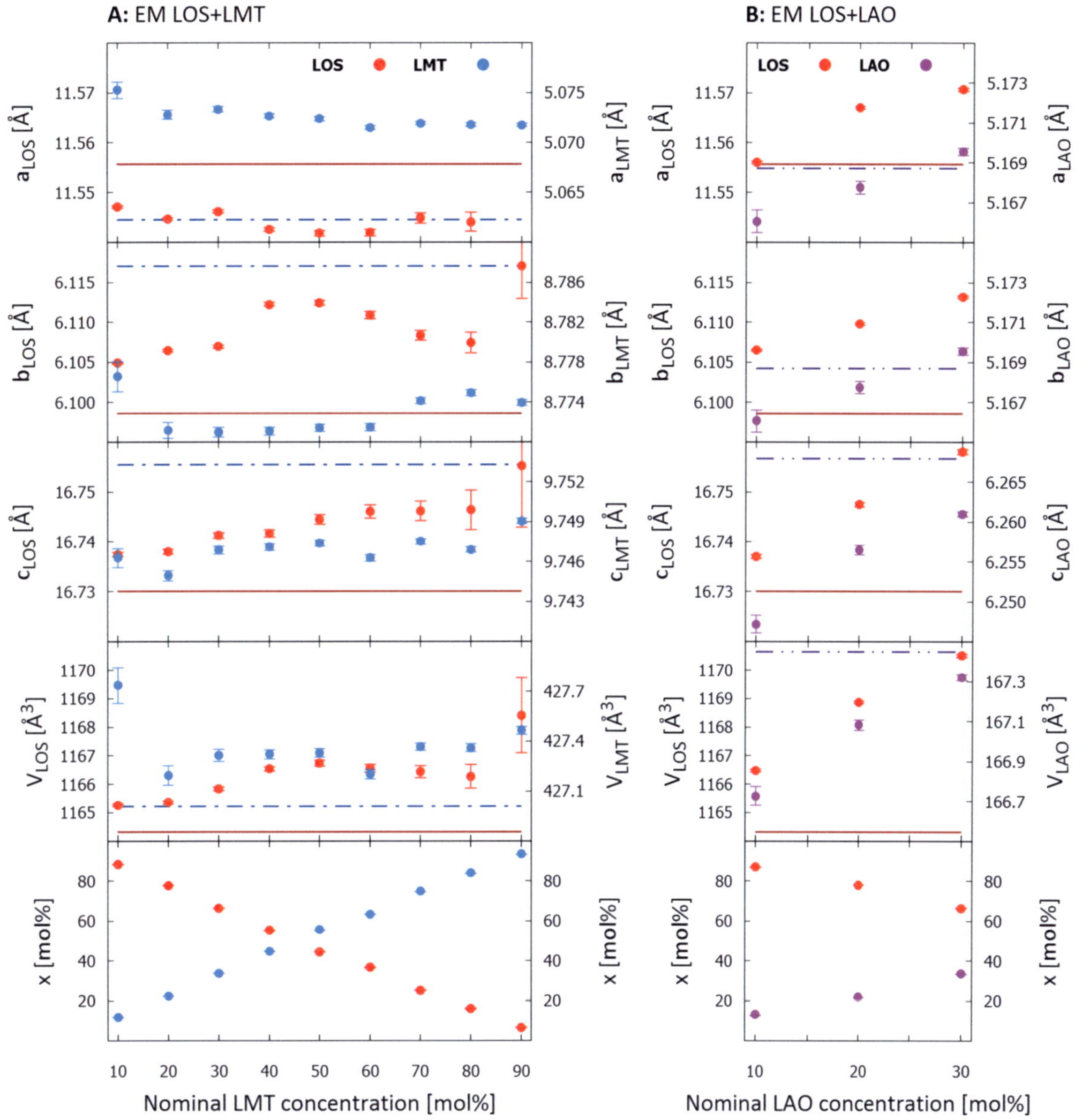

Figure 4.6: Resultant parameters from the refinements of the biphasic pebbles that were fabricated by the emulsion method. The results of the LMT containing samples are displayed in A, as well as the results of the LAO containing samples in B. The lines illustrate the reference lattice parameters and unit cell volume of LOS (solid lines), LMT (dash-dot lines) and LAO (double-dot dash lines) according to Deng et al. [56], Morales et al. [80] and Marezio [103], respectively. The error bars indicate the estimated standard deviation of each refined parameter. The determined parameter a_{LOS} of the nominally 90 mol% LMT containing sample is not displayed for the sake of visibility.

With the exception of the nominally 10 mol% LMT containing sample, the unit cell volume is slightly increased over the reference value and marginally increases with increasing LMT content. The nominally 10 mol% LMT containing sample shows a significantly larger unit cell of 427.74 Å^3, which is an increase of 0.17 % over the reference value. Significantly increased a- and b-parameter with respect to the other value that were determined for low LMT concentrations are apparently responsible for the increase in volume. For the other LMT containing samples, the increase in volume is due to a slight increase of the b- and c-parameters, while the a-parameter is slightly reduced. For all samples, the angle β is determined to about 99.84°, which is -0.37 % lower than the reference value.

The obtained phase composition of the LOS+LMT samples generally shows an overestimation of the LMT phase with respect to the nominal value, which ranges from about 1.6 % to 5.6 %. This overestimation is attributed to micro-absorption effects because of the significantly higher absorption of LMT (mass attenuation coefficient of 92.33 cm^2/g for Cu-Kα radiation) compared to LOS (mass attenuation coefficient of 21.18 cm^2/g for Cu-Kα radiation) [176].

For the LAO-containing samples, an increase of the unit cell volume of both phases LAO and LOS is clearly visible as a function of increasing LAO content (see Figure 4.6). For both phases all lattice parameters expand significantly. For the LOS phase, the unit cell volume is generally increased with respect to the reference value and is increases with increasing LAO content from 1166.47 Å^3 (increase of 0.19 %) to 1170.50 Å^3 (increase of 0.53 %). Similarly to the samples that were fabricated by the KA-LOS process, the largest strain is found in b-direction and the lowest in a-direction. The β angle almost insignificantly increases from a difference of -0.04 % to 0.02 % with respect to the reference value. In contrast to the LOS phase, the unit cell volume of the LAO phase is reduced with respect to the reference value (167.45 Å^3, from the values for γ-LAO in Table 2.5). With increasing LAO concentration the unit cell volume almost reaches the reference value, i.e. the LAO is virtually unstrained. Most prominently, the expansion of the unit cell originates from an expansion in c-direction, whereas the a-parameter and the b-parameter, accordingly, increase just slightly. Similarly to the LMT containing biphasic samples, also the LOS phase content is underestimated in the performed analysis. Also in this case micro-absorption possibly explains these findings.

4.2.3 Thermal expansion of lithium orthosilicate and lithium metatitanate

The thermal expansion of the two main constituting phases of the pebbles, lithium orthosilicate and lithium metatitanate, is determined by temperature varied XRD. The powders for this analysis were synthesized as described in section 3.2.3.2. The resultant diffractograms of the analysis are shown in Figure 4.7. None of the powder samples is free of impurities. Traces of lithium carbonate and lithium hydroxide can be found in the lithium orthosilicate sample, and the lithium metatitanate sample shows small amounts of lithium orthotitanate and lithium carbonate. The crystal structures of the three compounds are given by Idemoto et al. [172] (Li_2CO_3), Mair [177] (LiOH) and Gunawardane et al. [178] (Li_4TiO_4). Apparently, a slightly higher lithium content than stoichiometric is present in both samples.

During the temperature varied XRD experiment, the impurities expectedly show phase transformations, which is visible from Figure 4.7. A significant impact on the actual determination of the thermal expansion of LOS or LMT is not expected. At temperatures above 400 °C the lithium hydroxide impurities vanish and are also not detectable during cooling to room temperature. While the lithium carbonate impurities also vanish during heating (above 600 °C), it forms again during cooling below 700 °C. The vanishing of lithium hydroxide can easily be explained by reaching its melting temperature of 450-471 °C. At a temperature of 924 °C, lithium hydroxide decomposes as reported by its ICSC card. Therefore lithium hydroxide is not formed during cooling. Most probably the melting also explains the vanishing of lithium carbonate. Yet, the reported melting temperature of lithium carbonate is 723 °C (see its ICSC) and thus a melting of lithium carbonate is not expected at that temperature. But once the temperature falls below 750 °C, lithium carbonate is detectable again. As in contrast to lithium hydroxide, the decomposition temperature of 1310 °C (ICSC) is not surpassed in these experiments, lithium carbonate seems to persist in liquid form above its melting temperature.

For the lithium metatitanate sample, the detected lithium carbonate also vanishes at significantly lower temperatures as its reported melting point suggests (above 450 °C), but it crystallizes below 750 °C as for the lithium orthosilicate sample. The lithium orthotitanate impurities seemingly vanish when a temperature of 400 °C is reached. Yet, the reflexes reappear above 700 °C. Upon cooling, the lithium orthotitanate reflexes disappear again at temperatures below 850 °C. This behavior is not explained by the published phase diagram (see Figure 2.10) and no other satisfying explanation is evident from the available results.

A similar behavior is observed for the reflexes of the β-lithium metatitanate phase at about 20° to 25°. Theoretically, with the applied heating of the powder, the reflexes should become sharper and more pronounced as defects, like stacking faults (see section 4.2.2), are healed and the grain size increases during the more than 20 hours long measurement run. Yet, upon cooling some defects seem to be (re-)introduced into the material and thus the reflexes in this range blur as a result. The graph of Figure 4.7A clearly shows, that the lithium orthosilicate lattice changes continuously and reversibly with temperature. Yet, between 700 °C and 750 °C, the expansion behavior changes significantly. This is in accordance with the reported phase transformation at 723 °C (see Figure 2.9). From the simultaneous refinement of the lattice parameters the four coefficients of equation (3.11) were determined and are given in Appendix C. As shown in equation (3.12) the thermal expansion coefficient for each lattice parameter is therefore determined as a function of temperature.

In Figure 4.8 the thermal expansion coefficients for all four lattice parameters are shown as a function of temperature. Yet, the term thermal expansion coefficient is not well suited for the lattice parameter β, as it rather describes the distortion of the unit cell. Still the term is used for simplicity. It is evident from the graph, that there is a discontinuity at the temperature of the phase transformation and that as the transformation temperature is approached, the thermal expansion coefficient rises substantially.

At temperatures below 500 °C the thermal expansion coefficients are comparably constant. Yet, when compared to other ceramic materials, like rutile or alumina, the thermal expansion of lithium orthosilicate, taking only a, b and c into account, is at least two times higher. Above 500 °C, i.e. when the thermal expansion coefficients change drastically, the difference can reach up to 20 times near the phase transformation temperature of about 725 °C.

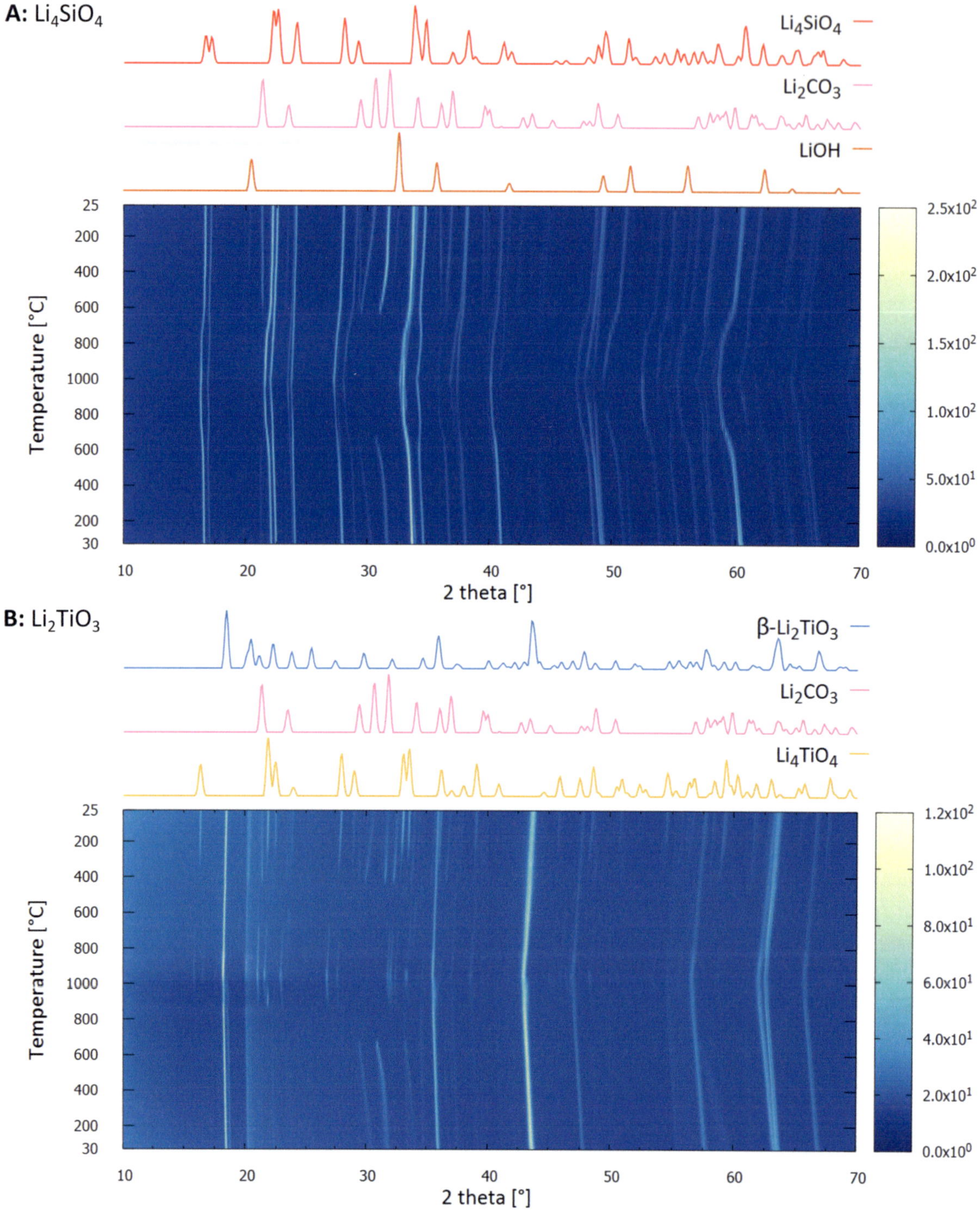

Figure 4.7: Temperature varied XRD of synthesized lithium orthosilicate (A) and lithium metatitanate (B), showing the square root of the intensity in arbitrary units as a function of the diffraction angle and the temperature. The temperature was changed from 25 °C to 1000 °C to 30 °C. Simulated diffractograms (at room temperature), also showing the square root of the intensity, of all identified phases are shown above each plot for comparison. The results are interpolated along the y-axis.

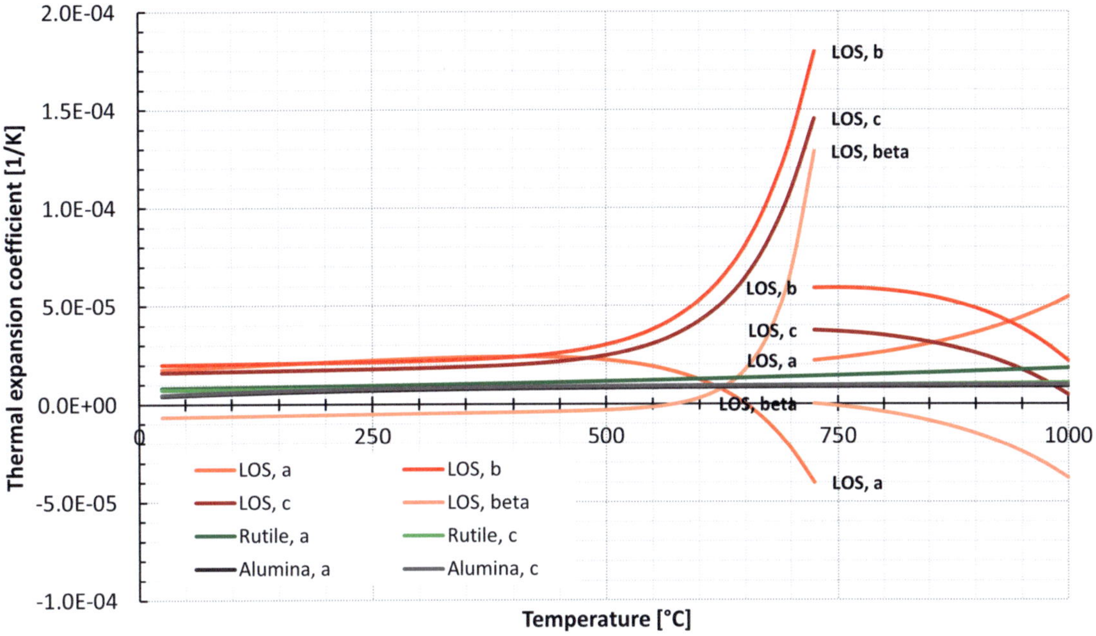

Figure 4.8: The determined thermal expansion coefficient of lithium orthosilicate as a function of temperature for all four lattice parameters. For comparison the thermal expansion of rutile (as reported by Rao et al. [179]) and the thermal expansion of α-alumina (as reported by Munro [169]) are also given in the graph.

Another significant difference between lithium orthosilicate and alumina as well as rutile is the degree of anisotropy or difference between the thermal expansions along the various axes of the crystal. Especially at temperatures above 500 °C, these differences are sizeable and even a contraction along the *a*-axis is observable while the crystals heavily expand in *b*- and *c*-direction. This is also visible in a converse way for temperatures above 725 °C, although not as strongly. The lattice of β-lithium metatitanate expands, as detailed in Figure 4.7B, as a function of temperature in a continuous way without an observable phase transformation, which is in agreement with the phase diagram of Li_2O-TiO_2 shown in Figure 2.10. Yet, the phase diagram suggests the phase transformation between the β- and the γ-modification between 1050 °C and 1155 °C depending on the composition, which is just slightly above the tested temperatures.

As shown in Figure 4.9, similarly to the behavior of lithium orthosilicate, the thermal expansion coefficients of the lattice parameters change significantly when the temperature of the phase transformation is approached. Apparently, for lithium metatitanate this is the case at temperature above 750 °C. At these temperatures, the expansion of the lattice seems to cease in [100]- and [001]-direction and expansion in [010]-direction slightly reduces. Below temperatures of 750 °C, the temperature dependence of the thermal expansion coefficients of β-Li_2TiO_3 resemble that of rutile, yet at about double the values.

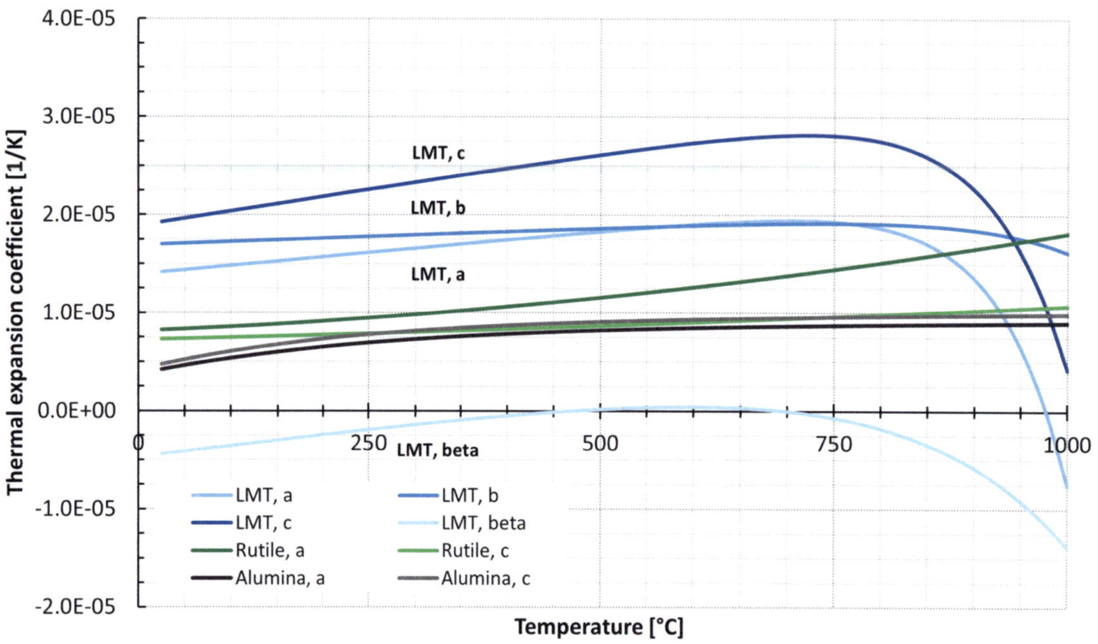

Figure 4.9: The determined thermal expansion coefficient of β-lithium metatitanate as a function of temperature for all four lattice parameters. For comparison the thermal expansion of rutile (as reported by Rao et al. [179]) and the thermal expansion of α-alumina (as reported by Munro [169]) are also shown.

From these measurements, the volumetric thermal expansion coefficient α_V can be calculated on the basis of equation (3.12). For a monoclinic unit cell of volume V it is

$$\alpha_V = \frac{1}{V}\frac{dV}{d\vartheta} = \frac{1}{V}\frac{d(a \cdot b \cdot c \cdot \sin\beta)}{d\vartheta} \cong 3\alpha_L. \tag{4.3}$$

Usually, for an isotropic material with mutually orthogonal axes, the volumetric thermal expansion coefficient is approximately three times the linear thermal expansion coefficient α_L. Of course, this criterion is not completely fulfilled for monoclinic unit cells, yet as the values of β for lithium orthosilicate and β-lithium metatitanate are very similar, a direct comparison between both phases is still reasonably possible. The resultant estimated linear thermal expansion coefficients of lithium orthosilicate and β-lithium metatitanate are shown in Figure 4.10 as a function of temperature. For comparison, based on Munro's [169] values, the equally estimated linear thermal expansion coefficient of α-alumina is also given. The so-determined thermal expansion coefficient of alumina is in very good agreement with reference data given by Touloukian et al. [180].

In principle, these values are well in line with the determined thermal expansion of the unit cells of the three phases. Of course, the linear thermal expansion coefficient follows the trend of the strongest expansions of the unit cells. Furthermore, the thermal mismatch between lithium orthosilicate and β-lithium metatitanate, which is simply the difference between both coefficients, is also shown. It is clear, that the thermal expansion of both phases does not differ greatly at low temperatures, but above

500 °C, when the phase transformation is approached, a substantial thermal expansion mismatch is visible. Beyond the phase transformation, the thermal mismatch is not as large, but still considerably high. The phase transformation of metastable γ-lithium metatitanate to β-lithium metatitanate at elevated temperatures was analyzed exemplarily for the LMT-richer KALOS pebbles (see Figure 4.11).

The increase of the LOS unit cell volume clearly shows a slight kink, i.e. a displacement of the data points towards lower cell volumes, at the measurements at 800 °C and 810 °C after which the thermal expansion continues by the same rate. This kink coincides with the observation of a substantial amount of the equilibrium β-lithium metatitanate phase. The determined strain of the LOS unit cell at room temperature before the temperature varied measurements was 0.68 % which was reduced to 0.38 % at room temperature after finishing the measurements.

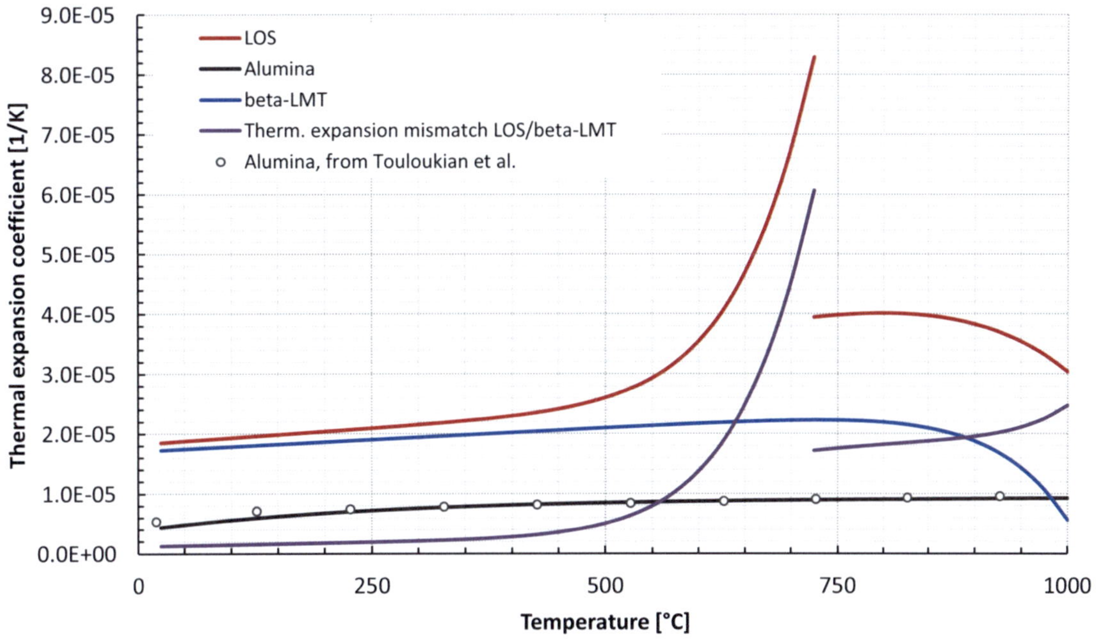

Figure 4.10: The estimated linear thermal expansion coefficients of lithium orthosilicate, β-lithium metatitanate and α-alumina as a function of temperature. The thermal expansion mismatch between lithium orthosilicate and β-lithium metatitanate is also shown. The linear thermal expansion coefficients were estimated from the measurements that are detailed in Figure 4.8 and Figure 4.9. The reference values for α-alumina were taken from Touloukian et al. [180].

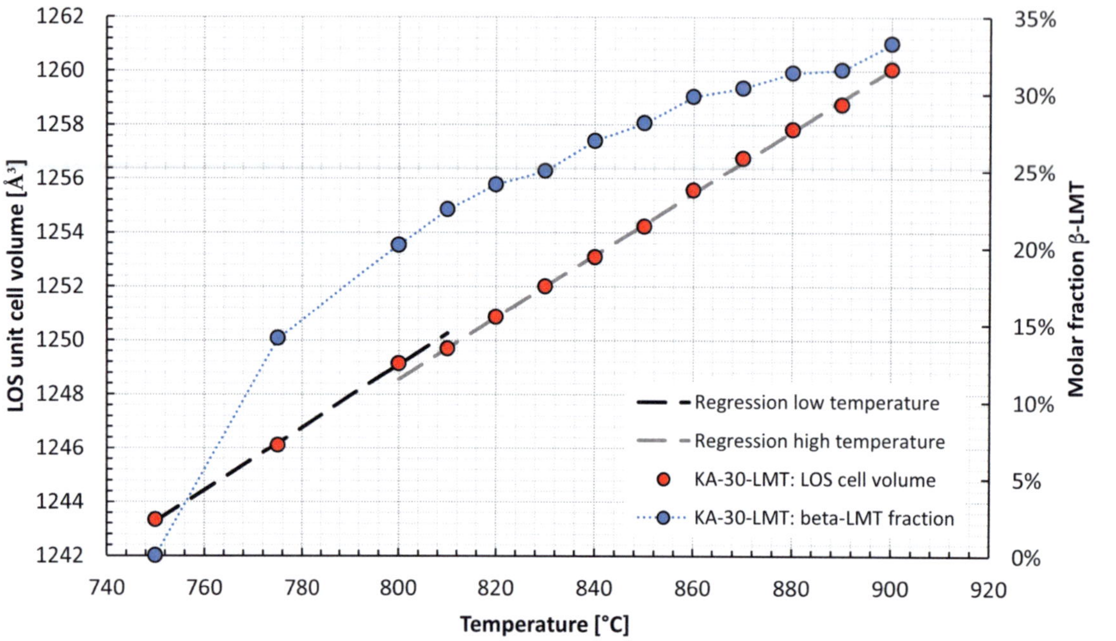

Figure 4.11: The thermal expansion of the LOS unit cell of the as-received KA-30-LMT pebbles as a function of temperature (left axis) is given as well as the evolution of the molar fraction of β-lithium metatitanate as a function of temperature (right axis). Two linear regression lines are shown for the low and high temperature part to illustrate the kink. Both regression lines show a coefficient of determination of 0.999, which is higher than using a single regression line. The dotted line is introduced to improve the visualization. The errors of the refinement are too small to be recognized in this plot.

4.3 Pebble morphology and microstructure

In the following section the surface morphology of the pebble samples as well as their microstructure are presented in a qualitative way by SEM investigations. In addition the determined pebble size distributions of the sintered as well as un-sintered biphasic pebbles that were fabricated by the emulsion method are displayed and the resulting shrinkage during sintering is presented.

4.3.1 Surface morphology

In the as-received state, the surface of the nominally 20 mol% LMT containing pebbles that were fabricated by the KALOS process is showing a lamellar structure of submicron spacing with primarily crystallized lithium orthosilicate dominating the structure (see Figure 4.12). The lithium metatitanate phase appears in a lighter grey than the lithium orthosilicate. The slight differences in contrast of surface images also show that a pebble is composed of a number of domains, which differ either in their surface relief and/or are inclined by an obtuse angle. For a number of pebbles, the vast majority of these domains obviously originate from a single point, which usually shows signs of an impact. For the other pebbles the domains are not ordered in a specific way.

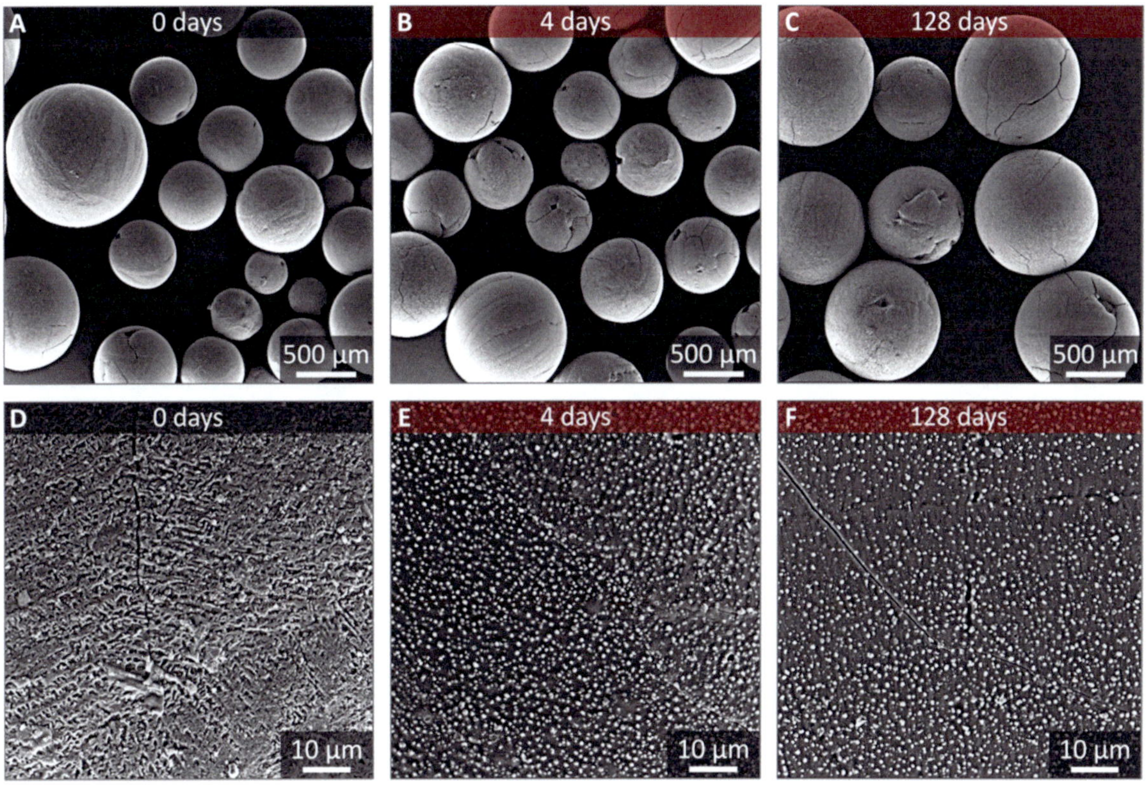

Figure 4.12: SEM surface images of the nominally 20 mol% LMT containing KALOS pebbles for different annealing durations. In the upper row, an overview of a number of pebbles is shown, while a detailed view of the surface is shown in the other images.

It is also visible from the images, that numerous cracks are formed on the surface of the pebbles. In addition to that the pebbles show mostly one major dent in an otherwise rather spherical body. From these dents the majority of cracks extend into the surrounding material. The cracks seem marginally deflected by the lamellar surface structure during transgranular fracture. Yet, major cracks may also follow boundaries between domains, which may be considered as a case of intergranular fracture. After 4 days of annealing, the opening of the cracks has apparently increased with respect to the as-received state. Therefore, the number of cracks in Figure 4.12B compared to Figure 4.12A appears to be increased as well. However, this may solely be caused by the higher visibility of the existing cracks due to the larger opening. Also the lamellar structure of the surface changes into seemingly large lithium orthosilicate grains in which a multitude of spheroid lithium metatitanate grains is dispersed. With continued annealing for more than 4 days, no significant changes in the surface morphology are visible. The opening of the cracks does not increase considerably and also the grain size of the lithium metatitanate spheroid within the lithium orthosilicate grains seems to be marginally increased at most. From the images that are shown in Figure 4.12 it is also clear that the LMT spheroids do not pose a significant resistance to crack propagation as they are easily bypassed.

For the nominally 30 mol% LMT containing pebbles that were fabricated by the KALOS process, two significantly different surface morphologies are visible (see Figure 4.13), which was not observed for the KA-20-LMT samples. One morphology is composed of primary crystallized lithium metatitanate dendrites which feature characteristic secondary arms that are inclined at an angle of 90° with the primary dendrite arm.

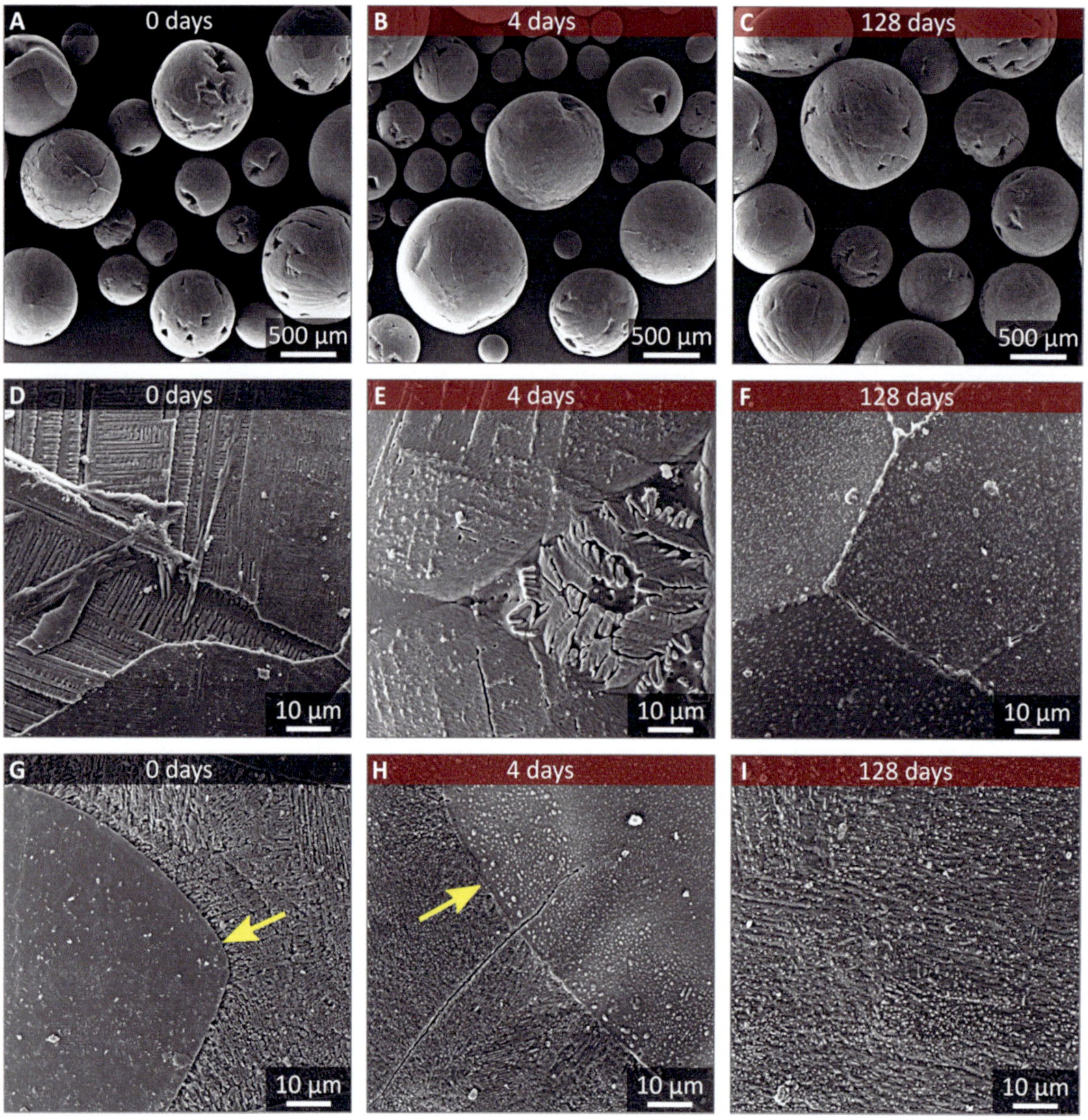

Figure 4.13: SEM surface images of the nominally 30 mol% LMT containing KALOS pebbles for different annealing durations. In the upper row (A-C), an overview of a number of pebbles is shown. Detailed views of the surface are shown in the other images. As two different morphologies exist within this set of samples, the two detail views are shown for each sample to illustrate the evolution of these morphologies. The yellow arrows point at the LMT 'platelets'.

In some cases, the secondary arm spacing can be so small and the secondary arms so thick, that an almost continuous structure is formed (see Figure 4.13D-F). The other morphology consists of a similar lamellar structure like the nominally 20 mol% LMT containing samples show. Yet, in addition to that, also almost single phase lithium metatitanate 'platelets' of about 10-100 μm size are found on the surface (see Figure 4.13G and H), which are separated by the lamellar structure. In the areas adjacent to the platelets, it often appears that the platelets nucleated the crystallization of the lamellar structure. Also contrasting the findings for the pebbles with lower LMT content, the KA-30-LMT pebbles clearly show more dents. Additionally, the size of the dents can be comparably large and the domains are usually more pronounced. However, large cracks are not visible as for the KA-20-LMT samples. There is one type of domains that is found in approximately half of the pebbles which is specifically angular. This type of domains is found in conjunction with numerous dents in the pebble surface. Nevertheless, the overall shape of all pebbles is still rather spherical.

After annealing the pebbles for 4 days, the overall appearance does not change visibly. Also a notable increase of the crack opening is not apparent. Even after 128 days of annealing, the changes are little, but the visibility of the cracks has clearly increased (also see Figure 4.14). The increased visibility of the surface cracks is also apparent for samples that were annealed for 64 days. Since a considerable number of the observable cracks do not appear to be rounded by the long time of annealing but rather show sharp flanks (see Figure 4.14D-F), it is self-evident that cracks are generated or advanced during the cooling prior to the sampling. In addition to that there are effects of the annealing on the structure of the surface. After 4 days the lamellar structure is, equally to the nominally 20 mol% LMT containing samples, coarsening and transforming into large lithium orthosilicate grains with lithium metatitanate grains within (see Figure 4.13H). The grains are usually still elongated in direction of the former lamellae and thus the initial structure is still guessable. This state is already achieved after 32 days (see Figure C.1 in Appendix C). Within the lithium metatitanate platelets and between the lithium metatitanate dendrite arms, a coagulation of lithium orthosilicate (unexpectedly appearing in light grey, most probably because of a number of pronounced edges and/or elevation over the other structures, see Figure 4.13 and Figure C.2) takes place after 4 days. Also the grain boundaries of the lithium metatitanate areas are locations, where lithium orthosilicate prominently concentrates (again appearing lighter than lithium metatitanate). With prolonged annealing the structures do not change considerably, with the exception of the lithium metatitanate rich areas in which pores are virtually eliminated after 128 days of annealing.

In Figure 4.15 the surface of the lithium excessive pebbles that were fabricated by the emulsion method are displayed. Overall the pebbles show no large defects as cracks or dents, independent of the annealing duration. Yet, the pebbles show in the as-received state a considerable open porosity with channels that seem to penetrate deep into the pebbles. Occasionally small intergranular cracks that usually span between a low monadic number of grains are visible (see Figure 4.15G). The observable grain size lies in the order of 1 μm. Also some debris from the porous surface can be identified on the surface. It presumably originates from the handling and transport of the pebbles.

After 4 days of annealing, the surface of the pebbles shows considerable densification and grain growth. The small cracks that were visible in the as-received pebbles' surfaces have almost completely vanished. The numerous channels that penetrate into the pebbles are considerably reduced in number and mostly also in diameter. Yet, a substantial amount of open porosity still exists. The grain size expanded to a value of approximately 3-5 µm. After 128 days of annealing, the surface of the pebbles shows no significant number of such channels and thus an open porosity that should lie close to zero. However, as the surface is smoothed in this way, the grain size is increased even more to values clearly exceeding 10 µm.

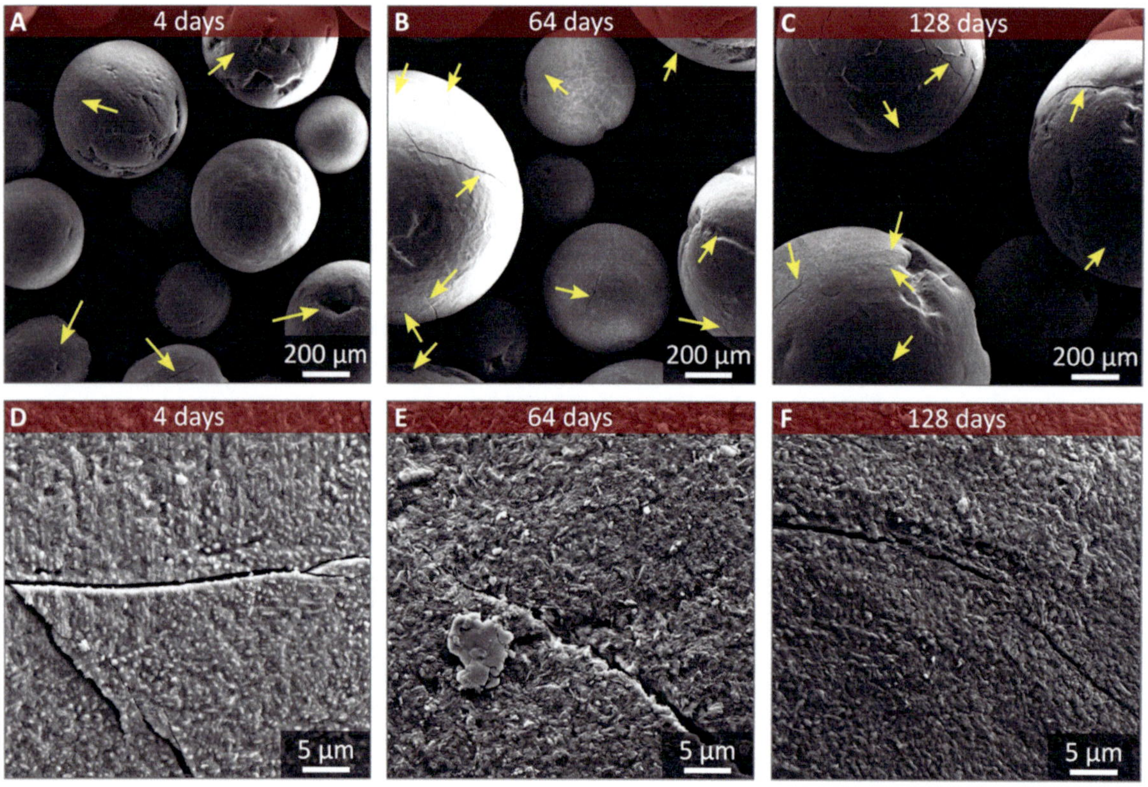

Figure 4.14: SEM surface images of the nominally 30 mol% LMT containing KALOS pebbles for different annealing durations. The yellow arrows in A-C indicate prominently visible surface cracks. In the bottom row, detail views of sharp-edged surface cracks are shown.

Most grains also show characteristic facets which already begin to emerge after 64 days of annealing (see Figure 4.15K). More importantly, however, is the observation that a considerable number of fine intergranular microcracks re-emerge after longer annealing periods (see Figure 4.15I-K). These microcracks first appear after 32 days of annealing as both intergranular and intragranular cracks. Yet not all pebbles show these cracks. After 64 days of annealing, and also after 128 days of annealing, virtually no intragranular cracks are visible.

Figure 4.15: SEM surface images of the lithium excessive LMT pebbles that were fabricated by the emulsion method for different annealing durations. In the images A-C, an overview of a number of pebbles is shown, while a detailed view of the surface is shown in the other images. The yellow arrows point at surface cracks.

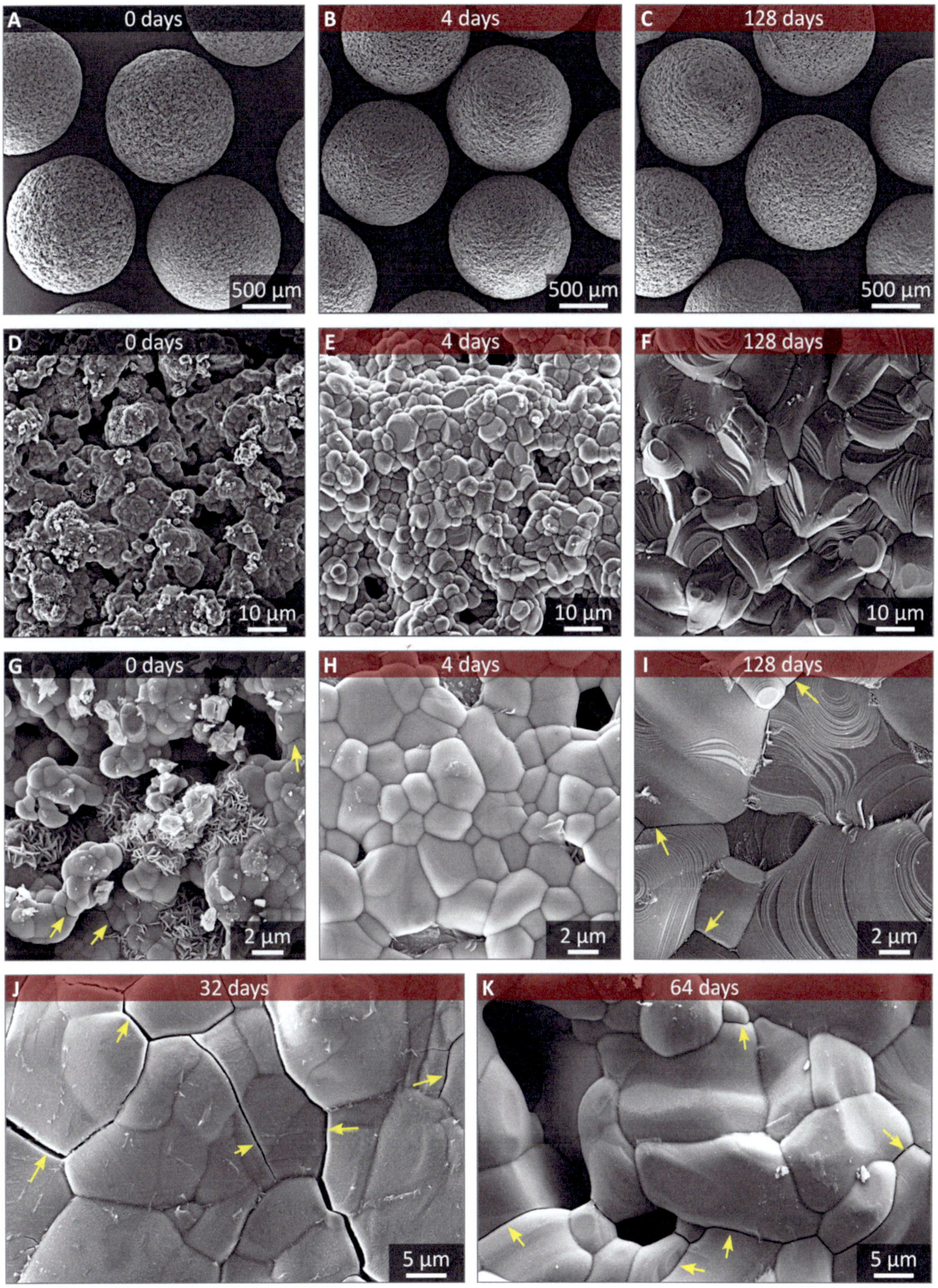
A
0 days
500 µm
B
4 days
500 µm
C
128 days
500 µm
D
0 days
10 µm
E
4 days
10 µm
F
128 days
10 µm
G
0 days
2 µm
H
4 days
2 µm
I
128 days
2 µm
J
32 days
5 µm
K
64 days
5 µm

In principle, the surface morphology of the lithium excessive pebbles that were fabricated by the sol-gel method is largely comparable to and develops similarly to the lithium excessive pebbles that were fabricated by the emulsion method (see Figure 4.16). Yet, some differences are visible. In general, the sol-gel derived pebbles are less spherical in comparison and the initially smooth appearance of the pebble surface roughens with prolonged annealing.

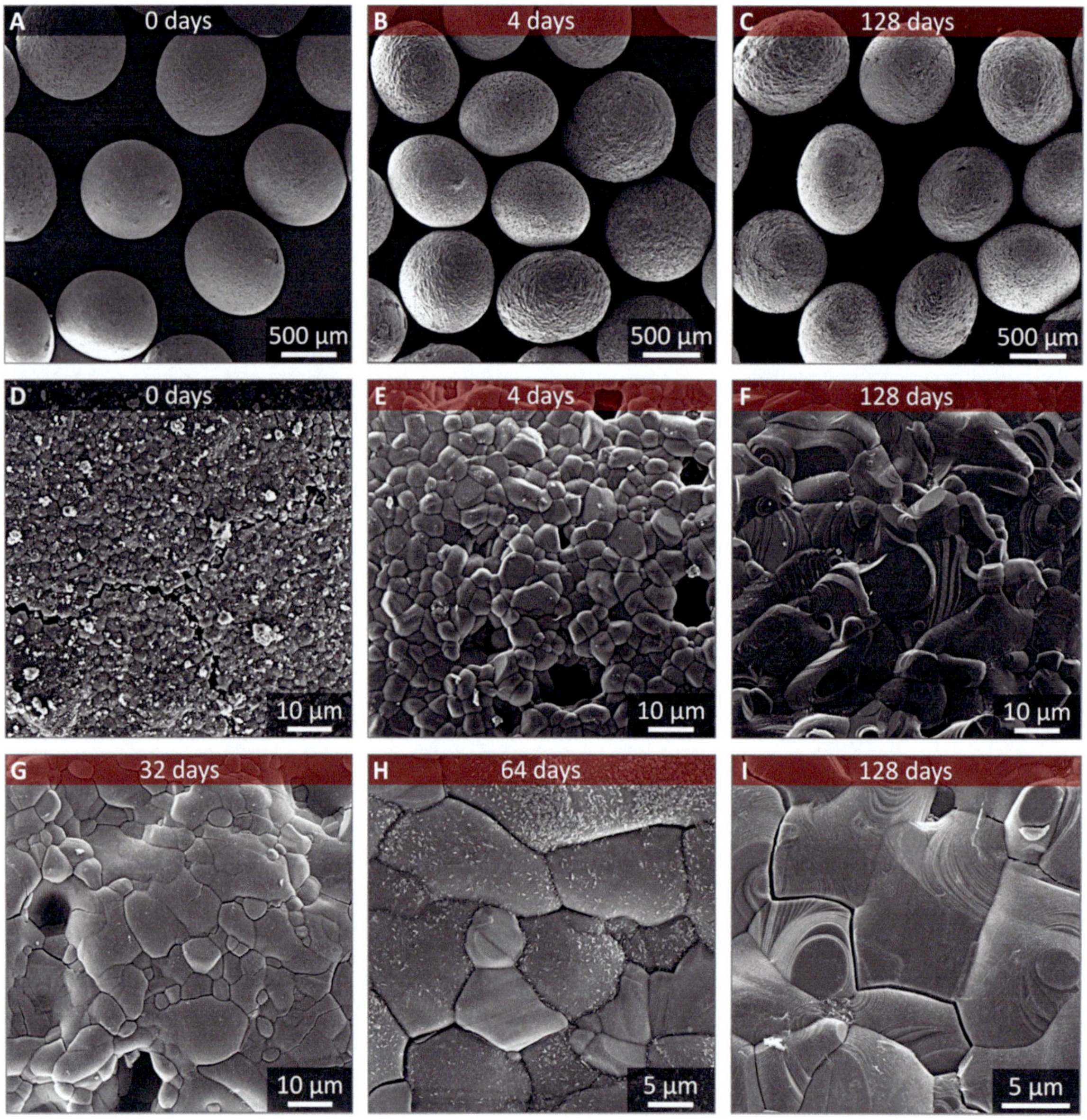

Figure 4.16: SEM surface images of the lithium excessive LMT pebbles that were fabricated by the sol-gel method for different annealing durations. In the upper row, an overview of a number of pebbles is shown, while a detailed view of the surface is shown in the other images.

But as for the pebbles that were fabricated by the emulsion method, large flaws are not apparent. Yet, a network of intergranular cracks that can span along a large number of grains as well as just a few grains is visible. As the impression of a smooth surface already suggests, there is virtually no open porosity visible for these pebbles as a relatively dense surface of grains of 1-2 µm diameter constitute the pebble surface. However, after 4 days of annealing, similar channels, in size as well as in number, appear as observed for the emulsion method pebbles. The grain size also increases to approximately 3-5 µm. After 128 days the grain size is increased expectedly to values greatly exceeding 10 µm. However, also the generated channels have vanished. The beginning of the vanishing of these channels is already visible after 32 days of annealing (see Figure 4.16G). Yet, also the formation of microcracks goes along with increasing annealing duration as observed for the pebbles that were fabricated by the emulsion method (see Figure 4.16G-I). The morphology of the sintered LOS/LMT containing pebbles that were fabricated by the emulsion method is displayed in Figure 4.17. Overall the pebbles show a rather sphere-like shape, although basically all pebbles show some sort of larger defects. Dents are the most frequently observable defect but also the accumulation of pores in a (curved) line, which resembles a sort of seam, is frequently found. Usually these seams have developed within the dents. However, the dents do not seem as large as those of the KALOS pebbles (see Figure 4.12 and Figure 4.13).

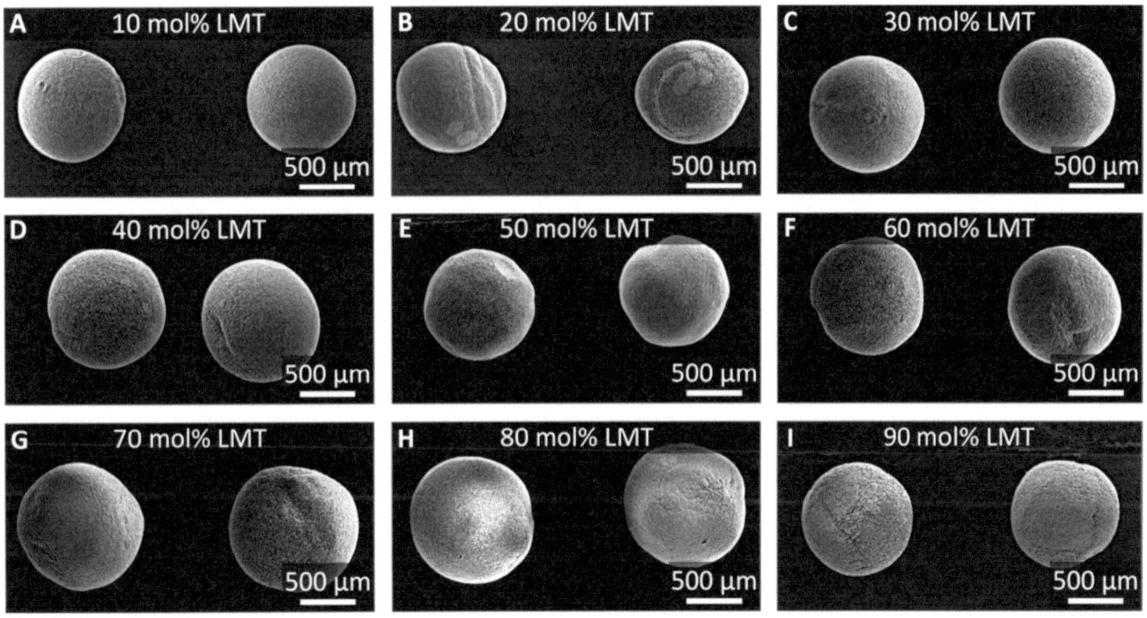

Figure 4.17: SEM surface overview images of LOS and LMT containing pebbles that were fabricated by the emulsion method as already published in [96]. For each picture, the nominal composition is given.

Additionally, some pebbles show wave-like structures on their surface. Yet, no palpable correlation between the nominal LMT content and the defects is visible. Apparently, large seams of pores are not visible for pebbles that contain 70 mol% LMT and 80 mol% LMT. The surface of all of these pebbles shows a considerable amount of open porosity (see Figure 4.18), which seems to increase with nominal LMT content. However, the complicated structures may be deceiving. Nevertheless the surface of the pebbles with high LMT concentrations seems less dense than for lower LMT concentrations. From the SEM images that are shown in Figure 4.18 it is clear, that the amount of LOS (the dark grey phase in the images compared to the brighter LMT phase) also determines the LOS grain size. For pebbles with a nominal content of 10 mol% LMT the LOS grain size is clearly largest with approximately 5 µm and diminishes with reduced LOS concentrations to about 2 µm. However, it is difficult to determine LOS grains on the surface of pebbles which contain 80 mol% or more LMT.

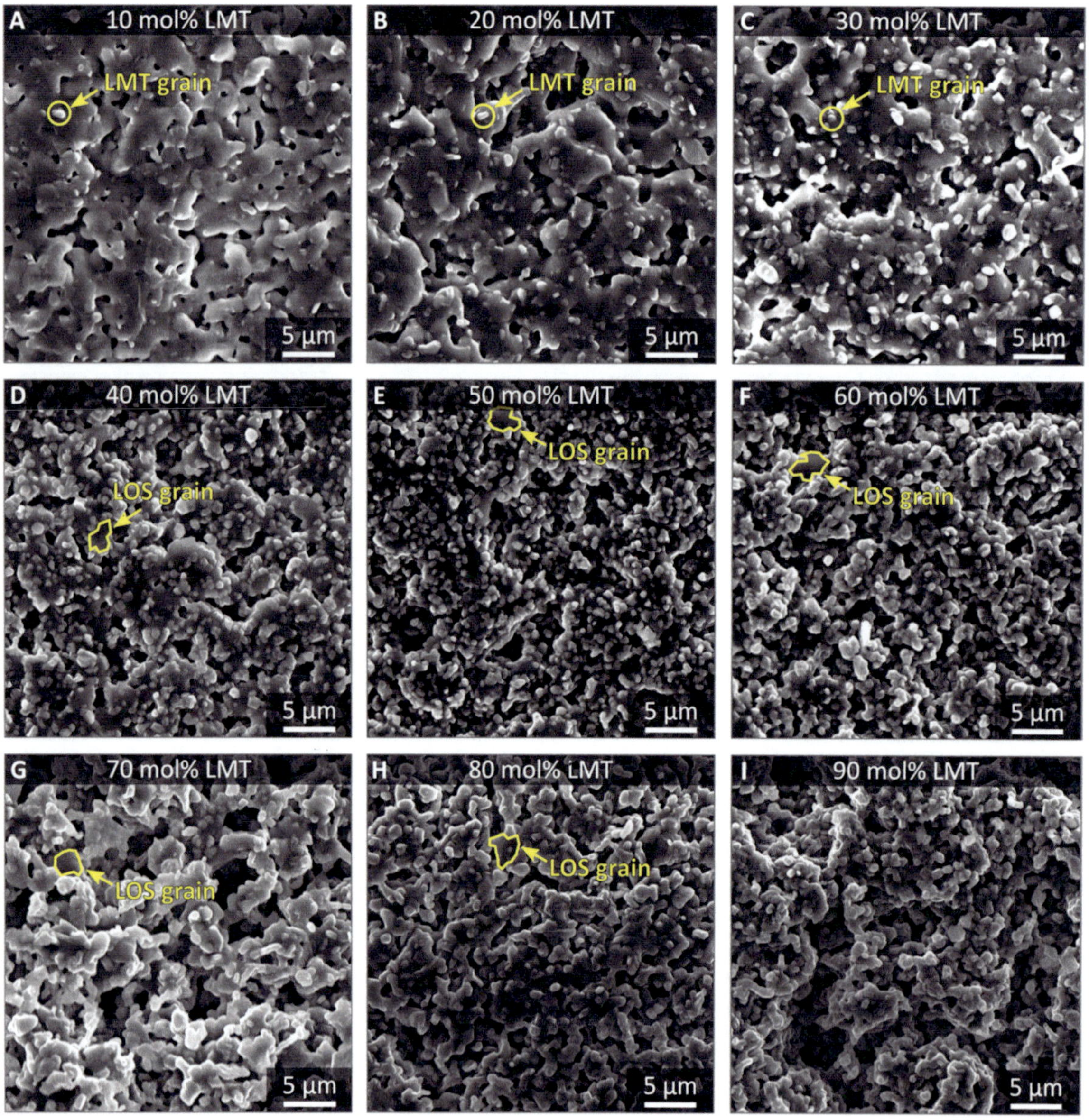

Figure 4.18: SEM surface structure images of LOS and LMT containing pebbles that were fabricated by the emulsion method. For each picture, the nominal composition is given. In the pictures one grain of LOS or LMT is exemplarily indicated.

For compositions lower than 60 mol% LMT (which equals a LMT content of about 50 vol.%), the lithium metatitanate grains are usually immersed within a matrix of LOS without notably contacting each other. This goes along with the seemingly less dense structure of the sample surfaces. These LMT grains show a size of approximately 1 µm. For higher LMT contents, the LOS grains span between several LMT grains, which then seemingly constitute the structural network of the pebbles instead of the LOS grains. At this point when the LMT grains share a high number of contacts, i.e. from nominally 70 mol% LMT to higher LMT contents, there is a slight increase of the LMT grain size with increasing LMT composition visible.

The non-sintered but dried green LMT containing pebbles show similar large defects as the sintered ones, as Figure 4.19 shows. The same wave-like structures are visible as well as forerunners of the seams that are equally prominently visible on the surface of the green pebbles as on the sintered pebbles. The wave-like structures that are found on every green pebble closely resemble drying cracks. Overall, the density of the surface seems relatively equal for all pebbles, except for the nominally 90 mol% LMT containing ones which show a notably rougher surface.

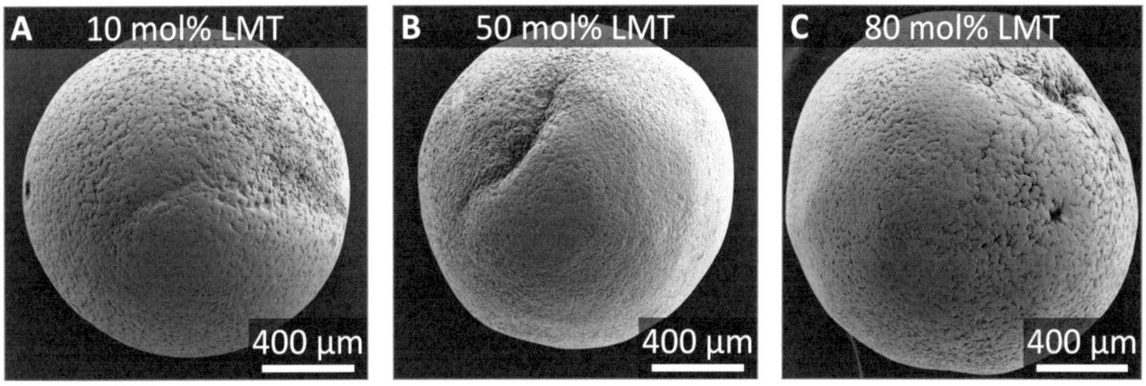

Figure 4.19: SEM surface images of selected green LOS and LMT containing pebbles that were fabricated by the emulsion method. For each picture, the nominal composition is given.

The lithium aluminate containing pebbles show in principle similar defects as the LMT containing pebbles that were fabricated by the emulsion method. All pebbles show dents, which can also be rather extreme as displayed by one pebble of nominally 10 mol% LAO in Figure 4.20. Apart from this pebble, all pebbles can be considered to be nearly spherical. Also the LAO containing pebbles show structures like the 'seams' that are visible for the LMT containing pebbles. Yet, in case of the LAO containing pebbles, the 'seams' do not correspond to an accumulation of pores. Also no wave-like structures are found on the surface of the LAO containing pebbles. The microstructure that is established on the surface of the pebbles shows a significantly different morphology as compared to the LMT containing pebbles that were fabricated by the emulsion method. There is no significant open porosity apparent and the grain size is considerably larger than for the comparable pebbles with LMT as second phase. For all pebble compositions, the lithium aluminate seemingly formed as a grain boundary phase between LOS grains with a size of approximately 10 µm.

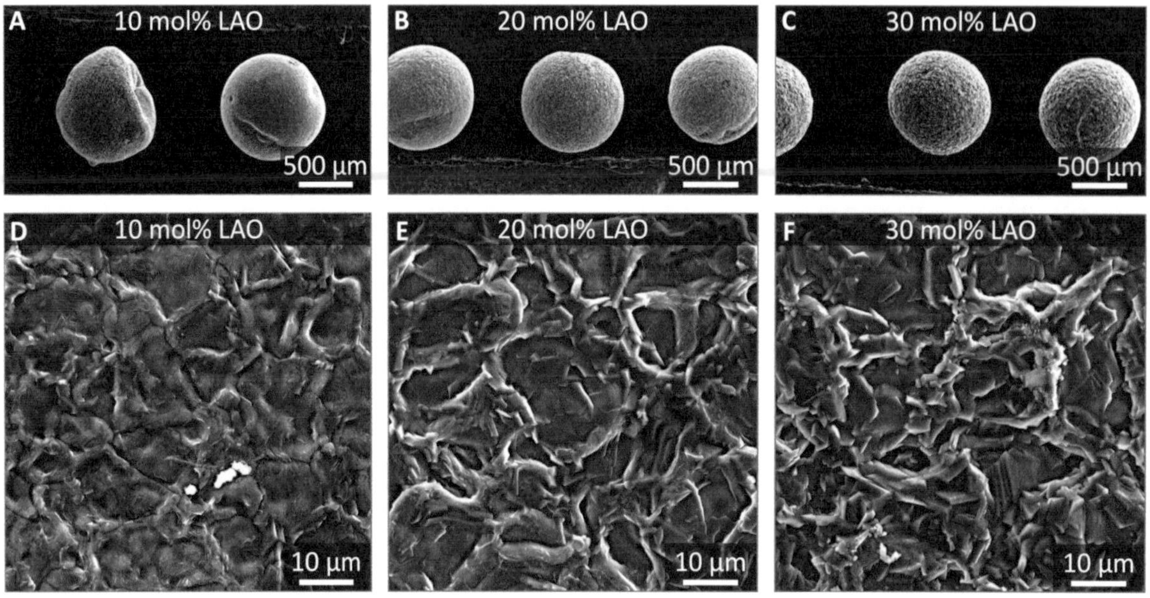

Figure 4.20: SEM surface images of LOS and LAO containing pebbles that were fabricated by the emulsion method as already published in [96]. For each picture, the nominal composition is given. In the upper row (A-C), an overview of a number of pebbles is shown, while a detailed view of the surface is shown in the other images.

It is almost impossible to identify either LOS or LAO by their brightness in SEM pictures as both phases show similar scattering of electrons. For the pebbles with a nominal composition of 10 mol% LAO, a considerable number of LOS grains share boundaries, while with an increase to 20 mol% LAO, LOS grains are seemingly enclosed by LAO. As the LAO concentration within the pebbles increases, the LOS grain size seems to decrease to about 5 µm. Yet, as the grain boundary LAO phase grows more salient with increasing LAO concentration, the estimation of the grain size is relatively difficult from surface images of 30 mol% LAO containing pebbles. Also, as the LAO phase shows a rather fissured surface structure, it is not possible to determine a meaningful value of its grain size.

4.3.2 Size distribution and sintering shrinkage

The determined pebble size distributions are given in Figure 4.21 with the frequency plotted as a function of the diameter and the second phase concentration in a color coded graph for both, the dried but non-sintered pebbles as well as the sintered pebbles. For the lithium metatitanate containing samples, a distinct concentration dependence of the dried pebble diameter is not observable. Yet, the size distribution is not as constant as desired and shows significant deviations for each prepared concentration. The majority of the samples (20 mol%, 30 mol%, 40 mol% and 60 mol% LMT) shows a mean pebble size of about 1530 µm. Some of the other samples (10 mol%, 50 mol% and especially 90 mol%) show smaller pebble sizes, while the samples with 70 mol% and most significantly 80 mol% LMT show larger sizes. All distributions seem to be multi-modal. After sintering, the size distributions considerably narrow and can be considered to be mono-modal. Overall minor deviations in dried pebble size do not seem to largely affect the final size of the sintered pebbles as most pebbles reach a mean pebble size of about 1030 µm and the multiple maxima of the dried pebbles are not observable for the sintered samples. Yet, the different pebble sizes of 70 mol% and 80 mol% LMT of the sintered pebbles can be related to the size of the dried samples.

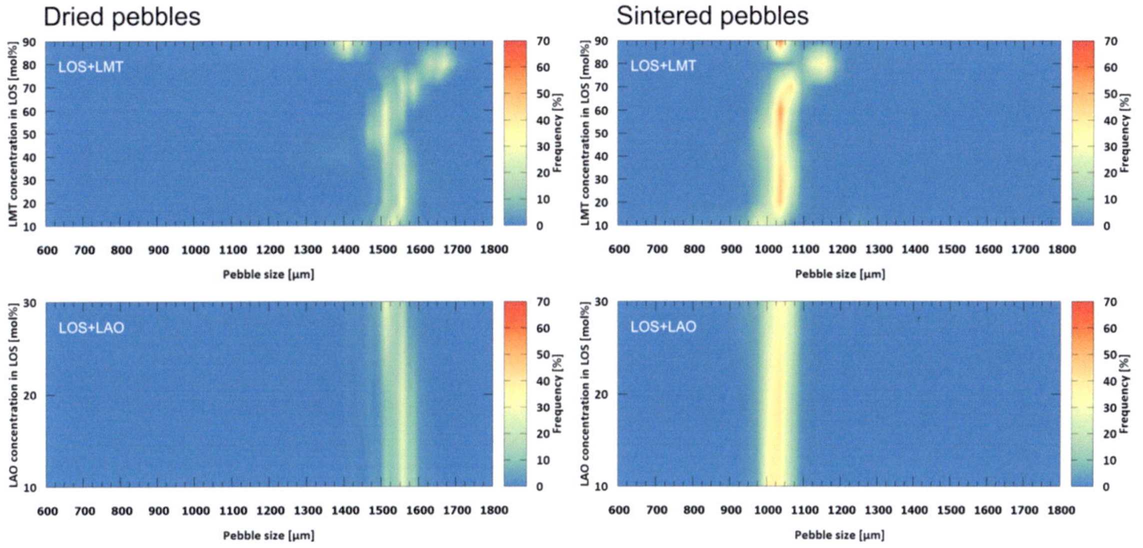

Figure 4.21: Optically determined diameter distribution of dried and sintered, biphasic emulsion method pebbles as a function of the nominal second phase content for the addition of LMT and LAO. The data has been interpolated between the data points along the x- and y-axes. These plots are an improved version of the ones that were already published in [96].

For the dried lithium aluminate containing pebbles, such variations are not visible. Yet, there is visible dependence on the nominal lithium aluminate content. From the lowest lithium aluminate content of 10 mol% to the highest of 30 mol% the pebble size distribution considerably shifts to lower pebble sizes for the dried pebbles, while the opposite is true for the sintered pebbles. The size distributions of the dried pebbles also clearly exhibit multiple maxima, while the sintered pebbles' distributions are more uniform, similar to the observations for the EM-X-LMT pebbles. Yet a slight broadening can be observed for the sintered samples with lower LAO content. Generally, the determined mean pebble sizes for dried and sintered pebbles are similar to the LMT containing pebbles. Overall the size distributions are still relatively narrow compared to the usual size distributions of pebbles fabricated by melt-based processes (compare to Figure 2.6). The exact mean pebble sizes of these samples are given in Appendix C. From the mean sizes of dried and sintered pebbles, the shrinkage during sintering can be obtained. All pebbles exhibit a significant densification during sintering (see Figure 4.22).

For both sets of samples, the shrinkage is similar with values between 30 % and 35 %. A slightly decreasing trend with increasing second phase content can also be observed. Only for 90 mol% LMT in LOS there is a significant decline in shrinkage to about 27 %. Figure 4.22 also shows the sum of the green and imaginary density of the pebbles, which seemingly mirrors the measured shrinkage as a function of the second phase content. As pebbles were produced by the same fabrication process and with very similar process parameters, it can be assumed that the fabricated pebbles show a similar degree of non-sphericity. Therefore, the value of imaginary density (see section 3.6) should be constant for these samples. A linear regression of the shrinkage (not shown in Figure 4.22) as a function of the nominal LMT content suggests a decrease of the shrinkage of 5.25 % over the fabricated composition range. The lowest sum of green and imaginary density is measured for the 70 mol% LMT containing sample with about 34 %, whereas the samples with 20 mol%, 30 mol% and 90 mol% show the highest values with about 39 %. As the values of the imaginary density are negative, the true values of green density have to be higher.

For the KALOS samples as well as the lithium excessive lithium metatitanate samples, the mean pebble diameter was obtained as a byproduct of the density determination, which means that much less pebbles were analyzed in contrast to a dedicated pebble size analysis. As a result, the values for the KALOS pebbles show a considerable scattering because of their broad pebble size distribution (see Figure 2.6), while the values that were obtained for the rather monodisperse EM-LMT and SG-LMT samples suffer significantly less from this issue.

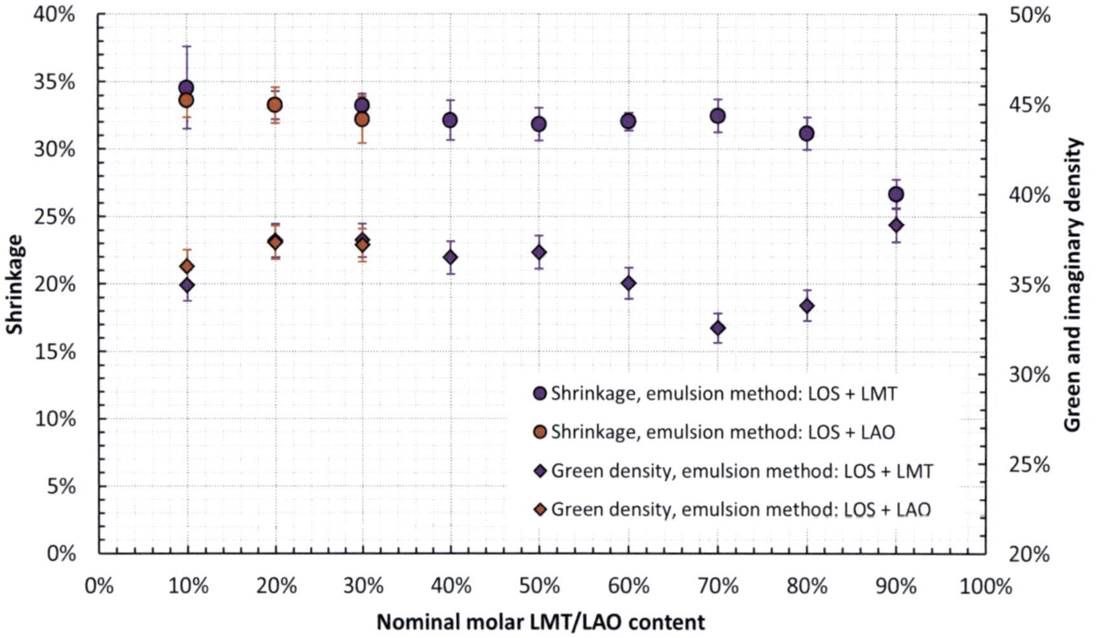

Figure 4.22: Mean shrinkage of the pebbles during sintering as a function of the nominal second phase content for the addition of LMT and LAO as already published in [96]. The error bars indicate the mean standard deviation of the individual measurements. Also the sum of the green and imaginary density as a fraction of the expected pebble density is given as a function of the nominal second phase content. The error bars for these values amount to ±2.5 % of the determined value.

The technical accuracy of this method however, is reasonably high as discussed in section 3.6. Similar to the observed shrinkage for the biphasic emulsion method pebbles during sintering (see Figure 4.21), the long-term heat treatment during annealing may also lead to a densification of the pebbles. The obtained values for all four grades of pebbles as a function of annealing duration are detailed in Table 4.6. The mean pebble diameters of the KALOS pebbles (KA20-LMT as well as KA-30-LMT), given the considerable scattering of the values, do not seem to follow a clear trend. Therefore, the pebble diameter of both sets of samples has to be considered as relatively constant as a function of annealing duration. For the EM-LMT samples, the pebble diameter is clearly reduced during annealing with respect to the as-received state, whereas the SG-LMT samples do not show such a trend. The obtained values of the size of these pebbles rather increases after 4 days of annealing and only decreases to a value close to the initial one after 128 days of annealing. The assumed uncertainty of the measurements mostly exceeds the determined change of the mean values as a function the annealing duration. This is particularly important for the EM-LMT samples, for which the determined mean pebble diameter decreases with increasing annealing duration. However, on the given basis it is neither possible to conclude that these values are statistically indistinguishable nor distinguishable [181,182].

Table 4.6: Mean pebble diameter as obtained by optical microscopy during the determination of the pebble density as a function of the annealing duration in He/H_2 atmosphere. For each value an uncertainty of ±11 µm is assumed.

	As-received	4 days	32 days	64 days	128 days
KA-20-LMT	1083 µm	889 µm	958 µm	1041 µm	1019 µm
KA-30-LMT	1132 µm	910 µm	889 µm	1106 µm	895 µm
EM-LMT	1449 µm	1406 µm	1403 µm	1395 µm	1388 µm
SG-LMT	1017 µm	1028 µm	1026 µm	1031 µm	1020 µm

4.3.3 Microstructure

The development of the microstructure of the pebbles is shown in this chapter. Therefore, SEM images of cross sections of a subsample of pebbles are presented in the following. In Figure 4.24 the microstructure of an as-received KALOS pebble that nominally contains 20 mol% LMT is shown after a special ion-polishing step by a Leica EM TIC 3X. By this ion-polishing step, the uppermost layer of the mechanically polished cross section is removed to reveal an almost undisturbed surface, which self-evidently provides a considerably higher surface quality than with conventional preparation and thus reveals finest details of the microstructure. Cross section images of conventionally prepared samples for different annealing durations are shown in Figure 4.24.

In the as-received state, the majority of the nominally 20 mol% LMT containing pebbles clearly shows a eutectic microstructure with very fine alternating parallel lamellae of lithium orthosilicate and lithium metatitanate, which form between primary crystallized lithium orthosilicate (see Figure 4.23). Therefore this microstructure is unambiguously assigned to a hypoeutectic composition. A single pebble is usually composed of a low number of domains with a size of more than 100 µm (see Figure 4.24D and much clearer after 4 days of annealing in see Figure 4.24F), in which the crystallized primary LOS dictates basic morphology (see Figure 4.24E). The primary LOS can crystallize in the form of dendrites, but it is not always clearly the case. A second type of pebbles, which is obviously hardly attacked by the water that is used for etching, constitutes the remaining part of the pebbles (see Figure 4.24D). Because of the difficulties during etching, only the large domains can be somewhat identified while the expected biphasic nature, which is displayed by the majority of the pebbles, is not visible.

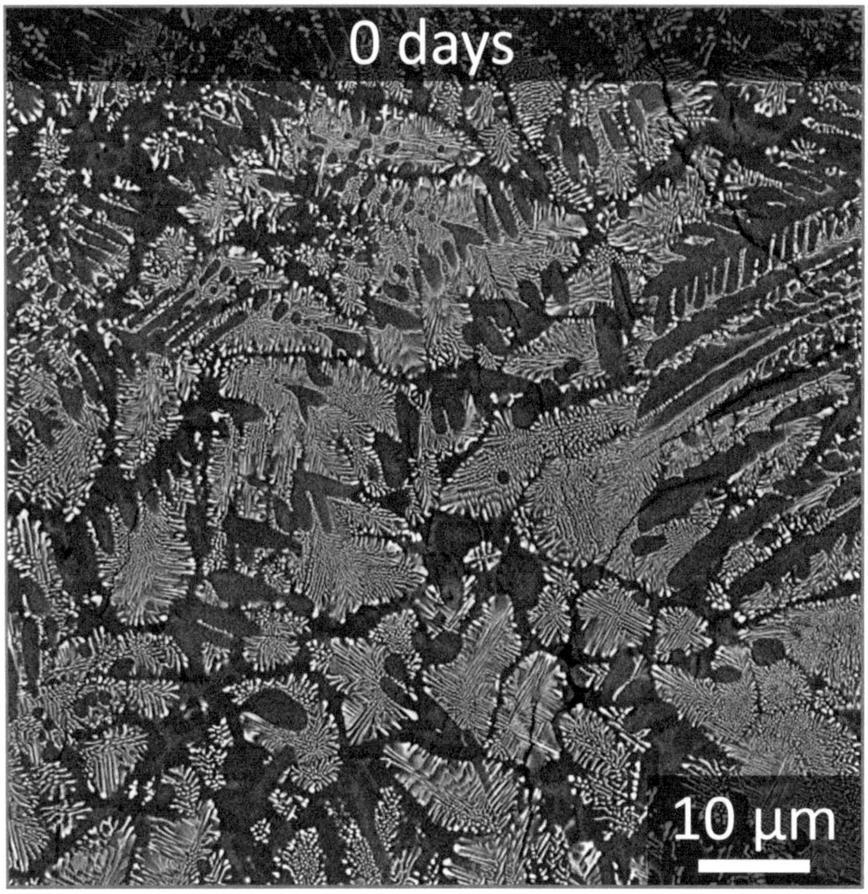

Figure 4.23: SEM micrograph of an ion-polished cross section of an as-received, nominally 20 mol% LMT containing KALOS pebble. For achieving a reasonable optical quality, the ASB detector for Z-contrast is used. Therefore, the bright phase represents lithium metatitanate while the dark phase represents lithium orthosilicate.

Yet, also these pebbles seem to crystallize dendritically. The lamellae are not necessarily planar and presumably follow strong thermal gradients that existed during the crystallization of the liquid melt. The thickness of the lamellae varies considerably. Yet it generally lies below 1 µm and can decrease to values of 10 nm or less (see Figure 4.23 and Figure 4.24E). These lamellae usually appear to be free of cracks, irrespective of the phase. For the vast majority of the pebbles that clearly show both phases, the cross section images reveal a slightly finer microstructure in terms of smaller dendritic domains within the enveloping region about 30 µm near the pebble surface (see Figure 4.24C). Such pebbles have obviously crystallized homogeneously.

Heterogeneous crystallization is no issue for these pebbles as there are only few pebbles that show hints of local heterogeneous crystallization. Within this complex microstructure, some small pores in the monadic micrometer range are also visible when the epoxy resin fills them (see Figure 4.24E). Very large pores are not present in these samples.

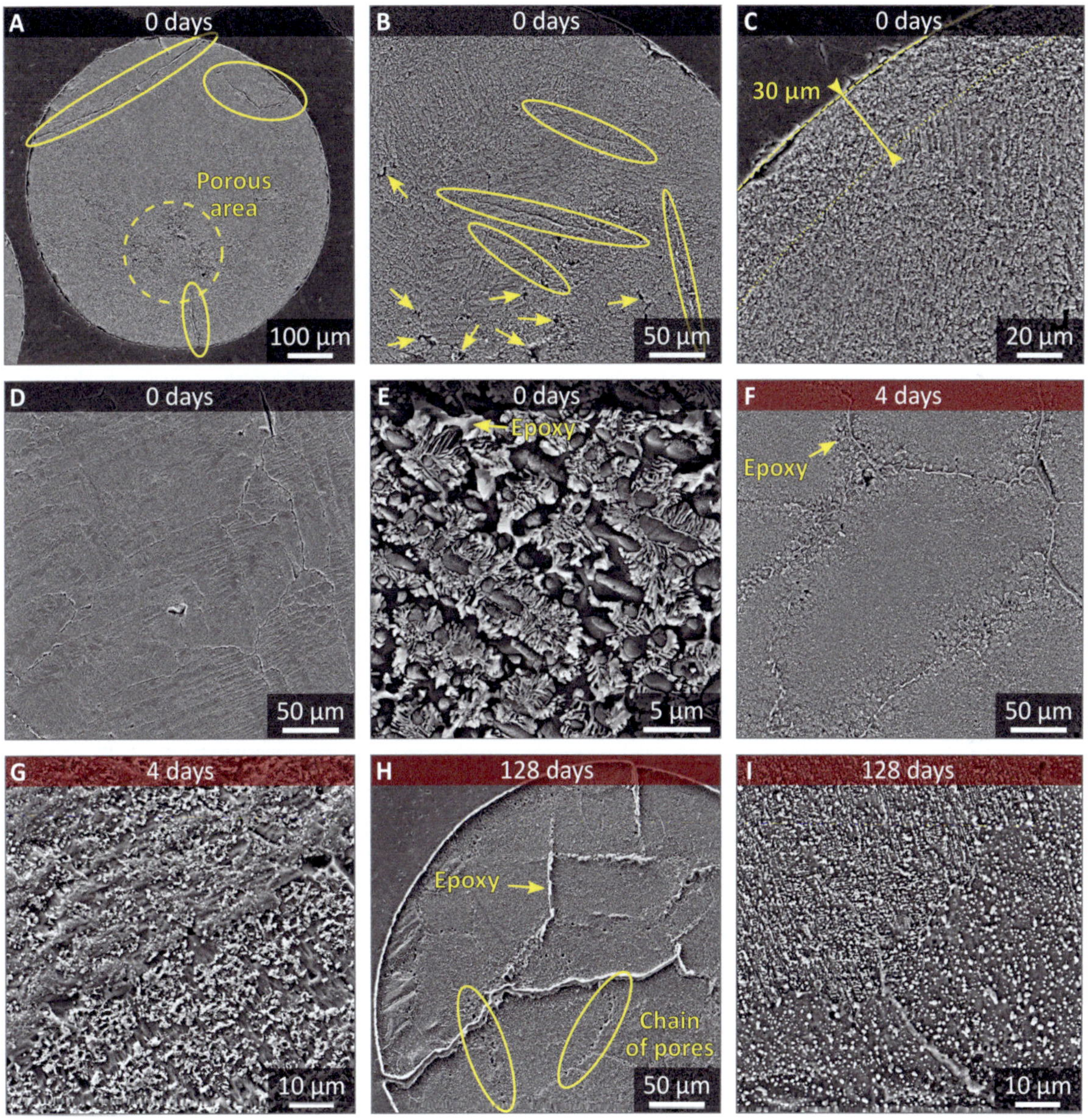

Figure 4.24: Etched cross sections of nominally 20 mol% LMT containing pebbles that were fabricated by the KALOS process imaged by the SEM. The images display the microstructure of the samples in as-received state and after annealing. The bright structures represent the lithium metatitanate phase, while the dark grey regions represent the LOS matrix. Along the larger cracks epoxy resin infiltrated the pebbles. For better visibility, some of these cracks are highlighted by yellow ellipses. The yellow arrows in image B point at pores that are also observed within the porous area in image A.

Yet, the actual porosity is difficult to judge from the etched cross section images is difficult because of the complex microstructure. Instead, it is easy to identify several large cracks within the pebbles, as the epoxy resin, into which the pebbles are embedded, is not etched and protrudes over the surface of the ceramic material (see Figure 4.24A and B). Many cracks start from the surface and either extend into the volume of the pebbles in a relatively straight manor or follow the boundaries of the eutectic domains in a branched manor. In contrast to the straight cracks, the branched cracks are far less gaping within the volume of the pebbles. As Figure 4.24B shows, in some pebbles cracks are observable which have no connection to the surface and have only formed within the pebble volume. With respect to porosity and cracks, the pebbles that are not attacked properly by water show no significant differences. Within the interior of the pebbles, pores that have formed along the borders of two eutectic domains are frequently observed (see Figure 4.24A and B). Figure 4.24F further shows the changes upon annealing for 4 days, as gaps between the eutectic domains grow significantly larger. These cracks also frequently interconnect with each other and also with the cracks that extend into the pebbles from the surface. The structure within the pebbles, i.e. the LOS and LMT lamellae, changes in morphology after 4 days of annealing. Yet, the degree of the changes usually differs considerably between different domains. While the lamella structure has almost vanished for some domains instead of spheroid LMT inclusions and a seemingly uniform LOS matrix, the lamella structure can still be relatively intact in different domains showing few signs of grain growth and coagulation (see Figure 4.24G). If spheroid LMT inclusions have formed, their size is usually in the order of 1 μm or less. A significant increase of the domain size is not unambiguously observable. The porosity, as far as it can be observed, does not seem overly large or increased over the as-received state, especially within the domains. Yet, the gaping cracks within the pebbles also have to be considered as a form of porosity and may be a result of pore coalescence along domain boundaries.

For the second type of pebbles, the etching by water seems to be more effective than for the as-received pebbles, although a thin layer still seems to cover the actual surface of the cross section for a number of pebble domains (see Figure 4.25A-C). Where this layer is removed, the underlying material appears in a notably darker grey. Furthermore, the areas that are eventually uncovered from preparation residues show angled etch pits and wavy surface structures that are anomalous for etched lithium orthosilicate (see Figure 4.25C). The microstructure of these pebbles consists, in contrast to the majority of the pebbles, of large lithium orthosilicate grains with some spherical precipitates of lithium metatitanate. These grains are approximately of the size of the domains that were also recognizable in the as-received state (compare Figure 4.24 and Figure 4.25). At the grain boundaries of these grains a thin layer of lithium metatitanate has formed which separates them (see Figure 4.25B). Along the so-decorated grain boundaries, cracks of substantial length are observable as a LOS grains can obviously be detached from the lithium metatitanate layer rather easily. Pebbles that show a microstructure of the second type feature notably larger pores compared to the pebbles with the ordinary microstructure of eutectic domains.

After an annealing for 128 days, the disparity of the transformation of the lamellar eutectic domains of the ordinary microstructure is still evident (see Figure 4.24I). Yet, the spheroid LMT grains seemingly grew slightly in size. In terms of the large cracks that extend through the pebbles, the overall appearance does not change significantly after 128 days of annealing, compared to the state after 4 days of annealing (see Figure 4.24H). Potentially the number of clearly visible cracks that formed along the boundaries between different domains increased slightly, which may also be a result of perpetual pore coalescence.

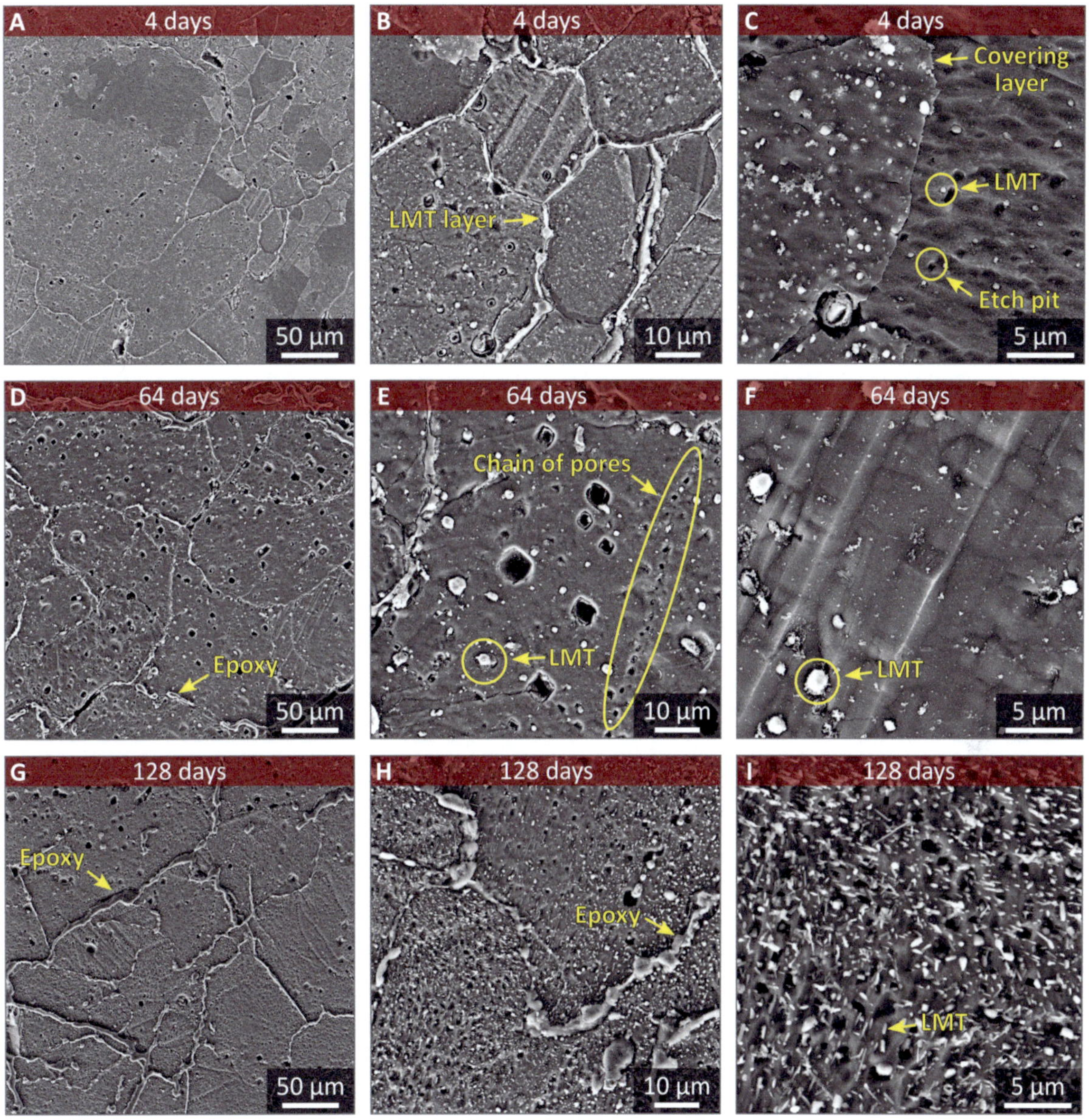

Figure 4.25: Cross sections of nominally 20 mol% LMT containing pebbles that were fabricated by the KALOS process imaged by the SEM. These pebbles are of the second type and do not show a microstructure which consists of eutectic domains and primary lithium orthosilicate. Along the cracks of the samples that were annealed for 64 and 128 days epoxy resin infiltrated the pebbles. Some of the pebbles that were annealed for 128 days (H-I) show the precipitation of LMT platelets.

Yet, it is also obvious that the lithium orthosilicate phase shows notable grain growth after long annealing durations as chains of pores intersect large domains although the matrix of lithium orthosilicate is in many cases unambiguously identically oriented on both sides of these chains (see Figure 4.24H). The pores are most probably remains of the former grain boundary and are seemingly unaffected by the growth of the matrix phase. Similarly, the lithium metatitanate grains are also not visibly affected and are not dragged along with the moving grain boundaries. The pores of these chains also grow in size to some degree, so that they are clearly visible after long annealing durations. It is, however, observed that cracks occasionally extend along these chains of pores as in Figure 4.24H.

For the second type of pebbles, the evolution of the microstructure after 64 days and 128 days of annealing is displayed in Figure 4.25D-I. The images clearly show that the thickness of the lithium metatitanate layer which surrounds the lithium orthosilicate grains grows and also the domain size significantly increases with respect to the state after 4 days of annealing (also visible from overview images in Figure C.3). Especially after 128 days of annealing, rather large lithium orthosilicate grains are observed, but apart from that there are insignificant difference between pebbles that were annealed for 64 days and 128 days. In further contrast to the samples that were annealed for 4 days, pebbles of this type show a significant, in all likeliness closed, porosity of rather uniform size. The high similarity between both samples (64 days and 128 days of annealing) also applies for the pore size, which is rather similar. In both samples also chains of pores are frequently observed, similarly to the samples that showed the ordinary microstructure in the as-received state and were annealed for 128 days (see Figure 4.24H and Figure 4.25E). For the LMT dispersoids that can be observed after 4 days of annealing, no significant changes in size are apparent as a function of annealing duration. Yet, for a significant fraction of pebbles that were annealed for 128 days one clearly distinguishing feature from the pebbles that were annealed for 64 days exists. These pebbles show the precipitation of small LMT platelets along certain planes within the LOS matrix (see Figure 4.25H-I). These platelets appear to be even thinner than the eutectic lamellae that were formed during the crystallization of the liquid droplets (compare to Figure 4.23). Apparently, these pebbles are supersaturated with titanium and may be the reason why the etching interacts visibly differently with the material, i.e. the observed anomalous surface structures are formed.

The cross section images of nominally 30 mol% LMT containing pebbles that were fabricated by the KALOS process are shown in Figure 4.26 and Figure 4.27, since there are two distinctly different microstructures observable for these pebbles. Both microstructures appear approximately equally often. One of these morphologies resembles that of the nominally 20 mol% LMT containing KALOS pebbles (see Figure 4.26) and so does the evolution of this microstructure during annealing. The pebbles that show this microstructure are mostly composed of eutectic domains (see Figure 4.26D), yet without the presence of significant amounts of dendritic primary lithium orthosilicate. Sporadically primary lithium metatitanate dendrites interrupt the otherwise eutectic microstructure within these pebbles (see Figure 4.26A). The eutectic domains are roughly equally sized as for the KALOS pebbles with less added LMT. However, the domain size varies considerably. Also similar to the nominally LOS-richer pebbles, an enveloping layer below the pebble surface is visible where the microstructure is notably finer than in the rest of the pebble. The size of this layer is about 30-50 μm (see Figure 4.26G). Furthermore, the vast majority of these pebbles has crystallized homogeneously.

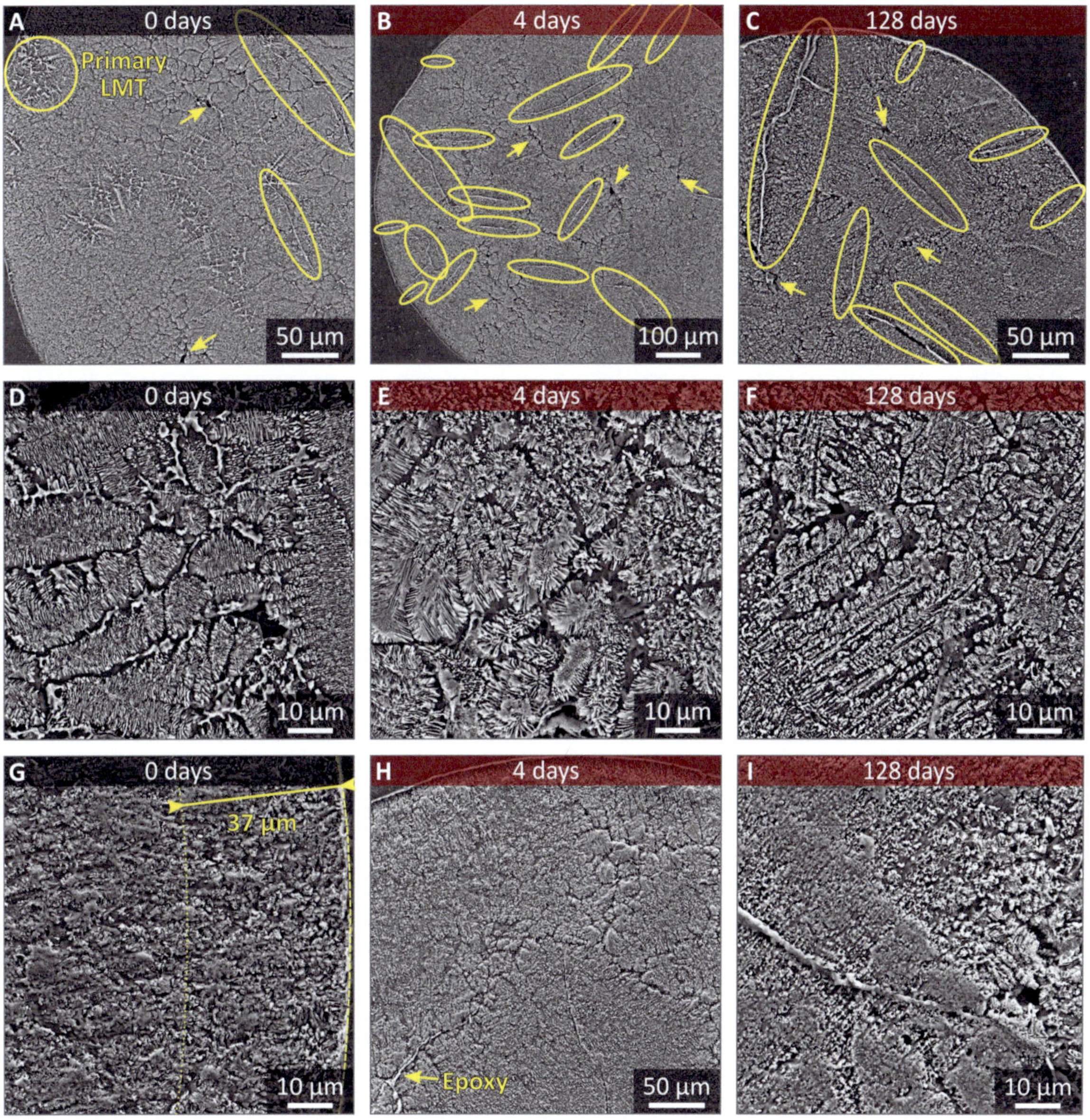

Figure 4.26: SEM images of etched cross sections of nominally 30 mol% LMT containing KALOS pebbles, which show the pearlite-like microstructure. Image A shows a pebble which also features some primary LMT dendrites. The samples are shown in the as-received state and after annealing. All cracks of significant size have been marked by yellow ellipse and yellow arrows point at pores of considerable size. Along the largest cracks epoxy resin infiltrated the pebbles.

The other microstructure that appears in the nominally 30 mol% KALOS pebbles exclusively shows primary lithium metatitanate dendrites which crystallize in a typical 90° pattern of dendrite arms (see Figure 4.27). Consequently this microstructure is unquestionably assigned to a markedly hypereutectic composition. This 90° pattern certainly reflects the cubic lattice in which LMT crystallizes at high temperatures (see also Table 2.3). Between the dendrite arms lithium orthosilicate crystallizes. There are, however, no clear indications, that a eutectic melt crystallizes between the LMT dendrites. Also these pebbles apparently crystallize homogeneously. While for the pebbles with pearlite-like microstructure a few heterogeneously crystallized pebbles are visible, not one pebble whose microstructure is dominated by lithium metatitanate dendrites was found that crystallized heterogeneously. In comparison to the pearlite-like pebbles, the second type of pebbles shows a notably thinner enveloping layer in which the microstructure is visibly finer than in the bulk (see Figure 4.27I). Usually the layer measures about 20 µm. Near the center of these pebbles pores of 5 µm to 10 µm size usually exist (see Figure 4.27J).

Generally, the LMT dendrite arm thickness and spacing is significantly larger than the thickness of the LOS and LMT lamellae, which lies in the same range as for the nominally 20 mol% LMT containing KALOS pebbles (compare Figure 4.24 and Figure 4.27F-H). The thickness of the LMT dendrite arms varies considerably from pebble to pebble, similar to the thickness of the LOS and LMT lamellae. The minimum LMT dendrite arm thickness lies at several hundred nanometers (see Figure 4.27F). During annealing, the 90° pattern persists in the microstructure and the thickness of the dendrite arms increases to about 1000 nm. In a number of cases a series of nearby lying fine dendrite arms is found to coagulate into a dense platelet with ongoing annealing (see Figure 4.27F-H). Other than that, the distribution of the two constituting phases is not changed significantly after 128 days of annealing.

For both microstructures, large cracks are an exception in the as-received state. Especially cracks that extend from the surface into the pebble are rare. Yet, for the pebbles that show the lamellar eutectic structure, small cracks of about 100 µm length as well as quasi-2D cavities along some of the domain boundaries, are observable within the pebble volume. The cracks mostly develop, unlike in the case of the nominally 20 mol% LMT containing ones, within the domains and occasionally span between them (see Figure 4.26A). Therefore, these transgranular cracks are relatively straight-lined in comparison to cracks that follow the domain boundaries which is often seen for the nominally 20 mol% LMT containing KALOS pebbles (see also Figure 4.24). The quasi-2D cavities can form real cavities when several domains meet. After annealing for 4 days, the number of cracks seemingly increases within the bulk of the pebbles with the lamellar eutectic microstructure (see Figure 4.26B). Also more surface cracks appear after 4 days of annealing in these pebbles, but they are usually outnumbered by the cracks that form within the bulk. Nevertheless, these surface cracks can occasionally reach comparable lengths as the cracks that form within the eutectic domains. The cracks along domain boundaries also seem to grow moderately, but the size of the other cracks is usually much larger (see Figure 4.26H).

With ongoing annealing, the number of cracks does not seem to increase and additional changes are relatively subtle (see Figure 4.26C). In some cases, the crack opening has increased after 128 days of annealing as well as the length of the cracks, which is most prominently visible for the surface cracks. From Figure 4.26I it is clear that also for these pebbles the transformation of the lamellar microstructure into a spheroidal microstructure strongly depends on the crystallographic orientation of the involved LOS and LMT grains.

Figure 4.27: SEM images of etched cross sections of nominally 30 mol% LMT containing KALOS pebbles, which feature the LMT dendrite dominated microstructure. All cracks of significant size have been marked by yellow ellipse. The cracks that extend from the surface into the pebble volume are additionally marked with a yellow arrow.

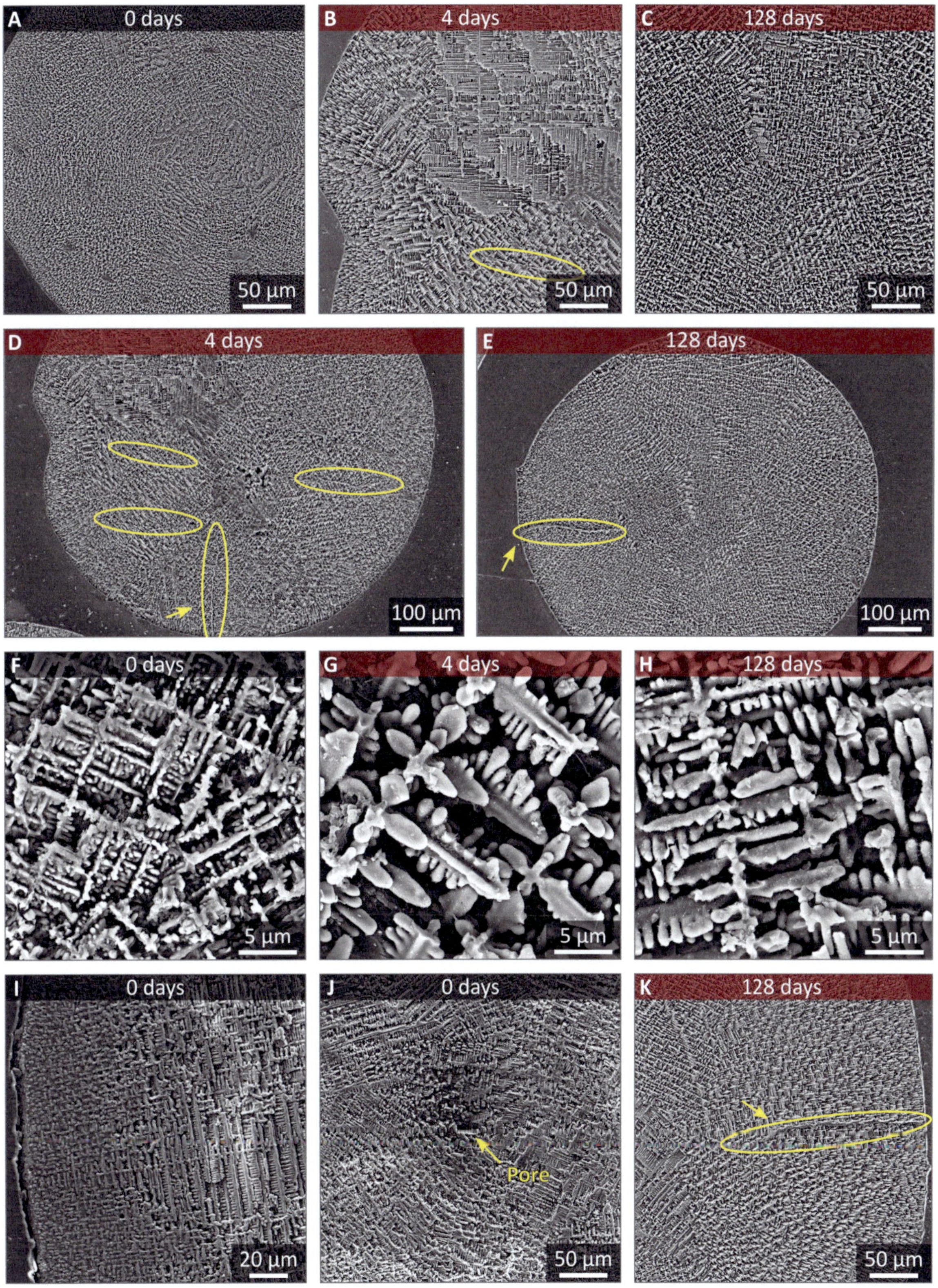
A
0 days
50 µm
B
4 days
50 µm
C
128 days
50 µm
D
4 days
100 µm
E
128 days
100 µm
F
0 days
5 µm
G
4 days
5 µm
H
128 days
5 µm
I
0 days
20 µm
J
0 days
Pore
50 µm
K
128 days
50 µm

For the pebbles with the LMT dendrite dominated microstructure, the few transgranular cracks that are present after the fabrication stay roughly unchanged during annealing. After 4 days of annealing, a very low number of surface cracks appear as well as in rare cases cracks within the volume of the pebble (see Figure 4.27B and D). Although the quantification is difficult, a visible increase of the amount of cracks with ongoing annealing cannot be observed and also the length of the cracks seems to remain constant in the order of 100 μm to 300 μm (see Figure 4.27E and K).

The microstructure of the as-received excessive lithium metatitanate pebbles that were fabricated by the emulsion method is built up of seemingly loosely interconnected, large blocks of about 100 μm diameter that consist of small, polyhedral LMT grains with diameters in the range of 1 to 4 μm (see Figure 4.28D). During the sample preparation a significant amount of these grains breaks from the surface of the cross section. Yet, the high amount of porosity that the pebbles show is not caused by this circumstance but an intrinsic feature of the samples. Other than the fragile interconnection between the denser blocks, no cracks are visible within the pebbles.

After 4 days of annealing, the microstructure shows clear signs of densification and grain growth. The grains grow to roughly 5 μm after these 4 days. On the majority of the pebbles, the block structure has disappeared and the microstructure appears to be more homogenous. Sporadically, the lack of interconnection between the blocks seems to lead to the formation of large cracks. If areas are isolated from the densification of the rest of the pebble, these areas are occasionally removed during the preparation of the cross section. As a result of the densification, a significant coalescence of the pores is observable within the pebbles, and thus an overall decreased porosity. In the later stages of annealing, the grains continue to grow to more than 10 μm after 128 days. The porosity is not reduced notably, yet it continuingly loses its spherical shape as a significant portion of grains slowly develops a jolted, elongated shape with ongoing grain growth (see Figure 4.28F). The contact between the individual grains also seems to increase as a lot less grains appear to breaking from the cross section. For all pebbles, no large cracks are observed after annealing the samples for more than 4 days. Yet, some small intergranular microcracks are visible.

Similar to the lithium excessive lithium metatitanate pebbles that were fabricated by the emulsion method, the as-received lithium excessive pebbles that were fabricated by the sol-gel method show small, polyhedral grains. Yet, the pebbles lack the block structure that the emulsion method pebbles show (see Figure 4.29A and J). Aside from that, the sol-gel-fabricated pebbles basically show two morphologies, with some variation, in the as-received state and also in the annealed state. One type of pebbles shows a reasonably dense bulk of 3 μm to 5 μm grains with numerous larger grains of a second phase of 10 μm to 15 μm (see Figure 4.29A and D). These grains appear in a slightly darker grey tone, which suggests that there is a difference in the chemical composition. In general, the grains seem to adhere reasonably well to each other. Obviously, these pebbles show a bimodal grain size distribution, but all in all, the microstructure is homogeneous.

For the second type of pebbles, the microstructure consists of small LMT grains of 1-2 μm size, without the presence of large grains of the second phase (see Figure 4.29G and J). However, the second phase seems to be relatively homogeneously distributed along between the grains. One probable effect of this distribution is a lowered cohesion between the grains of some pebbles, which can lead to the removal of large amounts of the pebbles during preparation, especially from the center of the pebbles (see Figure C.4 in Appendix C). With regard to the density of these pebbles, no definite conclusion can be drawn, yet it appears to be similar compared to the other type of pebbles. In contrast to this type of pebbles, the grain size distribution is rather monomodal.

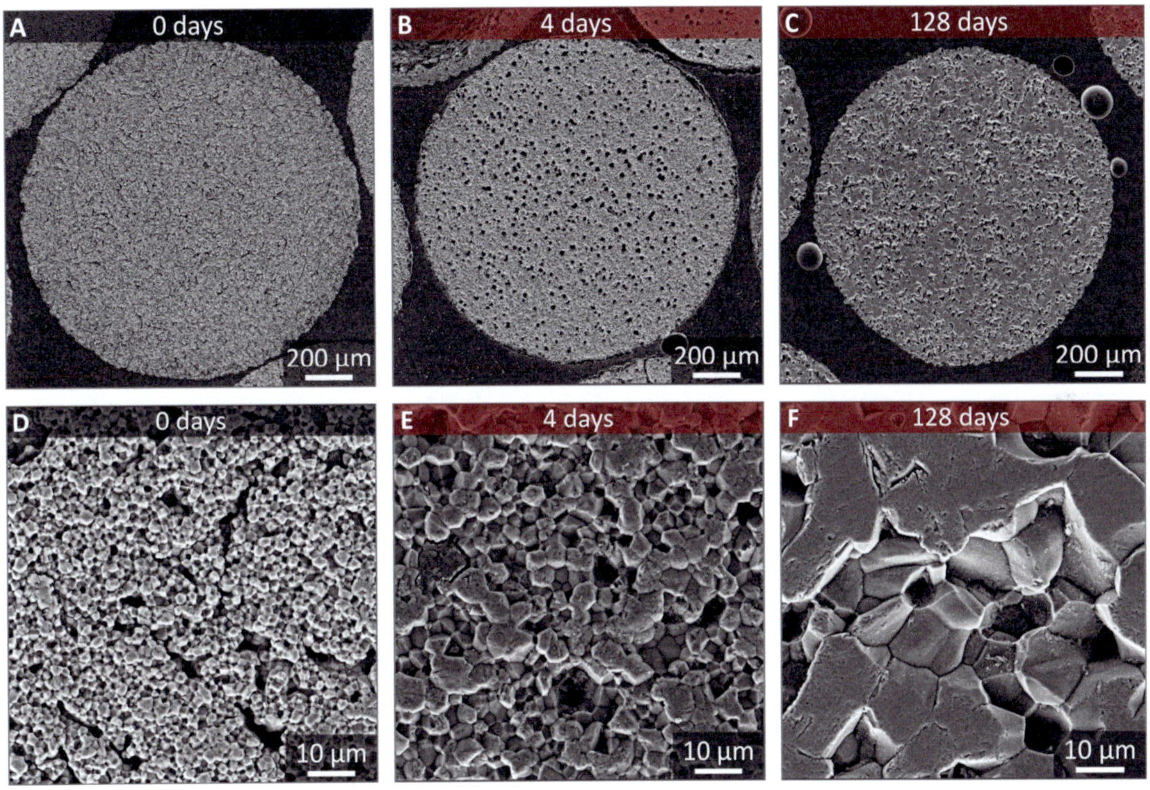

Figure 4.28: Cross section images of the lithium excessive samples are shown that were fabricated by the emulsion method. The samples are displayed in the as-received that as well as in different annealed states.

But, as already mentioned, for this sample in general, a monomodal grain size distribution has to be considered as one extreme of a spectrum, since the size of the large grains that constitute one mode differs from pebble to pebble and can ultimately approach the low size mode until indistinguishability. Yet, mostly the two types of microstructure are clearly developed. The formation of a shell of slightly higher density at the surface cannot be verified unambiguously and cracks are generally not found in the as-received samples of these pebbles.

After 4 days of annealing, a significant number of pebbles shows the formation of a distinct, relatively dense shell of varying thickness which appears to be denser than the pebble core (see Figure 4.29H and K and Figure 4.30A). The shell does not seem to be detached from the inner part of the pebble in most cases.

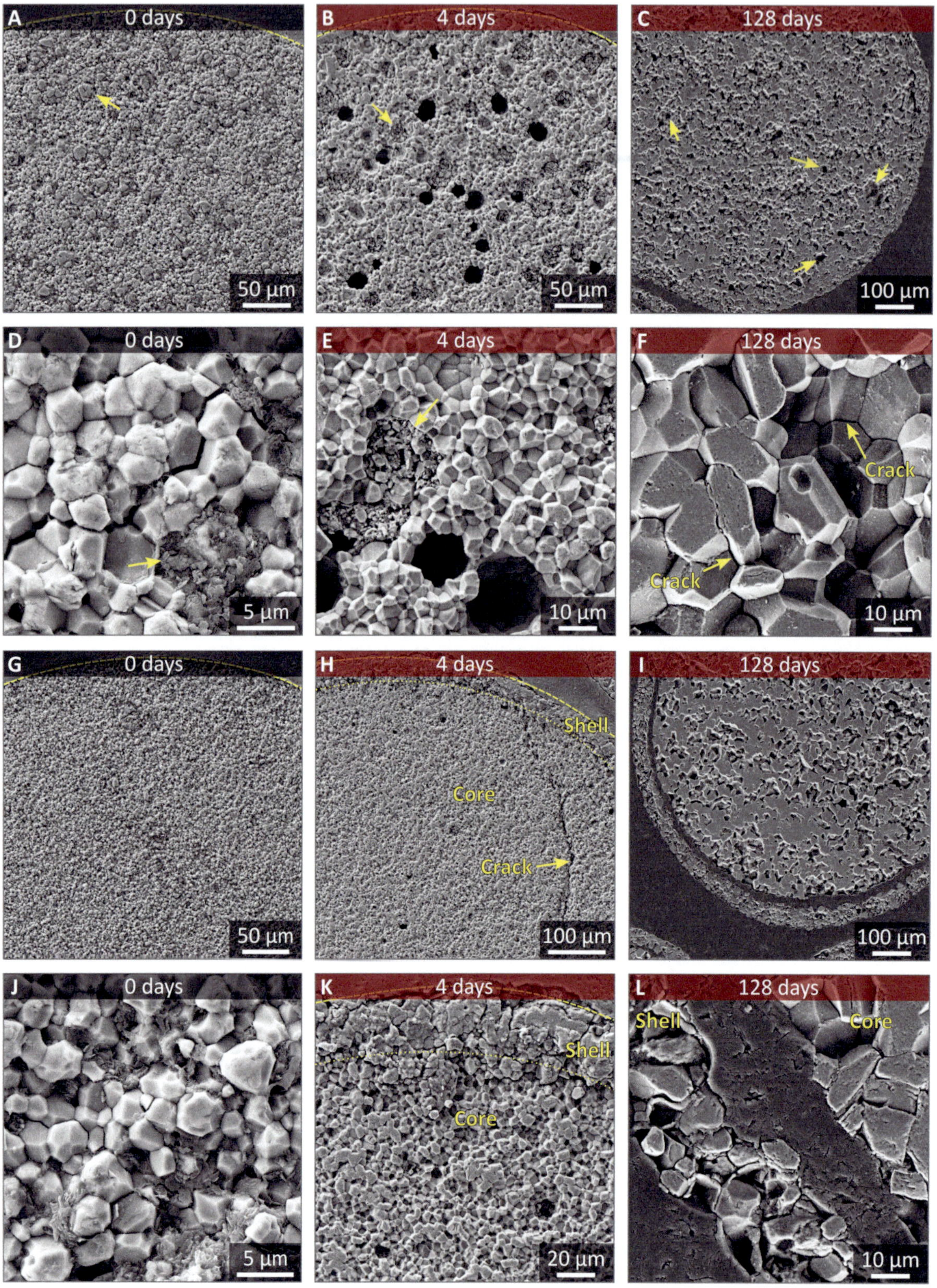
A
0 days
50 µm
B
4 days
50 µm
C
128 days
100 µm
D
0 days
5 µm
E
4 days
10 µm
F
128 days
Crack
Crack
10 µm
G
0 days
50 µm
H
4 days
Shell
Core
Crack
100 µm
I
128 days
100 µm
J
0 days
5 µm
K
4 days
Shell
Core
20 µm
L
128 days
Shell
Core
10 µm

The darker contrast in the SEM images suggests that the shell is composed of the same second phase that is found in the as-received samples. An EDX analysis finds a decreased titanium content in the shell relative to the amount of oxygen as well as a significant amount of carbon (see Figure C.5 and Table C.14 in Appendix C). The grain morphology of the shell is also significantly different from the core of the pebbles with large (more than 10 µm), angled grains which are intersected by numerous cracks (see Figure 4.29K and Figure 4.30A).

The microstructure of the core of the pebbles is mostly similar to that of the pebbles that were annealed for 4 days and were fabricated by the emulsion method with a grain size of about 5 µm. However, with the exception, that the presumably least dense pebbles show an aggravated version of the densification in blocks without proper cohesion between them. For such pebbles, large cracks dominate the core (see Figure 4.29H). For the pebbles that showed two distinct grain sizes in the as-received state no shell can be observed, but large pores of about 10-20 µm in diameter have formed homogeneously within the pebble volume. Apparently, these pores have formed from the second phase grains that were observed in the as-received state (see Figure 4.29A). Yet, only about half of these second phase grains have transformed into pores, while the other half is filled with a two phase material. Obviously, dispersoids that appear in a similarly light grey as the LMT bulk have formed within the second phase grains (see Figure 4.29E). As for all lithium excessive LMT pebbles the grains of the pebbles grow considerably after 4 days of annealing. In terms of grain size within the core/bulk of the pebbles, there seems to be no significant difference between both types, although some variation between individual pebbles exists.

When the annealing is extended to longer durations, the pebbles that were fabricated by the sol-gel method show an increasingly similar microstructure in the core to the lithium excessive pebbles that were fabricated by the emulsion method. However, as the grains within the core of the pebbles clearly grow larger, the microstructure of the shell changes relatively mildly with increasing annealing duration (see also Figure 4.30). Nevertheless, after 64 days of annealing a significant amount of small polyhedral grains has formed, which resemble the grains in the core of the pebble in shape (see Figure 4.30C). Eventually this shell develops a very similar microstructure, although finer, as the core after 128 days of annealing (see Figure 4.29L). In this case, the shell does not adhere well to the core and an occasional detachment of it is observed.

From Figure 4.30C it can be seen that this is rather a logical consequence of the development, because also for these samples the adherence between core and shell seems to be low. With continued annealing, it also appears that the darker contrast of the shell structure in the SEM gets brighter in comparison to the core (see Figure 4.29H and L). This effect may be associated with the constant decrease of the lithium content of the pebbles as the elemental analysis suggests (see Table 4.4).

Figure 4.29: SEM images of cross sections of the lithium excessive LMT pebbles that were fabricated by the sol-gel method are shown in the as-received state and after annealing. As the formation of a detached outer shell is observable for a number of pebbles, the microstructure of this shell is also displayed (H, I, K, L). The yellow arrows point at accumulations of a second phase, except for C, where a number of (once) spherical pores are highlighted. The surface of the pebbles is highlighted by a dashed line for better visibility in some images.

Figure 4.30: SEM detail images of the shell region of the sol-gel fabricated lithium excessive LMT pebbles for several annealing durations. The approximate position of the shell is indicated.

Similarly, the content of the second phase decreases with increasing annealing duration until it almost vanishes after 128 days of annealing. In the core of both types of pebbles, the grains grow to slightly larger sizes after 128 days of annealing than observed for the respective pebbles that were fabricated by the emulsion method.

Yet, the moderate differences in grain size between individual pebbles, which is already observed in early stages of annealing, is not leveled after 128 days of annealing as it is still perceivable. In this microstructure a number of fine intergranular microcracks can be observed with a considerable length of several grain diameters (see Figure 4.29F). The large pores that were observable in pebbles without shell after 4 days of annealing are still recognizable after 128 days of annealing. However, as large irregularly shaped pores develop because of the significant grain growth, the once spherical pores have lost their shape to some extent and are thus more difficult to identify (see Figure 4.29C).

The grain size of the pebbles that were annealed as well as that of the as-received ones that has been determined by the linear intercept method (see section 3.6) is given in Table 4.7. The given values are related to the bulk of the pebbles and do not consider the shell if the pebbles show one. Also, the values were not determined for both predominant morphologies of the pebbles that were fabricated by the sol-gel method, but only for the morphology which resembles that of the emulsion method pebbles the most. Yet, as already stated, the differences in microstructure between the pebbles that were fabricated by either method diminish considerably after 4 days of annealing.

In Figure 4.31 cross sections of green, dried pebbles are shown for which a broad region of slightly higher density is observable at the surface. The solid raw materials are apparently homogeneously distributed throughout the pebble volume. However, there seems to be an increased porosity within the core of the pebbles compared to the region at the surface (see Figure 4.31). Figure 4.31 also suggests that the dense surface layer increases with nominal lithium metatitanate content, but there is no clearly visible correlation.

Table 4.7: The grain size of the as-received as well as annealed lithium excessive pebbles as determined by the linear intercept method. The uncertainty of the mean value for the given measurements is estimated to be ±1 %. (*: Not representative for the whole sample as only the microstructure that resembles that of the as-received EM-LMT pebbles was analyzed.)

	As-received	4 days	32 days	64 days	128 days
EM-LMT: mean grain size [μm]	1.45 ± 0.01	3.27 ± 0.03	6.46 ± 0.06	8.16 ± 0.08	10.31 ± 0.10
SG-LMT: mean grain size [μm]	1.66* ± 0.02	3.46 ± 0.03	5.66 ± 0.06	7.79 ± 0.08	10.81 ± 0.11

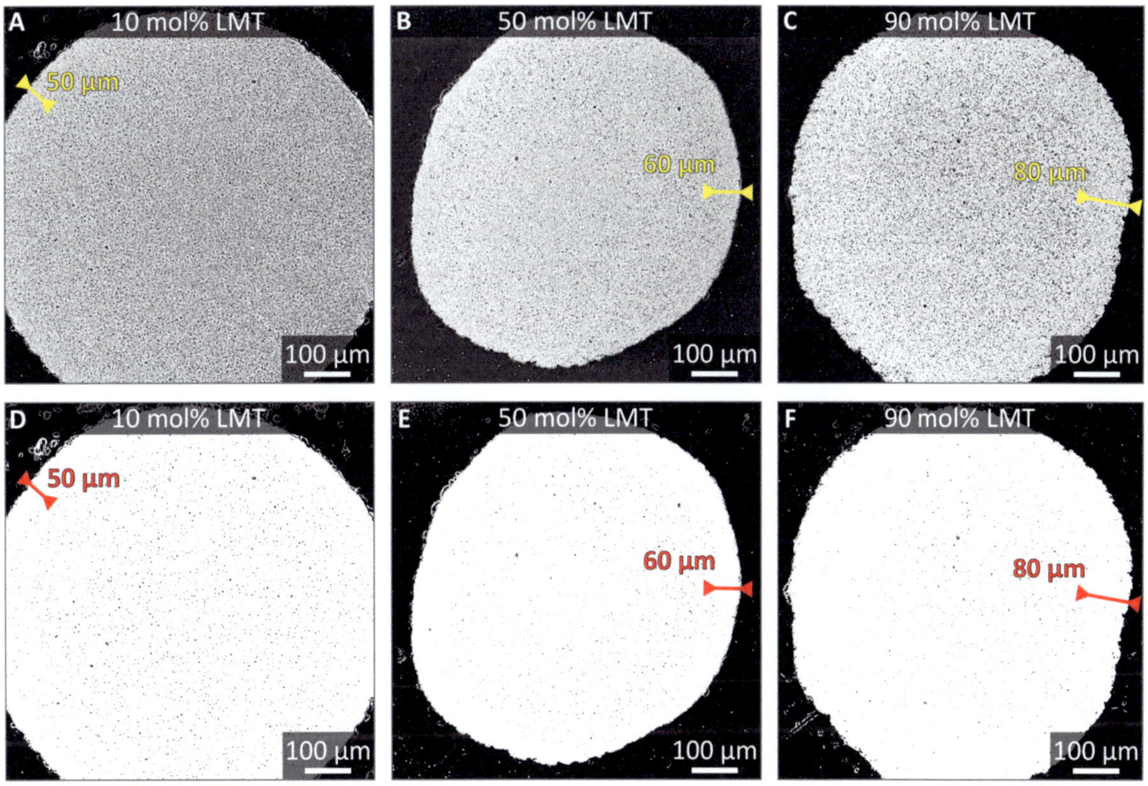

Figure 4.31: SEM image (Z contrast) of cross sections of green pebbles that were shaped by the emulsion method (A-C). The images D-F are binary images of A-C, that were created by applying a threshold to the greyscale value in order to emphasize the pores over the solid material in the samples. The respective sintered pebbles nominally contain 10 mol%, 50 mol% and 90 mol% lithium metatitanate. The approximate size of the denser surface region is indicated in the images.

The cross section images of the sintered biphasic, LMT-containing samples that were fabricated by the emulsion method show that these pebbles are virtually free of cracks of significant size (see Figure 4.32). Yet, the microstructure changes significantly as a function of nominal LMT concentration. With the exception of the 10 mol% LMT containing pebbles, the porosity of the pebbles continuously increases from nearly zero for low LMT contents to a significant amount with increasing nominal LMT content (see Figure 4.33). However, the nominally 80 mol% containing pebbles seem to show the maximum porosity of this series. It should be noted that there is a considerable increase in porosity observable for the 70 mol% and 80 mol% LMT containing samples with respect to the 60 mol% LMT containing ones (see Figure 4.33F-H). The pore size of the samples with a nominal LMT of 70 mol% to 90 mol% lies at about 5 µm.

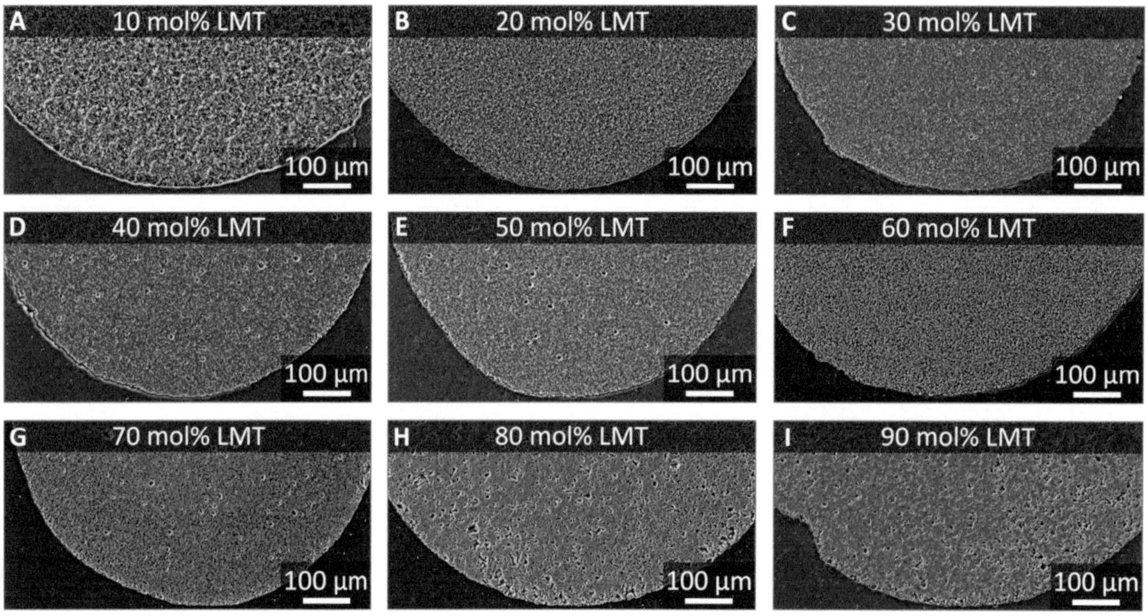

Figure 4.32: SEM overview images of cross sections of the biphasic, emulsion method fabricated LMT containing pebbles are shown for different nominal LMT contents.

Yet, sporadically several pores connect to form larger cavities which easily exceed lengths of 10 µm. For these samples with high LMT concentrations, the porosity seems to originate from incomplete densification of the lithium metatitanate grains, which leads to the angled shape of the pores. In contrast, the pebbles with lower LMT concentrations show pores that are predominately of ellipsoidal shape, which is usually associated with the formation of gaseous species that are sealed in during sintering. For the nominally 10 mol% LMT containing samples this kind of pores appear in much higher frequency and size (about 3 µm) than for all other samples (see Figure 4.33A). With a varied concentration of LMT, the distribution of the phases LOS and LMT within the pebbles also changes. Basically two regimes can be determined, one in which LMT is embedded in a matrix of LOS and vice versa. At about 60 mol% LMT, which coincides with an almost equal volume fraction of both phases, the transition from one structure to the other is observed. For the samples with a LOS matrix, the size of the LMT grains increases with increasing LMT content. The increase in grain size is especially strong between the samples with a nominal LMT content of 10 mol% and 20 mol%, from sub-micron size to 1 µm to 2 µm, respectively (see Figure 4.33A and B). The morphology of the LMT grains is generally plate-like, when they are embedded in the LOS matrix.

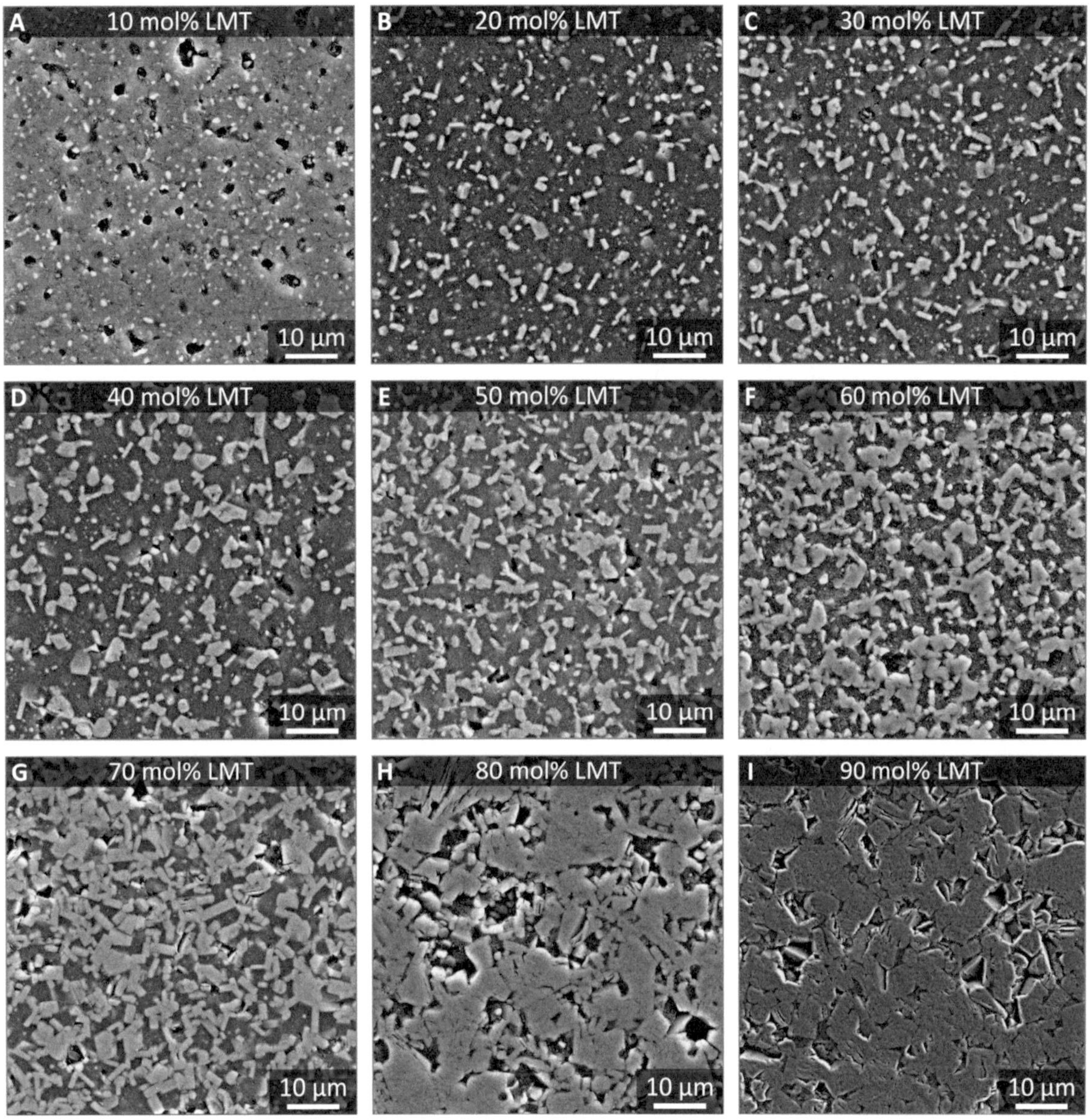

Figure 4.33: SEM detail images of cross sections of the biphasic, emulsion method fabricated LMT containing pebbles are shown for different nominal LMT contents as already published in [183]. The bright phase corresponds to LMT.

The grain size of the LOS phase cannot be estimated from these pictures. For samples with a matrix of LMT, the size of the LMT grains seems to be just slightly affected by the LMT content. Also the LMT grains in this regime show a plate-like structure, yet mostly not as pronounced, i.e. with a lower aspect ratio, as when the grains are dispersed within an LOS matrix. The observable grains show a size between about 5 µm to 10 µm. For high contents of LMT, the LMT grains tend to adopt a less plate-like morphology. It is least pronounced for the sample that nominally contains 90 mol% LMT. The LOS grains for these samples are located at the triple points of the LMT grains, and therefore their grain size is determined by the LMT content and decreases with increasing LMT content from up to 5 µm for the nominally 70 mol% LMT containing samples to about 2 µm for the maximum LMT content (see Figure 4.33G). Whether the so-determined grains show additional grain boundaries, cannot be determined.

The microstructure of the LAO containing samples shows a significantly different morphology (see Figure 4.34). All three fabricated samples show a significant porosity, which increases considerably with increasing nominal LAO concentration. The pores are of near ellipsoidal shape. Their size ranges from about 10 µm for the lowest LAO content to about 18 µm for the medium LAO content and reaches about 30 µm for the nominally 30 mol% LAO containing samples. Generally the pores are homogenously distributed throughout the pebble volume, except for the area close to the surface, which are virtually free of pores. None of the samples contains any notable cracks.

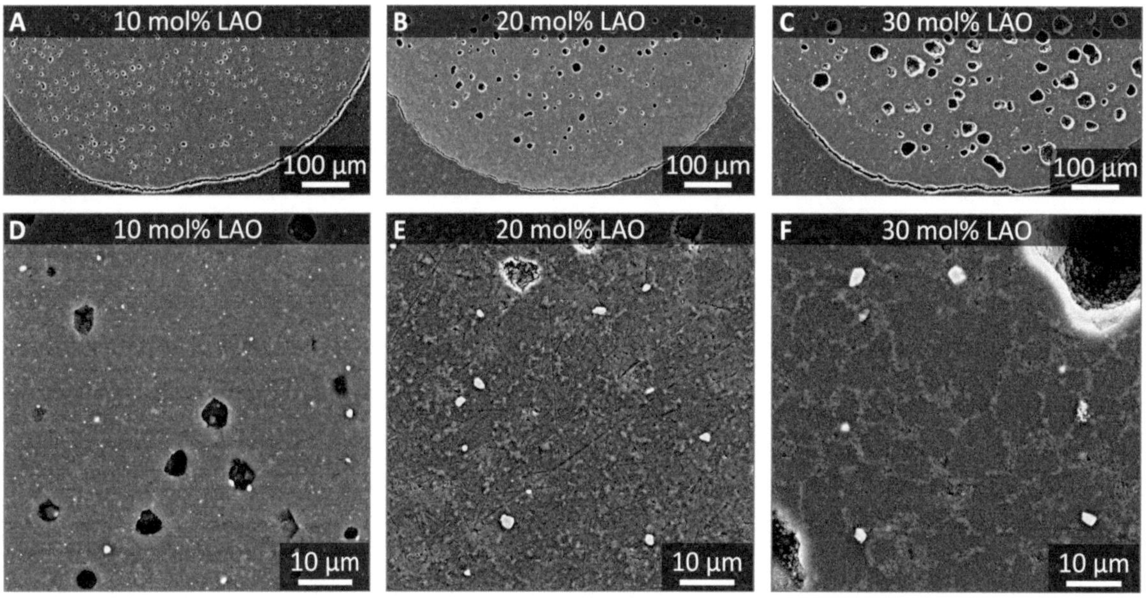

Figure 4.34: SEM images of cross sections of the biphasic LAO containing pebbles that were fabricated by the emulsion method are shown with different nominal LAO content as already published in [183]. The brightest phase is attributed to the lithium zirconate while the medium bright phase is attributed to lithium aluminate.

Within the microstructure of all three samples, very bright particles can be observed, which increase in size with increasing LAO content. These particles were identified as zirconium rich phase by means of EDX (see Figure 4.35). As the XRD analysis showed that samples of this series contain a small but significant amount of Li_2ZrO_3 (see Figure 4.3) these particles are unambiguously attributed to this phase. In contrast to the XRD analysis, the SEM images show that lithium zirconate is present in all three LAO containing samples. It is located at the grain boundaries and its size is seemingly increased with increasing LAO content. The absence of a Li_2ZrO_3 phase in the XRD investigations is in all likelihood explainable by a too low concentration in conjunction with a relatively low particle size in case of the 10 mol% LAO sample for detection by this technique.

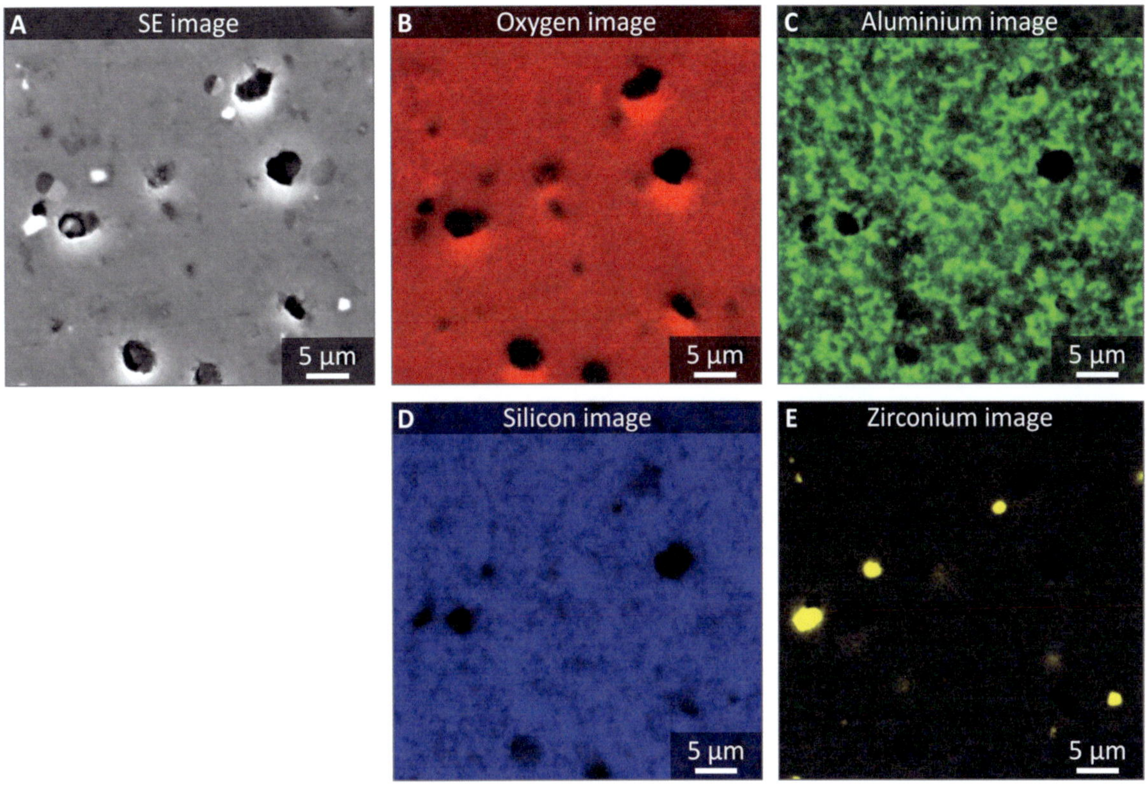

Figure 4.35: EDX elemental maps showing oxygen, silicon, aluminium and zirconium on a cross section of a sintered LOS sample with 30 mol% LAO (nominal concentration). These results were already published in [183].

Presumably, the abrasion of zirconia during ball milling is increased with increasing LAO concentration and thus more and larger particles are introduced into the samples. As lithium aluminate apparently shows an almost similar electron backscattering cross section as lithium orthosilicate, the two phases are hard to separate in the SEM cross sections. Still, the lithium aluminate phase appears in a slightly lighter grey. For the samples with nominally 20 mol% and 30 mol% LAO, lithium aluminate is clearly located at the grain boundaries of the LOS grains. The LAO phase seems to be composed of grains that are generally smaller than 1 µm. An effect of the LAO content on the size of the LAO grains is not obvious from the SEM images. As the LAO phase separates the LOS grains, the size of these grains can be estimated. While the pebbles that nominally contain 20 mol% LAO show a LOS grain size of about 5 µm, the grain size is significantly increased for the LAO richer pebbles with approximately 10 µm. For the sample that contains the lowest LAO content, the LAO grains are difficult to observe.

Yet, also these pebbles show a comparable microstructure with respect to the other two samples of the series, but with an apparently smaller LAO grain size and fraction. For the nominally 10 mol% LAO containing sample the grain size of the LOS grains cannot be determined. Yet, especially for the sample with 30 mol% of lithium aluminate, the EDX elemental map (see Figure 4.35) of this sample suggests a very fine grained microstructure, although the grain size seems to exceed 10 µm in the SEM image. Furthermore, it can also not be precluded that LOS and LAO show as negligible miscibility as LOS and LMT [73], because a significant aluminum signal is also found for silicon-rich grains.

4.4 Porosity

For the calculation of the porosity, the theoretical density of the sample is necessary as reference. The following results on the pebble porosity take the obtained elemental composition (see section 4.1) into account to estimate the phase composition. As the elemental analysis was not performed for all biphasic pebbles that were fabricated by the emulsion method, the missing values were interpolated from the available data. Yet, the differences to the nominal values are low. Also the data of the quantitative phase analysis by X-ray diffraction (see section 4.2.2) is taken into account to use the correct density of each present phase in equation (3.13) to obtain the theoretical density of the samples. Of course, any amorphous or seemingly amorphous phase cannot be addressed in this way. Therefore, the density was calculated as if no amorphous phases were present and the pebbles are only constituted by the detected crystalline phases. The closed porosity of the pebbles is then calculated from the difference of the theoretical density to the determined pebble density by helium pycnometry.

With regard to the open porosity of the samples, in this section only the results that were obtained by the optical method, i.e. the sum of the open and the imaginary porosity, are presented, because this method was applicable for all samples. Yet, referring to the considerations in section 3.6, it is not surprising that the obtained absolute values are significantly higher than those determined by mercury intrusion porosimetry (see Table 4.8) because of significant deviations from ideal spherical shape, which is called 'imaginary porosity' here.

Moderate deviations from the ideal spherical shape do not interfere with the mercury intrusion measurements and therefore the so-determined value of total porosity of the pebbles is used for calculating the open porosity by subtracting the determined closed porosity. Consequently, range of these values, which is about one fourth in mean, is much more realistic in comparison to the ones that are obtained by the optical method. Yet the obtained trends of both methods are similar, which is expectable as the macroscopic deviations from ideal sphericity, i.e. the 'imaginary porosity', are constant within one set of samples as a function of annealing duration. Since the fabrication process of the biphasic emulsion method samples is identical and their macroscopic shape relatively similar, except for the 10 mol% LAO containing samples (see Figure 4.17 and Figure 4.20), the amount of 'imaginary porosity' of these pebbles can be considered to be approximately the same for all samples, irrespective of their nominal second phase content. Consequently, also for these samples, the obtained values by the optical method qualitatively represent the changes of the open porosity very well. For all samples, the total porosity values are listed in Appendix C. The closed porosity of the as-received pebbles that were fabricated by the melt-based process is close to zero. As reported by Knitter et al. [73], the closed porosity increases with increasing LMT concentration for biphasic LOS/LMT pebbles that were fabricated by the KALOS process. This conforms to the results of the KALOS samples in this study.

Table 4.8: The total porosity as well as the open porosity as determined by mercury intrusion porosimetry of the KALOS samples and the lithium excessive LMT samples in their as-received state as well as annealed state.

	As-received	4 days	32 days	64 days	128 days
KA-20-LMT					
Total porosity	5.02 %	6.20 %	6.90 %	7.10 %	8.43 %
Open porosity	4.84 %	2.61 %	4.96 %	4.66 %	7.31 %
KA-30-LM					
Total porosity	3.52 %	4.26 %	3.55 %	2.60 %	5.89 %
Open porosity	2.83 %	2.30 %	3.05 %	2.04 %	4.22 %
EM-LMT					
Total porosity	12.90 %	10.10 %	9.90 %	8.53 %	9.06 %
Open porosity	8.76 %	0.59 %	7.67 %	8.34 %	8.26 %
SG-LMT					
Total porosity	14.15 %	10.65 %	11.64 %	10.91 %	10.86 %
Open porosity	8.39 %	3.38 %	6.98 %	8.44 %	7.73 %

The samples that were fabricated by the emulsion method and the sol-gel method show a significantly higher closed porosity with about 4 % and 6 %. As a function of annealing duration, in principal all samples behave similarly as Figure 4.36 shows. Initially, the closed porosity significantly increases after 4 days of annealing. The increase ranges from about 1 % to 6 % in the following order: KA-30-LMT < SG-LMT < KA-20-LMT < EM-LMT. With increased annealing duration the determined closed porosity decreases again and overcompensates the initial increase, except for the KA-20-LMT sample. After 64 days of annealing, a constant level of closed porosity seems to have established, which lies below 4 % for the SG-LMT pebbles and below 2 % for all other samples. Yet, except for the KA-20-LMT pebbles, all other samples seemingly show an increase in the closed porosity after 128 days of annealing.

The change of the open porosity with annealing duration is illustrated in Figure 4.37, since the imaginary porosity is considered to be constant. The values of the sum of the open and imaginary porosity of the non-annealed samples lie in the interval 17 % to 28 % and increase in the following order: KA-20-LMT < EM-LMT < KA-30-LMT = SG-LMT. The values of the KA-30-LMT pebbles and the SG-LMT pebbles differ only insignificantly.

As a result of the annealing, the open porosity values of all samples but the SG-LMT pebbles sharply reduce after 4 days to their respective minimum in this study. This finding constitutes the opposite of the observations in Figure 4.36. The absolute reduction ranges from about 6 % to about 8 %. From this minimum, the values of all samples increase again for 32 days of annealing, but the initial values are not exceeded, except in case of the SG-LMT samples. After this increase, the open porosity can be considered to stay relatively constant as a function of annealing duration although the KA-30-LMT and SG-LMT samples show an additional increase after 64 days of annealing. Yet, this increased open porosity does not seem to persist after 128 days of annealing and values close to those that were determined after 32 days of annealing are observed. At the end of the annealing, the open porosity values fan out more than in the as-received state.

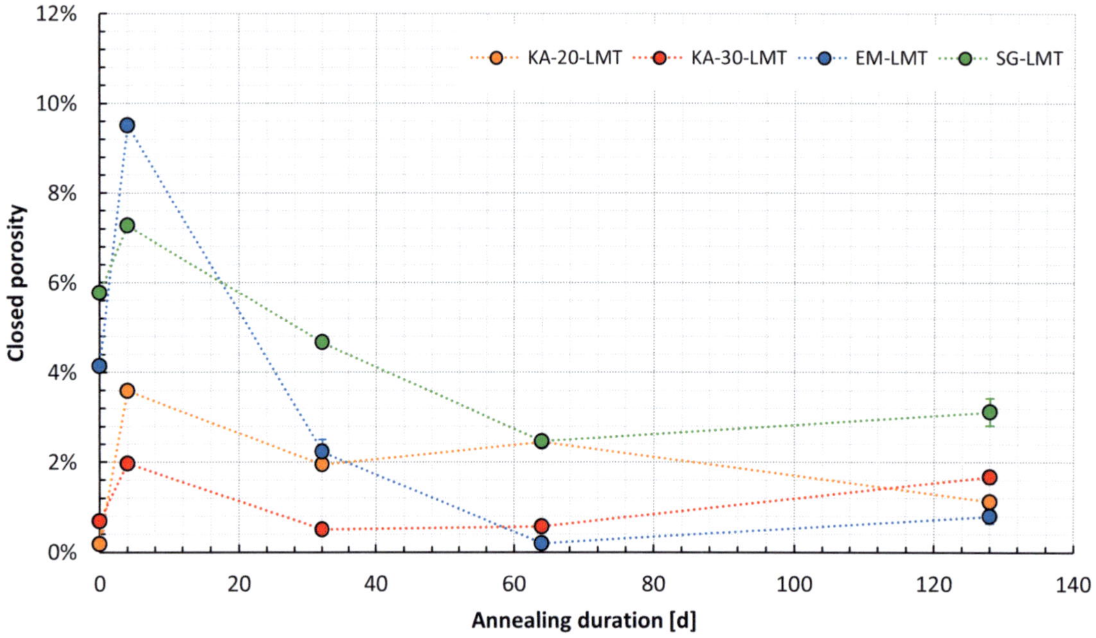

Figure 4.36: Closed porosity of the samples that were annealed at 900 °C as a function of annealing duration. The uncertainty of the measured values that is calculated from the average absolute deviation of the measurements is marginal and thus rarely visible in this graph. Between the measured values dotted links are introduced to improve visualization.

Because of the ellipsoidal shape of the SG-LMT samples (see Figure 4.16), the imaginary porosity value may be significantly increased with respect to the EM-LMT samples. The values that were obtained by mercury intrusion porosimetry do not suffer from this circumstance and lead to virtually the same values of open porosity for the EM-LMT and SG-LMT samples, also after annealing (see Table 4.8). There is only one exception for the samples that were annealed for 4 days where a considerable difference is determined. For this SG-LMT sample also a significant drop of the open porosity is observed after annealing for 4 days. Apart from that, the so-determined values follow the same trend that is shown in Figure 4.37 for the porosity values that were obtained by the optical method. For the biphasic samples that were fabricated by the emulsion method, the determined closed porosity values are illustrated in Figure 4.38. Apart from the sample with the lowest LMT concentration, a fitful but steady increase in closed porosity from about 2 % to more than 4 % with increasing LMT concentration is observed. The observed 4.3 % of closed porosity for the 90 mol% LMT containing sample are only insignificantly different from the value obtained for the as-received EM-LMT pebbles. The 10 mol% LMT in LOS sample shows about 6 % of closed porosity. The samples that contain LAO instead of LMT also show an increase of closed porosity with increasing second phase content. Yet, the observed increase, from about 7.5 % to about 11 %, is much larger.

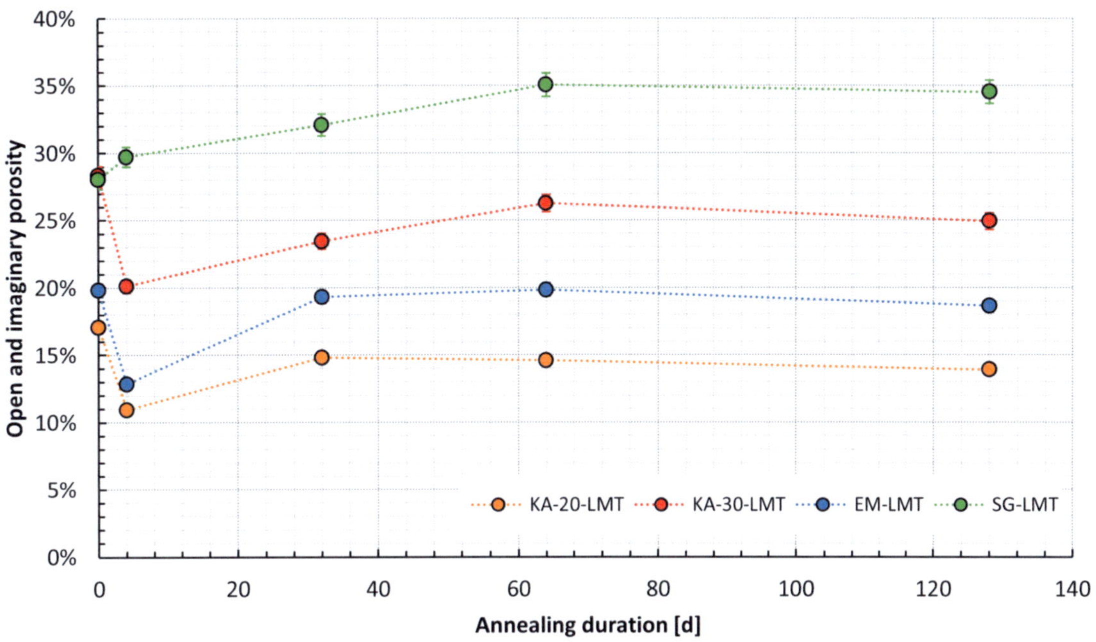

Figure 4.37: The sum of the open and imaginary porosity of the samples that were annealed at 900 °C is shown as a function of annealing duration. For each sample the determined closed porosity is considered (see Figure 4.36). As uncertainty ±2.5 % of the determined values are considered as a reasonable estimate. However, compared to the scale the uncertainty is small and thus barely visible in this graph. The dotted links between the measured values shall improve visualization.

The biphasic samples that were fabricated by the emulsion method show open porosity values in the same order as the samples that were annealed, except for the nominally 30 mol% LAO containing sample. In Figure 4.39 the obtained open porosity values are displayed as a function of the nominal second phase content. The LMT containing pebbles apparently show significantly higher open porosity values with respect to the LAO containing ones. While the LAO containing pebbles show values similar to the ones of the KA-20-LMT samples, the LMT containing samples are more similar in this regard to the KA-30-LMT or SG-LMT samples. Yet, the measured open porosity of the 90 mol% LMT containing pebbles is almost identical to the non-annealed pebbles (EM-LMT) that were also fabricated by the emulsion method. With the exception of the 70 mol% and 80 mol% LMT containing pebbles, similar to the observations for the closed porosity, a volatile, but steady, trend with the nominal LMT concentration is visible.

Yet, the open porosity decreases with increasing LMT content. Also for the LAO containing pebbles, a contrary trend compared with the determined closed porosity exists. For these samples the open porosity reduces significantly to less than 5 % with increasing LAO content.

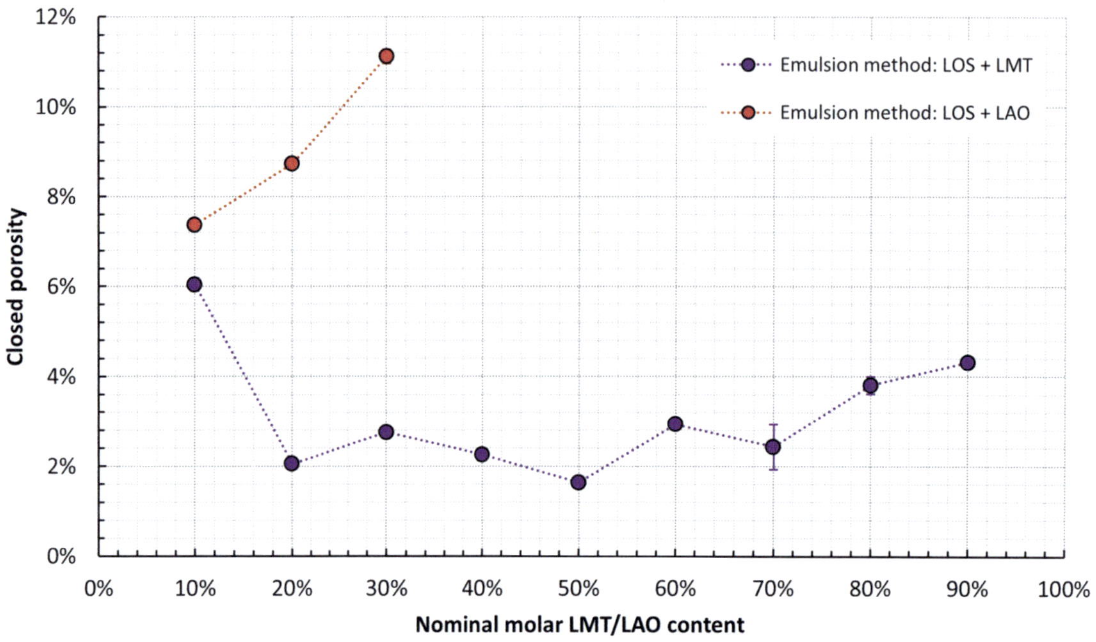

Figure 4.38: Closed porosity of the biphasic samples that were fabricated by the emulsion method. The uncertainty of the measured values that is calculated from the average absolute deviation of the measurements is marginal and thus rarely visible in this graph. Between the measured values dotted links are introduced to improve visualization. These results were already published in [183].

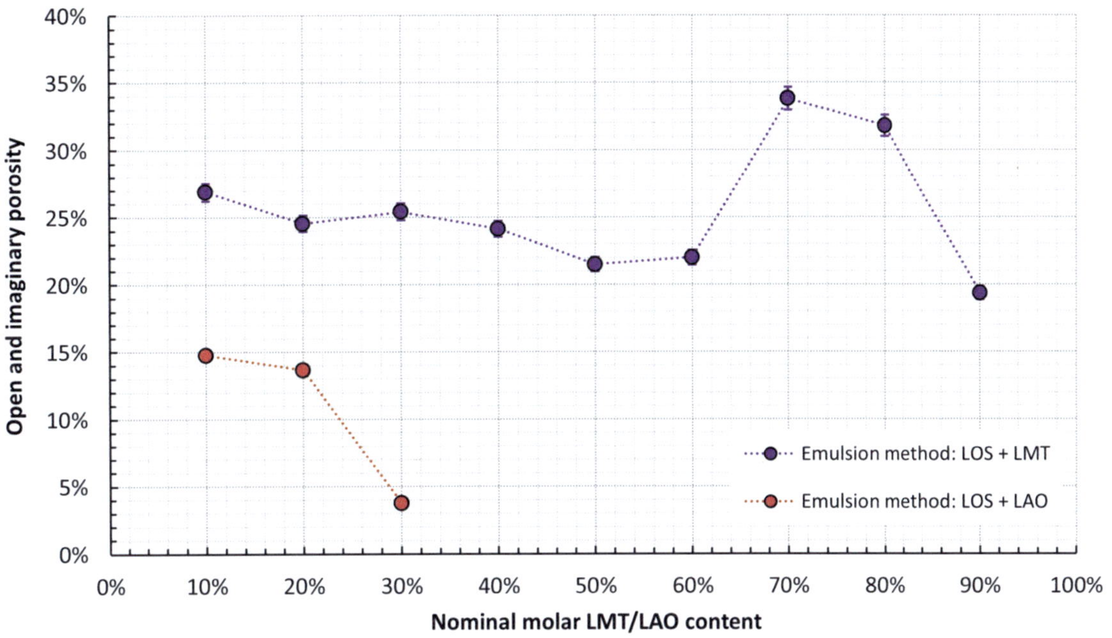

Figure 4.39: The sum of the open and imaginary porosity of the biphasic samples that were fabricated by the emulsion method is shown as a function of nominal composition. The determined closed porosity (see Figure 4.38) is considered. As uncertainty ±2.5 % of the determined values are considered as a reasonable estimate. However, compared to the scale the uncertainty is small and thus barely visible in this graph. Between the measured values dotted links are introduced to improve visualization. An earlier interpretation of these results was already published in [183].

4.5 Mechanical properties

In this chapter the results of the uniaxial compression experiments that were detailed in section 3.4 are shown. The obtained crush load values of the individual pebbles were converted to the mean contact pressure by applying equation (2.5), in order to be able to compare the individual measurements. For the sapphire plates a Young's modulus of 362 GPa, which is the mean of the values given by Shipway and Hutchings [184], and a Poisson's ratio of 0.26 [184] were used. As explained in section 3.4, the mechanical properties of the materials were estimated by taking the closed porosity into account but not the open porosity. The resultant elastic properties of the samples are given in Table B.2 in Appendix B. For each sample 40 to 50 (successful) uniaxial compression tests were performed to obtain a reliable measure of the mechanical strength. The exact number of pebbles that were tested in each individual case is shown in Table B.1 in Appendix B. The chronological evolution of the mechanical strength is presented, followed by a Weibull analysis of all tested samples. To foster the findings of the Weibull analysis, the SEM analysis of selected fragments of failed pebbles is presented to conclude this section.

4.5.1 Evolution of the mechanical strength

The average mean contact pressures at which the pebbles fail are displayed as a function of either the annealing duration or the nominal second phase concentration in this section. For the pebbles of 500 μm diameter that were fabricated by the KALOS process, the strengths of the two types of samples develop in significantly different ways as Figure 4.40 shows. From an initial mean contact pressure of about 3.85 GPa the KA-20-LMT pebbles can only withstand pressures of about 2.98 GPa after 4 days of annealing. With prolonged annealing, no clear trend establishes and the average pressure at which the pebbles fail stays relatively constant. After 128 days of annealing the observed average mean contact pressure of 3.16 GPa is only marginally different from the value observed after 4 days of annealing. In contrast to the KA-20-LMT pebbles, the nominally 30 mol% containing KALOS pebbles do not lose their strength as a function of annealing. The obtained mean contact pressure rather increases after 4 days of annealing from the initial value of 3.68 GPa to 4.17 GPa. The strength of these pebbles seemingly increases with annealing duration almost linearly from this value to about 4.56 GPa after 64 days of annealing. Yet, after an additional annealing of 64 days, the strength of the samples drops to a value of 4.21 GPa which is just slightly higher than the value that is obtained after 4 days of annealing.

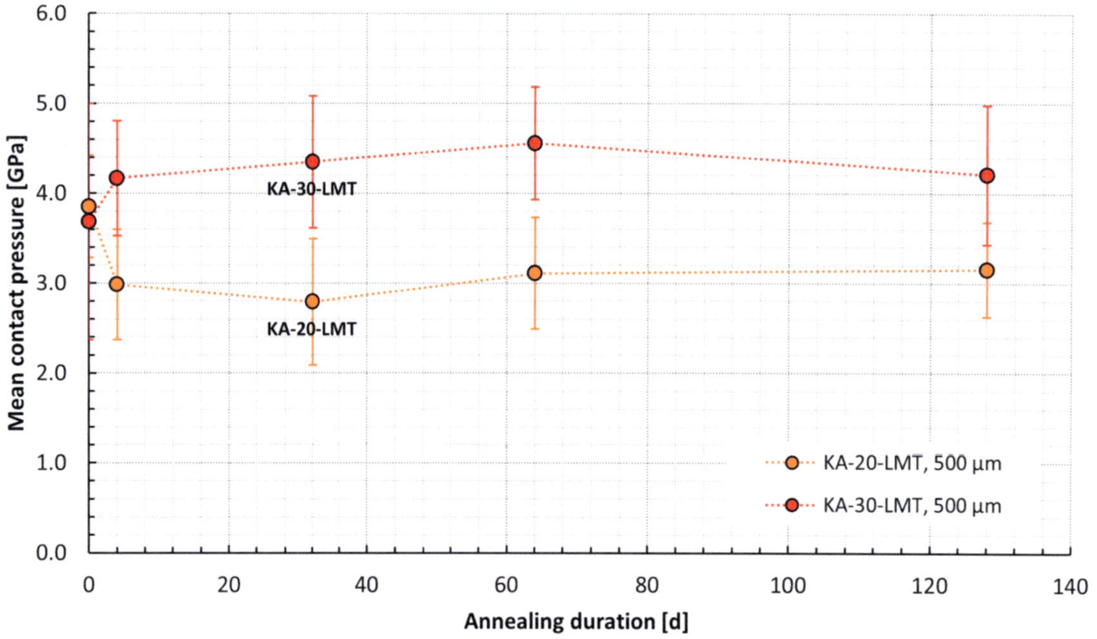

Figure 4.40: The average mean contact pressures (see equation (2.5)) that the 500 μm pebbles, which were fabricated by the KALOS process, withstand during uniaxial compression are displayed as a function of the annealing time. The uncertainty of the measurements, considering the quality of the load cell (see section 3.4), is too low to be displayed in this plot. Instead, the standard deviation of the determined mean contact pressures is given.

For the 1000 µm pebbles of the KA-20-LMT and KA-30-LMT samples the development of the mechanical strength as a function of annealing duration is basically identical to that of the 500 µm pebbles as Figure 4.41 shows. Also for the nominally 20 mol% LMT containing pebbles the strength drops considerably after 4 days of annealing. The 1000 µm pebbles of the KA-20-LMT sample withstand an average mean contact pressure of 3.08 GPa in the as-received state which is decreased to about 2.15 GPa for any annealed state.

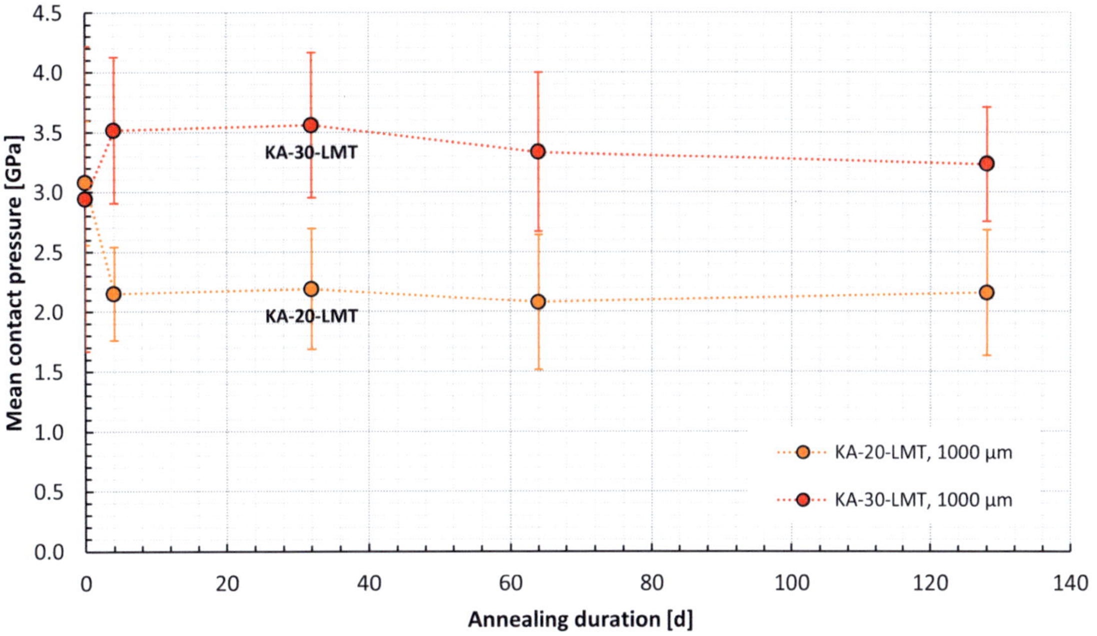

Figure 4.41: The average mean contact pressures (see equation (2.5)) that the 1000 µm KALOS pebbles withstand during uniaxial compression are displayed as a function of the annealing time. The uncertainty of the measurements, considering the quality of the load cell, is too low to be displayed in this plot. Instead, the standard deviation of the determined mean contact pressures is given.

Also the development of the mechanical strength of the 1000 µm KA-30-LMT pebbles resembles that of their 500 µm counterparts to a certain degree. The average mean contact pressure that can be withstood by the pebbles increases from 2.94 GPa to 3.52 GPa after 4 days of annealing. However, this increase does not continue at a lower but steady slope until 64 days of annealing as for the smaller pebbles, but stalls at about the level of strength that is observed after 4 days of annealing. The decrease of strength after 128 days of annealing that is observed for the 500 µm pebbles seemingly sets in after 64 days of annealing and continues until an average mean contact pressure of 3.23 GPa is reached after 128 days of annealing which is still notably higher than the value that is measured for the as-received 1000 µm KA-30-LMT pebbles. In general the mechanical strength of the 500 µm pebbles is significantly higher than that of their 1000 µm counterparts as they withstand about 10-33 % higher mean contact pressures. As a function of the annealing duration, the mechanical strength of the lithium excessive pebbles that were fabricated by the emulsion method as well as the sol-gel method show a significantly different behavior with respect to the LOS containing pebbles that were fabricated by the KALOS process as Figure 4.42 clearly shows. In principle, both samples show a decelerating decrease of the average mean contact pressure that can be withstood by the pebbles with increasing annealing duration.

Solely the strength of the as-received pebbles that were fabricated by the sol-gel method do not fit into this description as it is significantly lower than the value that is measured after 4 days of annealing with 4.23 GPa and 4.76 GPa, respectively. The as-received pebbles that were fabricated by the emulsion method show the highest average mean contact pressure for these two sets of samples with 5.52 GPa, and in general the pebbles that were fabricated by the emulsion method show higher strength compared to the pebbles that were fabricated by the sol-gel method. As a result of the annealing, the mechanical strength continuously decreases to very similar values of 3.36 GPa and 3.53 GPa after 128 days of annealing for the emulsion method and the sol-gel method pebbles, respectively.

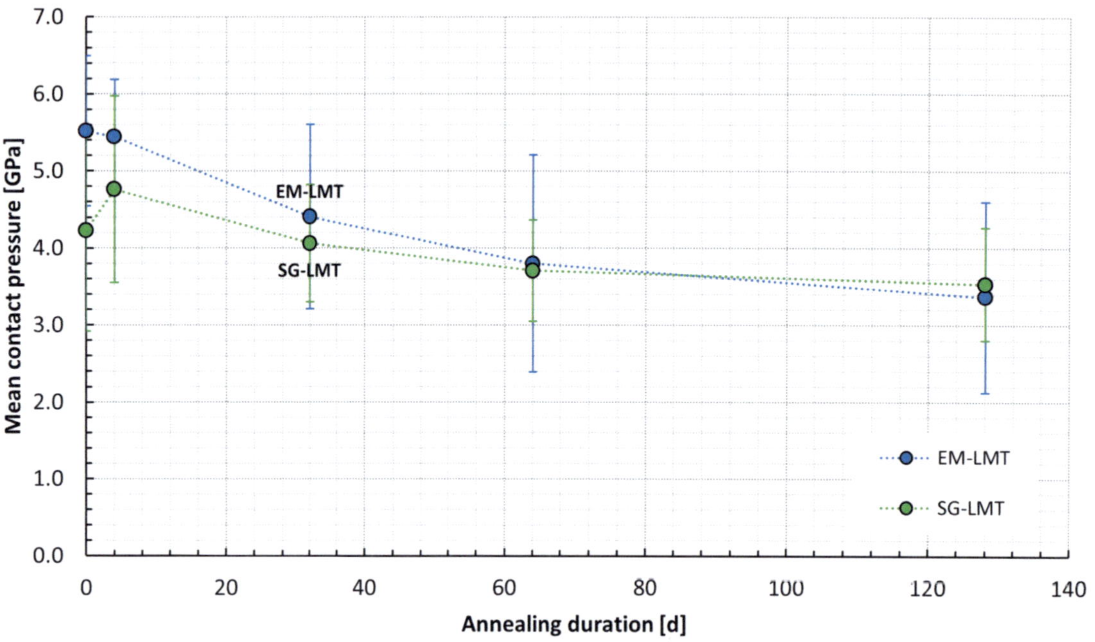

Figure 4.42: The average mean contact pressures (see equation (2.5)) that the lithium excessive pebbles (fabricated by the emulsion method and the sol-gel method) withstand during uniaxial compression are displayed as a function of the annealing time. The uncertainty of the measurements, considering the quality of the load cell, is too low to be displayed in this plot. Instead, the standard deviation of the determined mean contact pressures is given.

In Figure 4.43 the average mean contact pressures of the biphasic LOS containing pebbles that were fabricated by the emulsion method are detailed as a function of the nominal second phase concentration. It is evident, that the addition of LAO instead of LMT only marginally influences the mechanical strength of the pebbles.

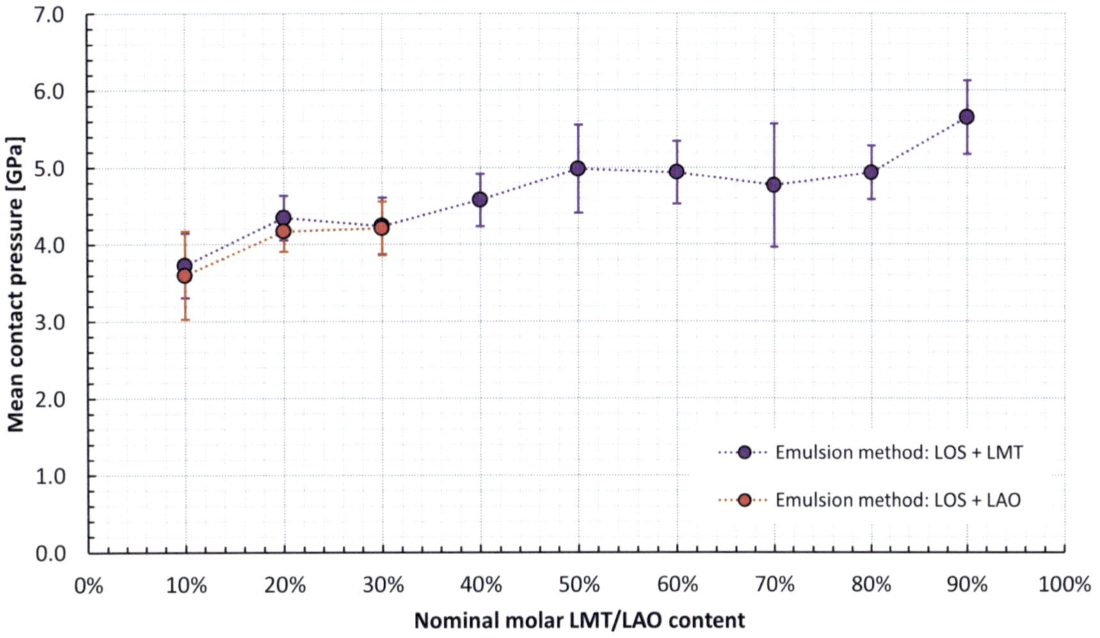

Figure 4.43: The average mean contact pressures (see equation (2.5)) that the biphasic pebbles (fabricated by the emulsion method) withstand during uniaxial compression are displayed as a function of the nominal second phase concentration. The uncertainty of the measurements, considering the quality of the load cell, is too low to be displayed in this plot. Instead, the standard deviation of the determined mean contact pressures is given.

However, the obtained values for the average mean contact pressure of the LAO containing pebbles are generally slightly lower compared to the respective LMT containing ones. For the LMT containing pebbles the average mean contact pressure that the pebbles can withstand principally increases with increasing LMT concentration from the minimal value of 3.73 GPa that is determined for the nominally 10 mol% LMT containing pebbles. Yet, the measured values fluctuate significantly about the general trend. Especially the pebbles with nominally 50 mol% LMT seemingly show a local maximum of strength with 4.98 GPa. This value is only exceeded by the 90 mol% LMT containing pebbles that are able to withstand a mean contact pressure of 5.65 GPa. Between these two values the measured average mean contact pressures stagnate at slightly lower values.

4.5.2 Weibull analysis

The Weibull analysis was performed according to chapter 3.4. It was found that for many samples a single Weibull line sufficiently describes the failure behavior. The so-determined parameters can thus be used in a fracture mechanical analysis, because a single defect size distribution determines the failure behavior of these samples. But, as the results of the fractography in section 4.5.3, demonstrate, the determination of the defect (and its size) which led to failure cannot be determined with appropriate certainty. Consequently a quantitative fracture mechanical analysis of any of the pebble samples is not possible in this study. In most other cases, when a single Weibull line insufficiently models the data points, the failure behavior can be approximated by a maximum of two Weibull lines. However, it is important to note that such a description is rarely suitable for a fracture mechanical analysis. Only in the exceptional case that the two relevant defect size distributions are reasonably approximated by the Weibull parameters, a quantitative fracture mechanical analysis can sensibly be carried out.

Nevertheless, the approximation of the data by two Weibull lines is well suited for describing the failure probability as a function of the mean contact pressure, which is of high value for the engineering of the breeder blanket and the necessary modeling of the ceramic breeder pebble beds within it [133,135]. In some cases however, also this descriptive approach does not appropriately model the data points of a sample. As a consequence a number of data points are censored, i.e. ignored in terms of mathematical description. There are numerous reasons why censoring may be necessary, but especially if single data points clearly deviate from a Weibull line and strongly deteriorate the description of the experimental data, it is often difficult to find a satisfying explanation for the deviation of this data point. In all other cases, any necessary censoring is explained and discussed in this section or eventually in section 5, so that no data point is actually ignored.

Of course, also the censoring of data points leads, like the use of multiple Weibull lines, to an approximation of the failure behavior, since in the ideal case the complete failure behavior has to be modelled. Nevertheless, the failure behavior of the overwhelming majority of pebbles and its development as a function of annealing can be comprehended in this way. All so-determined Weibull parameters are presented as likelihood contour plots for visual clarity, with the characteristic mean contact pressure $p_{m,0}$ plotted on the horizontal axis and the Weibull modulus m on the vertical axis.

In Figure 4.44 the determined Weibull parameters are displayed in this way. It is evident, that the samples in their initial state show a far higher characteristic contact pressure ($p_{m,0}$: 4.1 GPa) than the other samples with the same composition and size. After annealing, the Weibull parameters of all those samples cluster around the values that were found after 4 days of annealing (m: 5.6, $p_{m,0}$: 3.2 GPa). This constitutes a reduction of the Weibull modulus as well as the characteristic mean contact pressure. Although there is no overlap of the confidence boundaries of all four sets of samples, at least three overlap so that the failure modes of these samples may be indistinguishable.

For the 1000 µm pebbles of the same KALOS sample, the Weibull likelihood contour plot looks relatively similar to the one for the smaller pebbles (compare Figure 4.44 and Figure 4.45). Yet, overall the determined characteristic mean contact pressures and the Weibull moduli appear to be higher for the smaller pebbles in comparison. In the initial state, the pebbles show a significantly higher characteristic mean contact pressure (m: 7.3, $p_{m,0}$: 3.3 GPa), which is lowered significantly after 4 days of annealing (m: 5.3, $p_{m,0}$: 2.3 GPa). For both sets of samples (as-received and annealed for 4 days), 4 and 3 apparent failures were observed, respectively. These failures happen for the lowest observed pressures for the as-received samples while these values are scattered between the lowest 50 % of the observed mean contact pressures of the 4 days sample.

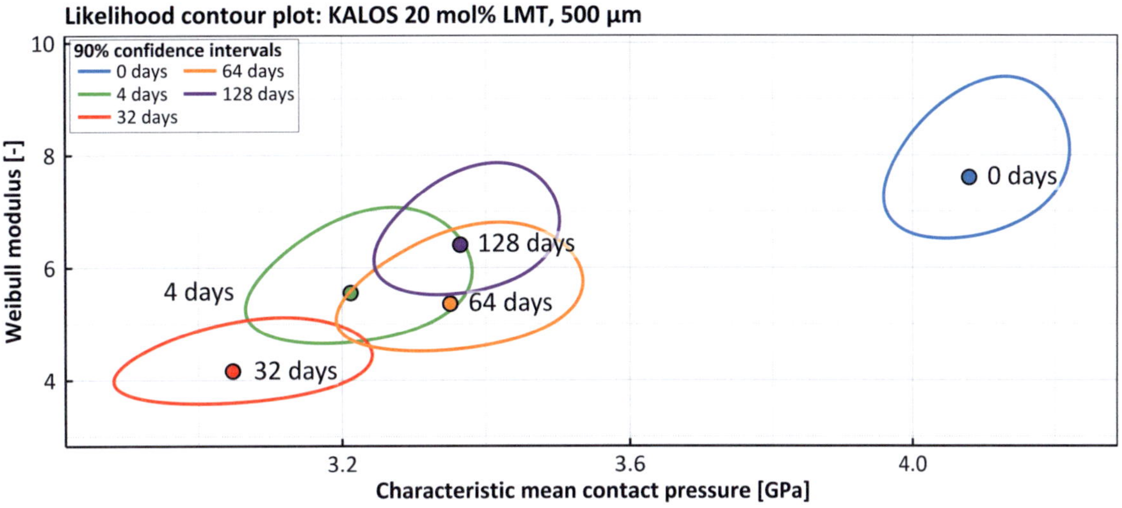

Figure 4.44: Weibull likelihood contour plot of the 500 µm, 20 mol% LMT containing pebbles that were fabricated by the KALOS process and that were annealed for different durations. The points mark the determined Weibull parameters (i.e. the maximum of the respective likelihood function), while the ellipsoidal curves mark the confidence boundaries of the respective parameter estimation.

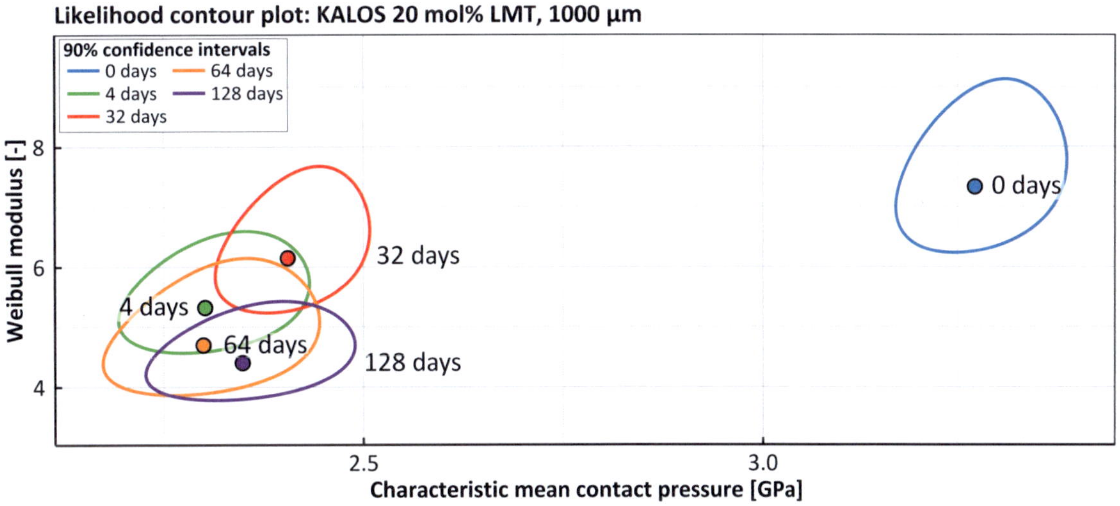

Figure 4.45: Weibull likelihood contour plot of the 1000 µm, 20 mol% LMT containing pebbles that were fabricated by the KALOS process and that were annealed for different durations. The points mark the determined Weibull parameters (i.e. the maximum of the respective likelihood function), while the ellipsoidal curves mark the confidence boundaries of the respective parameter estimation.

Yet, also a few quasi-continuously failing pebbles are observable in the annealed sample. Similar to the 500 µm pebble samples, the Weibull modulus is reduced as a consequence of the annealing and the likelihood ratio boundaries of the Weibull parameters overlap with the ones that are determined for longer annealing durations. The annealing of these pebbles for 32 days leads to a slight increase of the Weibull modulus as well as the characteristic mean contact pressure. For the samples that were annealed for 64 and 128 days, the characteristic mean contact pressure is rather similar while the Weibull modulus is determined to be lower. It must be noted that for both datasets of the samples that were annealed for 32 days and for 64 days, 8 data points of the lowest strength of 50 and 40 data points, respectively, were censored.

For both datasets, these data points lie above the determined Weibull line and follow a curved line which is clearly no random deviation from the Weibull line (see Figure 4.46). Furthermore, both sets of censored data points do not lie within the confidence boundaries of the determined Weibull line. Therefore, the representation of both datasets by a single Weibull line is just an approximation which may represent the majority of the data points reasonably well. For these samples, 13 and 16 non-ideal failures are detected, i.e. failures which do not lead to the immediate disintegration of the tested pebble (see section 3.4). These failures also constitute the censored range and are, as discussed in section 5.3.1, the reason for the observed behavior. The apparent failures are, however, also scattered between regular failures at the lower mean contact pressures of each sample, which is a sign for the presence of two independent failure mechanisms in the samples. Also for the other samples of this set, a considerable number of non-ideal failures is observable.

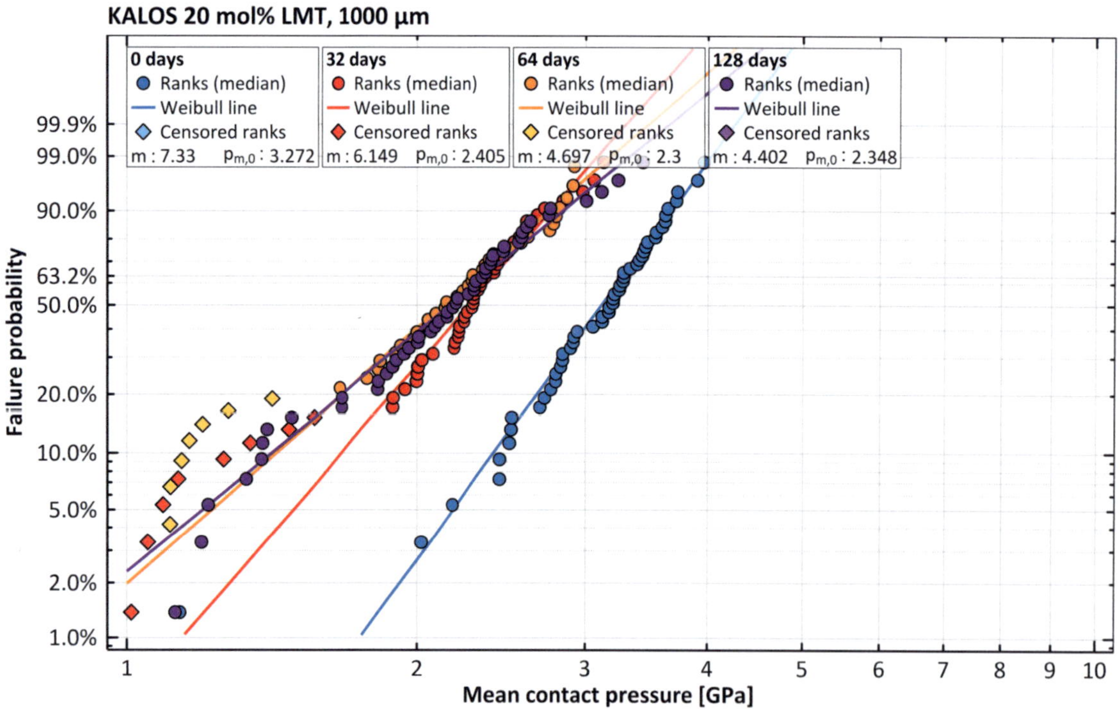

Figure 4.46: Weibull plots of selected datasets of the 1000 µm, 20 mol% LMT containing KALOS pebbles.

As for all of these non-ideal failures an unusually large traverse displacement is obtained, these failures appear to be quasi-continuous failures. In tendency the fraction of non-ideal failures increases with increasing annealing duration from about 10 % in the as-received state to about 40 % after 128 days of annealing. For the 500 µm pebbles of the same grade, non-ideal failures are similarly found, but just about two thirds of the amount of the 1000 µm pebbles. The 500 µm pebbles that were fabricated by the KALOS process and nominally contain 30 mol% LMT show a different behavior than the comparably fabricated pebbles that nominally contain 20 mol% LMT (see Figure 4.47). In the as-received state, the LMT-richer samples show the lowest Weibull modulus of this subgroup of samples with 3.0. However, a single Weibull line does not describe the experimental data well (see Figure 4.48), but it appears to be an at least moderate approximation of the data. Like in the previous cases, the determined Weibull parameters are statistically significantly different from all other sets of parameters.

When the samples are annealed, the Weibull moduli as well as the characteristic mean contact pressures increase until an annealing duration of 64 days (m: 8.5, $p_{m,0}$: 4.8 GPa). Also the likelihood contours of these sets of Weibull parameters overlap in sequence. When the pebbles are annealed for 128 days, the determined Weibull parameters are both reduced again to an intermediate level (m: 6.3, $p_{m,0}$: 4.5 GPa) between the initial pebbles and the pebbles that were annealed for 64 days.

For the pebbles in the as-received state, the 3 (out of 50) data points of the low strength were censored (see Figure 4.48). These data points as well as the fourth lowest data point were obtained at basically the same load, which is just above the detection limit of the measurement system at 0.5 N (see section 3.4). At this experimental boundary, the condition of random drawing of pebbles from the sample is not fulfilled anymore, since no pebbles with lower failure loads can be measured. In consequence the failure probability is incorrectly assigned for the three samples with the lowest loads (see also section 2.4.1) or to put it differently, for the assigned failure probability the determined failure loads are too high and proper values have to be smaller but are unknown. Instead of censoring these data points, they could have been combined into a single data point, i.e. effectively deleting the measurements. By the censoring of data points, the assigned failure probability of all other data points is maintained and with it the benefits of a larger sample size.

For the pebbles of 1000 µm size of the same type of pebbles the determined Weibull parameters show a different development with annealing as well as generally different values. Compared to the 500 µm pebbles, all determined characteristic mean contact pressures show clearly lower values of about 75 % to 80 %. For the determined Weibull moduli the picture is less clear as the spread of the Weibull moduli for the 1000 µm pebbles is large.

The determined Weibull parameters of the 1000 µm pebbles as well as the likelihood contours are displayed in Figure 4.49. The Weibull plot that is obtained for the data of the 1000 µm pebbles in the as-received state is conspicuously different from the Weibull plots of the annealed samples, but it resembles the Weibull plot of the as-received 500 µm pebbles (see Figure 4.48). The data is also moderately described by one Weibull line (m: 2.4, $p_{m,0}$: 3.3 GPa) and compared to the annealed samples, both Weibull parameters are the lowest (see Figure 4.49).

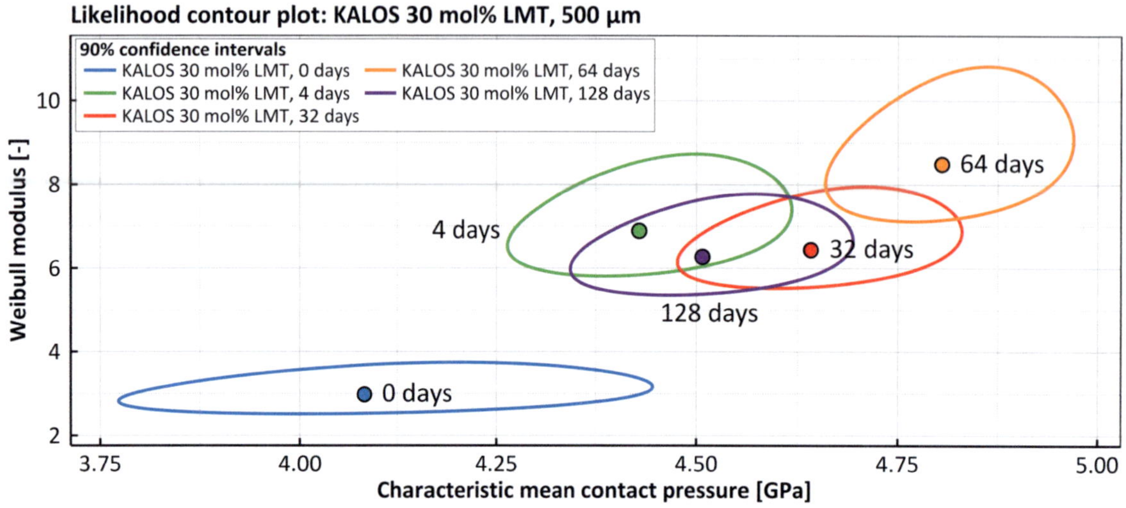

Figure 4.47: Weibull likelihood contour plot of the 500 µm, 30 mol% LMT containing pebbles that were fabricated by the KALOS process and that were annealed for different durations. The points mark the determined Weibull parameters (i.e. the maximum of the respective likelihood function), while the ellipsoidal curves mark the confidence boundaries of the respective parameter estimation.

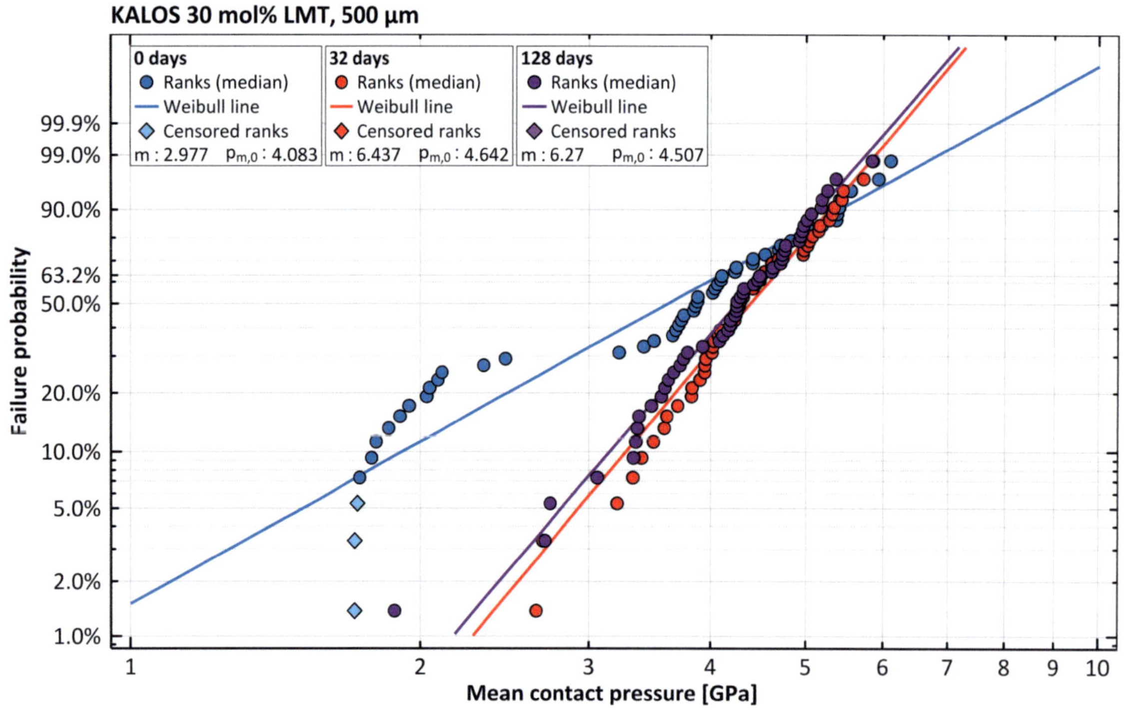

Figure 4.48: Weibull plots of selected datasets of the 500 µm, 30 mol% LMT containing KALOS pebbles.

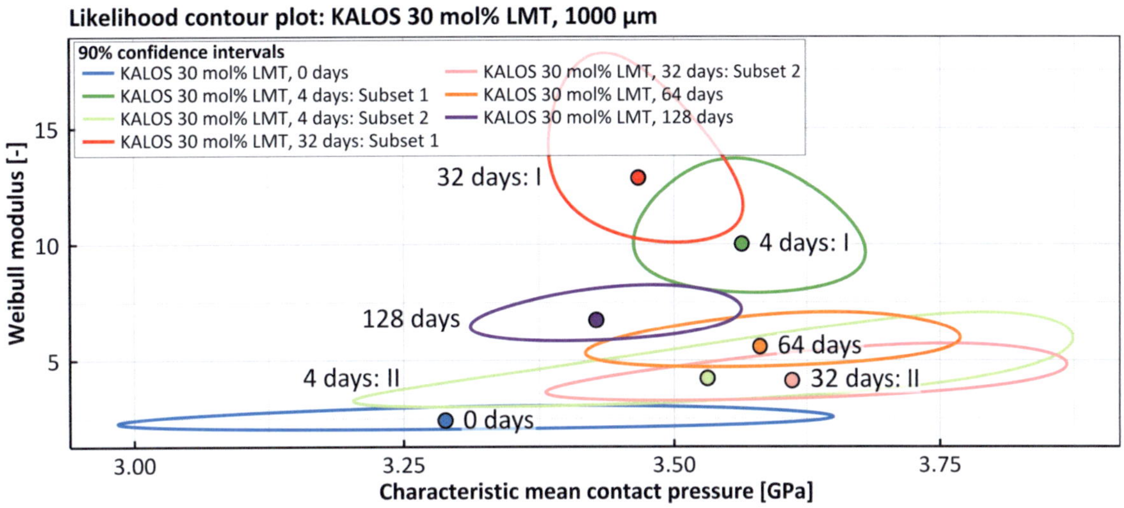

Figure 4.49: Weibull likelihood contour plot of the 1000 µm, 30 mol% LMT containing pebbles that were fabricated by the KALOS process and that were annealed for different durations. The points mark the determined Weibull parameters (i.e. the maximum of the respective likelihood function), while the ellipsoidal curves mark the confidence boundaries of the respective parameter estimation.

Furthermore similar to the 500 µm pebbles of this composition, the 3 (out of 50) data points of the lowest strength were censored, as these values are identically influenced by the experimental detection limit (see Figure 4.50 and compare with Figure 4.48). In contrast to the smaller pebbles, the Weibull plots of the samples that were annealed for 4 days and 32 days cannot be described well with a single Weibull line. In these cases, the use of two Weibull lines leads to a far better result (see Figure 4.50). The reason for this finding is discussed in section 5.3.1.

Generally, the low strength subsets, i.e. subsets I (30 and 28 out of 40 and 50 pebbles for 4 and 32 days of annealing, respectively), resemble the behavior of the 500 µm pebbles which also see a significant increase in the Weibull modulus and a moderate increase of the characteristic mean contact pressure. Also the confidence boundaries of both subsets clearly overlap for the 1000 µm pebbles. The confidence boundaries of the two high strength subsets, i.e. subsets II, also overlap and show considerably lower Weibull moduli, while the characteristic strengths are relatively similar to the ones of the low strength subsets. The necessity to use two Weibull lines vanishes after annealing the pebbles for 64 days and 128 days. In both cases, an intermediate position in the contour plot between the high and low strength subsets is determined for both samples. The confidence boundaries of these sets of Weibull parameters overlap with each other. Yet, there is also a considerable overlap of the confidence boundaries of the values for 64 days and the two high strength subsets of the samples that were annealed for 4 days and 32 days.

From Figure 4.50 the high similarity of high strength part (i.e. above 3.8 GPa) of the Weibull plots of the as-received and the samples that were annealed for 32 days is obvious. This is also true for the pebbles that were annealed for 4 days as both high strength subsets (II) of these samples are very similar as Figure 4.49 suggests. Furthermore, the Weibull line of the low strength subset (I) of the pebbles that were annealed for 32 days almost perfectly extends into a group of data points of the as-received pebbles (between 3 GPa and 3.8 GPa) and obviously describes their failure behavior very well. It must also be noted, that within the low strength section of the Weibull plot of the as-received pebbles, i.e. the flat part below 2.5 GPa, a large portion of the pebbles show non-ideal apparent failures (see section 3.4). While the number of pebbles that show an apparent failure is negligibly low (2 for each sample) for annealing durations of 4 days and 32 days, it is slightly increased to 6 for an annealing duration of 64 days. For all three samples, the cases of apparent failure are found at about the lowest determined mean contact pressures and, similar to the observations for the as-received samples, a considerably higher value is found for the maximum mean contact pressures that the pebble can withstand before disintegrating completely. The same number of pebbles shows an apparent failure after annealing for 128 days as the pebbles that were annealed for 64 days. However, in contrast to the other samples, a significant number of these apparent failures is also found for high strength values. In contrast to the KA-20-LMT samples, no significant number of quasi-continuous failures is observed for the nominally 30 mol% LMT containing pebbles.

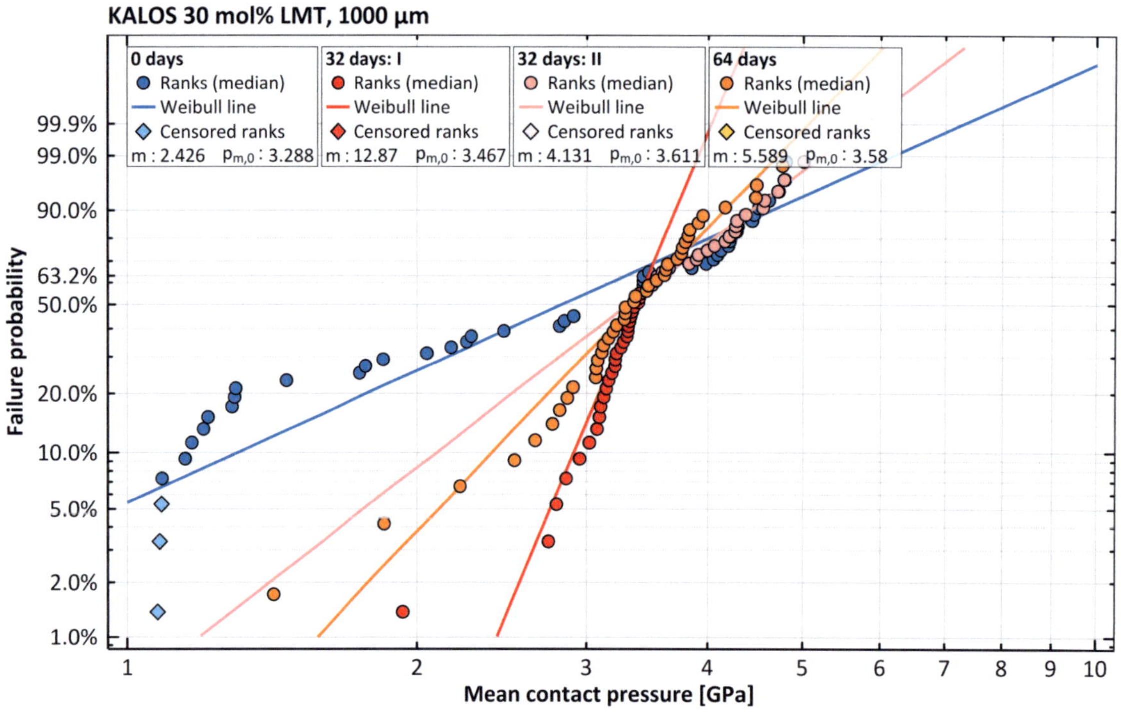

Figure 4.50: Weibull plots of selected datasets of the 1000 µm, 30 mol% LMT containing KALOS pebbles.

The results of the Weibull analysis of the lithium excessive pebbles that were fabricated by the emulsion method are detailed in Figure 4.51. For all samples a single Weibull line was sufficient to represent the majority of data points. The as-received pebbles show the highest characteristic mean contact pressure of this set of samples, while showing one of the lowest Weibull moduli (m: 10.1, $p_{m,0}$: 5.9 GPa). In principle, the annealing leads to a significant reduction of the characteristic mean contact pressure of the samples, while the Weibull modulus is just slightly increasing for these samples (m: 11.7, $p_{m,0}$: 4.2 GPa for the 128 days sample). An exception of this trend mark the pebbles that were annealed for 4 days, as they show a significantly higher Weibull modulus, while the characteristic mean contact pressure fits the overall trend. As Figure 4.51 further shows, all of the lithium excessive samples that were fabricated by the emulsion method behave significantly different considering their mechanical failure behavior.

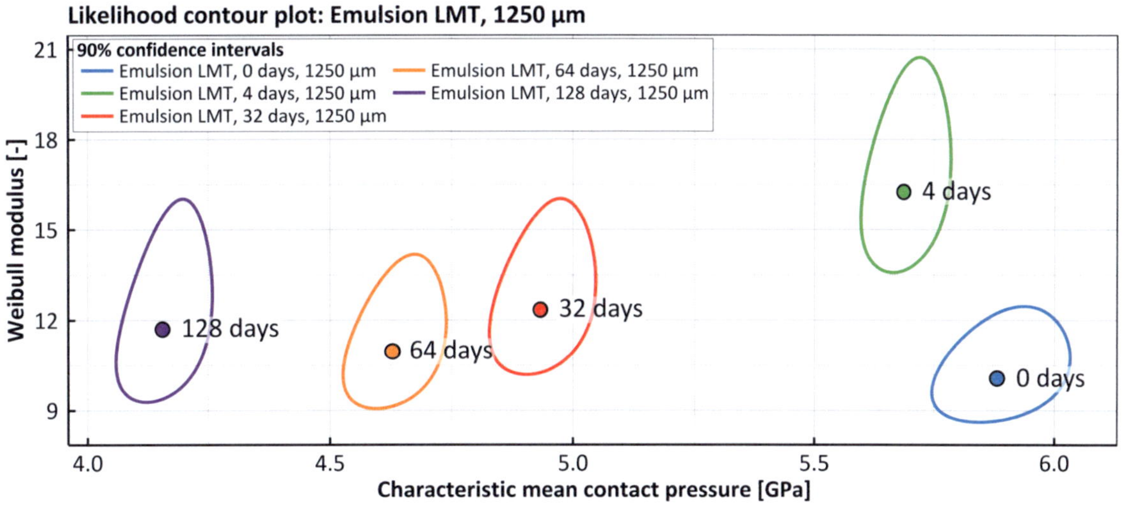

Figure 4.51: Weibull likelihood contour plot of the lithium excessive LMT pebbles that were fabricated by the emulsion method and that were annealed for different durations. The points mark the determined Weibull parameters (i.e. the maximum of the respective likelihood function), while the ellipsoidal curves mark the confidence boundaries of the respective parameter estimation.

However, as Figure 4.52 shows the determined Weibull parameters only approximate the high strength part of the obtained data, because all Weibull plots show strong deviations to lower strengths than the Weibull line suggests. Considering the shape of the Weibull plots, the presence of two strength determining failure distributions seems likely (see also section 5.3.2). From this graph it is also obvious, that the number of pebbles that cannot be assigned to the high strength part increases with increasing annealing duration. From 2 out of 50 data points for the as-received pebbles, 1 out of 40 pebbles that were annealed for 4 days to 6 data points out of 40 for pebbles that were annealed for 32 days are not represented by the given Weibull parameters. These numbers increase to 15 from 50 and 14 from 40 data points for the 64 days and 128 days samples, respectively. All low strength data points can clearly be attributed to 'apparent failures', like the ones for the mentioned KALOS samples.

The lithium excessive pebbles that were fabricated by the sol-gel method show a different failure behavior compared to the pebbles that were fabricated by the emulsion method. The respective likelihood contour plot is shown in Figure 4.53.

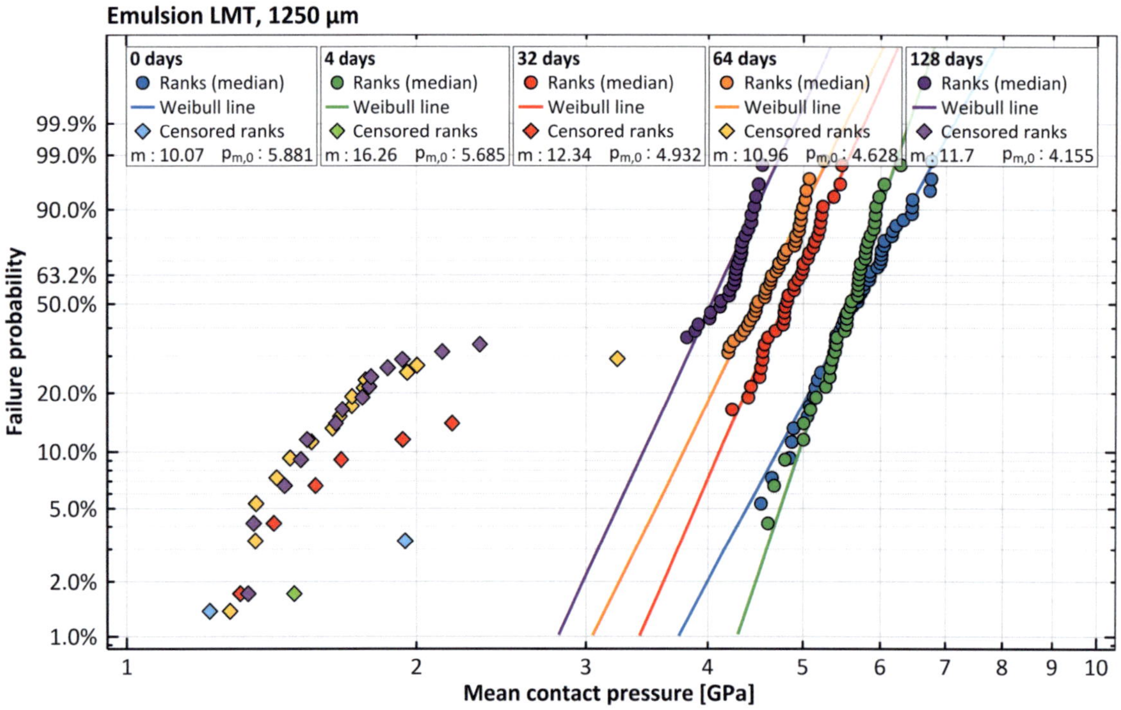

Figure 4.52: Weibull plots of selected datasets of the lithium excessive LMT pebbles that were fabricated by the emulsion method.

Also in this case, the determined Weibull modulus slightly increases with increasing annealing duration while the characteristic mean contact pressure considerably decreases. Yet, in contrast to the samples that were fabricated by the emulsion method, the Weibull modulus is generally lower with values between 6.5 and 8.8 for the 4 days and 64 days annealed samples, respectively. The determined characteristic mean contact pressures for the sol-gel samples are moderately lower than those of the emulsion method samples. Additionally, the as-received samples behave differently. The data of these mechanical tests can only be represented by two Weibull lines which show either a very low characteristic mean contact pressure, while showing a noteworthy high Weibull modulus (subset I) or a noteworthy low Weibull modulus, while featuring an intermediate characteristic mean contact pressure at a value between the samples that were annealed for 32 days and 64 days. This characteristic of the as-received pebbles hints at the presence of two independent strength determining failure populations, i.e. two different materials (see also section 5.1.2). Also for this set of samples, the pebbles that were annealed for the longest time (i.e. 64 days and 128 days) are statistically not clearly divisible as for many KALOS samples.

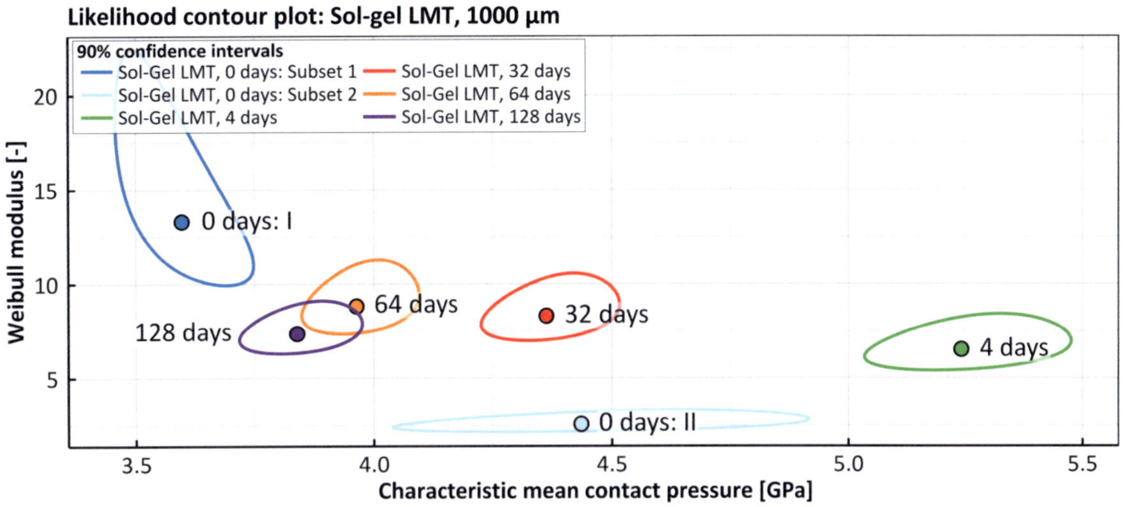

Figure 4.53: Weibull likelihood contour plot of the lithium excessive LMT pebbles that were fabricated by the sol-gel method and that were annealed for different durations. The points mark the determined Weibull parameters (i.e. the maximum of the respective likelihood function), while the ellipsoidal curves mark the confidence boundaries of the respective parameter estimation.

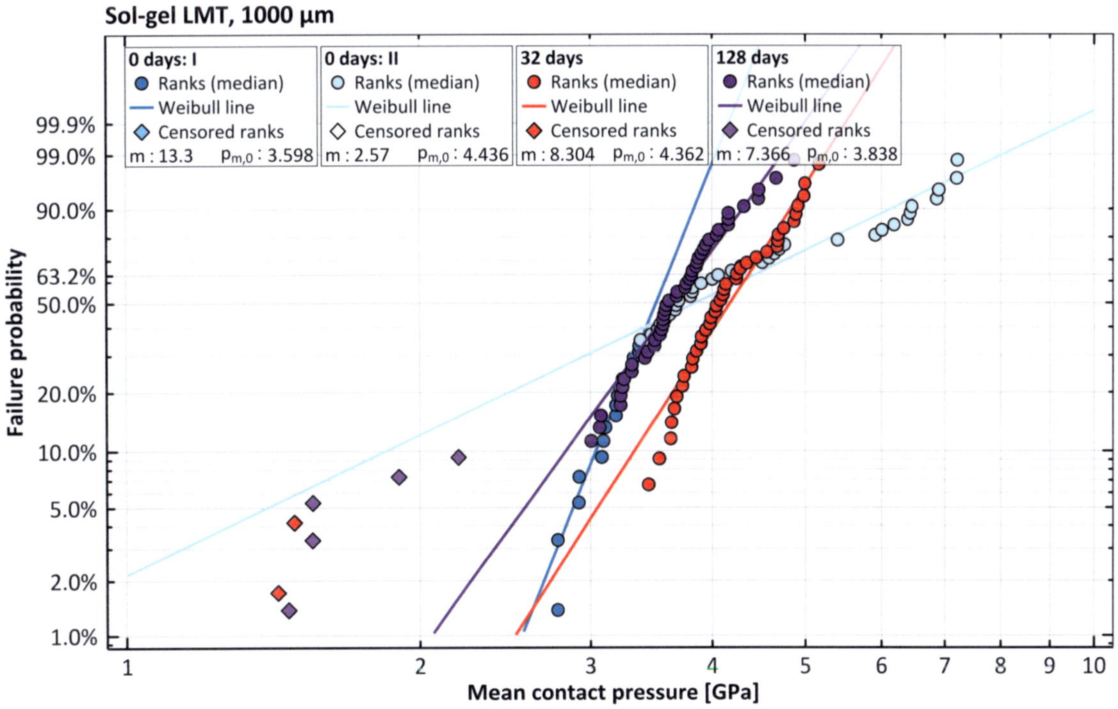

Figure 4.54: Weibull plots of selected datasets of the lithium excessive LMT pebbles that were fabricated by the sol-gel method.

When the Weibull plots of the individual data sets of these samples are plotted (see Figure 4.54), it is evident, that some (2 to 5) low strength data points significantly deviate from the majority of data. Although by far not as many as for the lithium excessive LMT samples that were fabricated by the emulsion method, these data points also form a curve on the left side of the respective Weibull line for all data sets, except for the as-received samples. In order to reasonably approximate the failure behavior of the majority of the data points, these low strength points were censored.

For the biphasic LOS/LMT pebbles that were fabricated by the emulsion method, the summarizing likelihood contour plot of the determined Weibull parameters is given in Figure 4.55. It is obvious, that the samples with 70 mol% LMT and 80 mol% LMT show considerably higher Weibull moduli than most of the other samples. The two sets of Weibull parameters are also very similar and thus the likelihood contours largely overlap. Especially the samples of the series 10 mol%, 30 mol%, 60 mol% and 90 mol% LMT show a relatively constant Weibull modulus at 11.6 to 13.5, while the determined characteristic mean contact pressure significantly increases with increasing LMT content. In comparison to the lithium excessive pebbles that were fabricated by the emulsion method (see Figure 4.51), the determined characteristic mean contact pressures, with the exception of the 10 mol% LMT sample, are similar.

Within the series of biphasic LMT containing pebbles, the samples with 20 mol% and 40 mol% LMT show visibly higher Weibull moduli than the 4 samples from 10 mol% to 90 mol% mentioned before. While the characteristic mean contact pressure that was determined for the 40 mol% sample fits moderately well into the series, the one for the 20 mol% sample exceeds, if only slightly, the value for the 30 mol% LMT containing sample. As a result, the likelihood contours of both sets of Weibull parameters overlap.

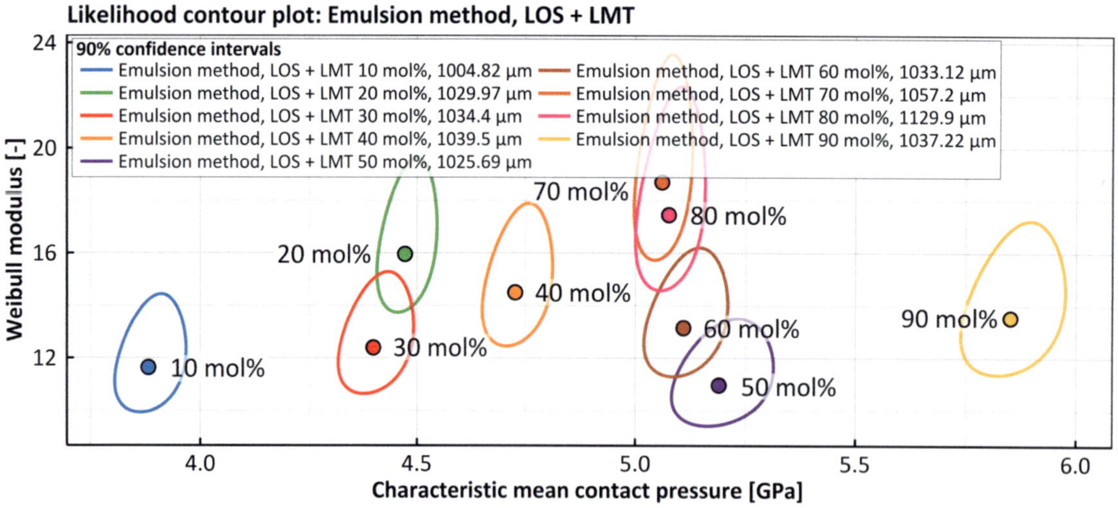

Figure 4.55: Weibull likelihood contour plot of the biphasic LOS/LMT pebbles that were fabricated by the emulsion method. The points mark the determined Weibull parameters (i.e. the maximum of the respective likelihood function), while the ellipsoidal curves mark the confidence boundaries of the respective parameter estimation.

Also the likelihood contour of the 60 mol% LMT sample overlaps with that of another sample, namely the 50 mol% LMT containing one. This sample shows a slightly lower Weibull modulus than the 60 mol% LMT sample but a considerably larger characteristic mean contact pressure. The 60 mol% LMT sample also shows some overlap with the 80 mol% LMT containing one. Other than that, all samples can be considered to be statistically distinguishable with a level of confidence of 90 %. For the 70 mol% LMT containing sample, 4 low strength data points deviate significantly from the Weibull line and form a curve on the left side of it, like the lithium excessive LMT samples that were fabricated with the emulsion method (see Figure 4.56).

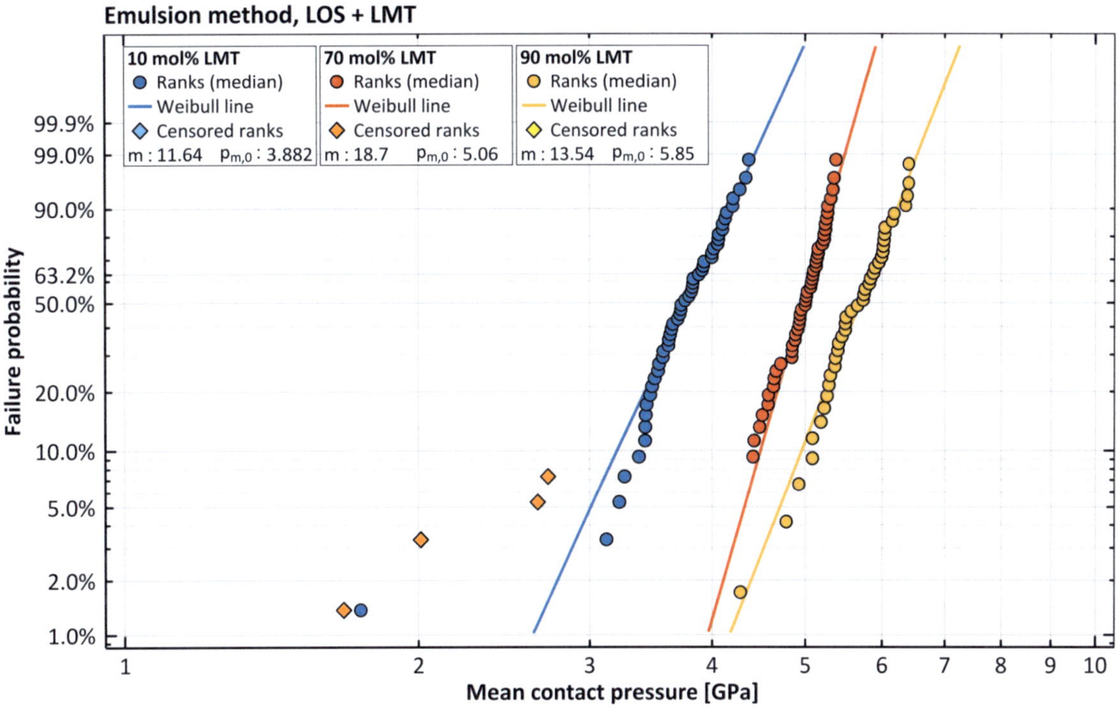

Figure 4.56: Weibull plots of selected datasets of the biphasic LOS/LMT pebbles that were fabricated by the emulsion method.

When LAO instead of LMT is added to the LOS pebbles, the determined Weibull parameters also change as shown in Figure 4.57. For the 10 mol% LAO containing sample, two Weibull lines were necessary to describe the data. The determined Weibull modulus of subset I is notably low with a value of 5.1 compared to all other samples that were fabricated by the emulsion method, but the determined characteristic mean contact pressure is higher than that of the 10 mol% LMT containing sample. For the other subset of data of the 10 mol% LAO containing sample, the Weibull modulus is comparably high and at roughly the same level as the one of the 70 mol% and 80 mol% LMT containing samples that were also fabricated by the emulsion method.

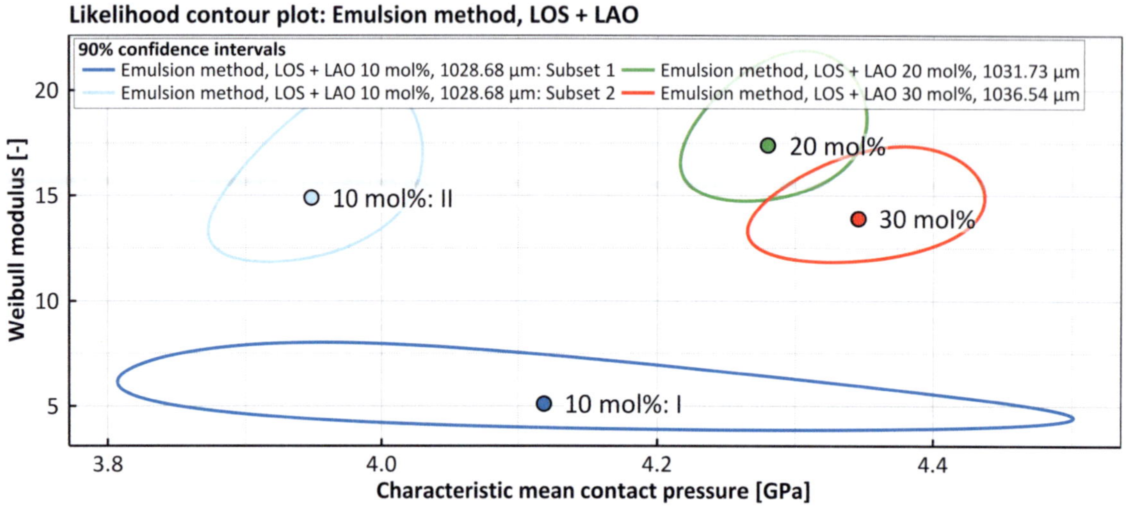

Figure 4.57: Weibull likelihood contour plot of the biphasic LOS/LAO pebbles that were fabricated by the emulsion method. The points mark the determined Weibull parameters (i.e. the maximum of the respective likelihood function), while the ellipsoidal curves mark the confidence boundaries of the respective parameter estimation.

The Weibull moduli of the 20 mol% and 30 mol% LAO containing samples lie moderately higher than that of the 10 mol% LAO containing sample with a value of 17.4 and 13.9 for the 20 mol% LAO and 30 mol% LAO containing ones, respectively. These values are close to those of the 70 mol% and 80 mol% LMT containing samples. Also similar to these two samples, the determined Weibull parameters are very similar and therefore the likelihood contours of both data sets largely overlap. The characteristic mean contact pressures of both samples are, however, significantly lower than those of the LMT containing samples that share similar Weibull moduli and lie in the same range as LMT containing samples with comparable second phase additions.

4.5.3 Fractography

The examination of fragments of pebbles which failed at a known load is shown in this section. Yet, the classical fractographic analysis of the samples is not possible, since the microstructure of the pebbles features considerable porosity (see also section 4.4) as well as for the majority of the samples two distinct phases. Also the presence of a number of fabrication induced cracks complicates the fracture surface considerably. As a result the commonly searched for notable features like mirror, mist or hackle are either not clearly visible, or a number of areas on a given sample is equally probable for one of these features. Nevertheless, a number of features can be identified on the fragments. Therefore, the analysis was not carried out for all samples, but for a number of samples to showcase these findings.

In Figure 4.58 two fragments of the 500 µm, nominally 20 mol% LMT containing pebbles are shown. Both pebbles failed at comparably high loads with respect to the data set of this sample. Yet the failure behavior of these samples can be described by a single Weibull line. In both images the estimated location of 80 % of the pebble diameter is indicated. As detailed in section 2.3.1, within this region the pebbles experience tensile stress along the axis of compression. The maximum of these tensile stresses is expected to lie at about these 80 % of pebble diameter. Yet, the origin of the fracture does not seem to involve volume defects. Instead, the fracture seemingly begins at or near the contact points of the loading. This is much more evident for the pebble that failed at 4.51 GPa, for which the large cavity at one side of the pebble apparently leads to a non-Hertzian contact. Such a large cavity is not visible for the pebble that failed at 3.70 GPa, however, it also shows considerable porosity in several grains, especially in a broad surface-near mantle (see Figure 4.58C). Around the Hertzian contacts also the formation of inner cone cracks can be observed. These inner cone cracks are much more pronounced on the pebble that failed at higher load (see Figure 4.58D).

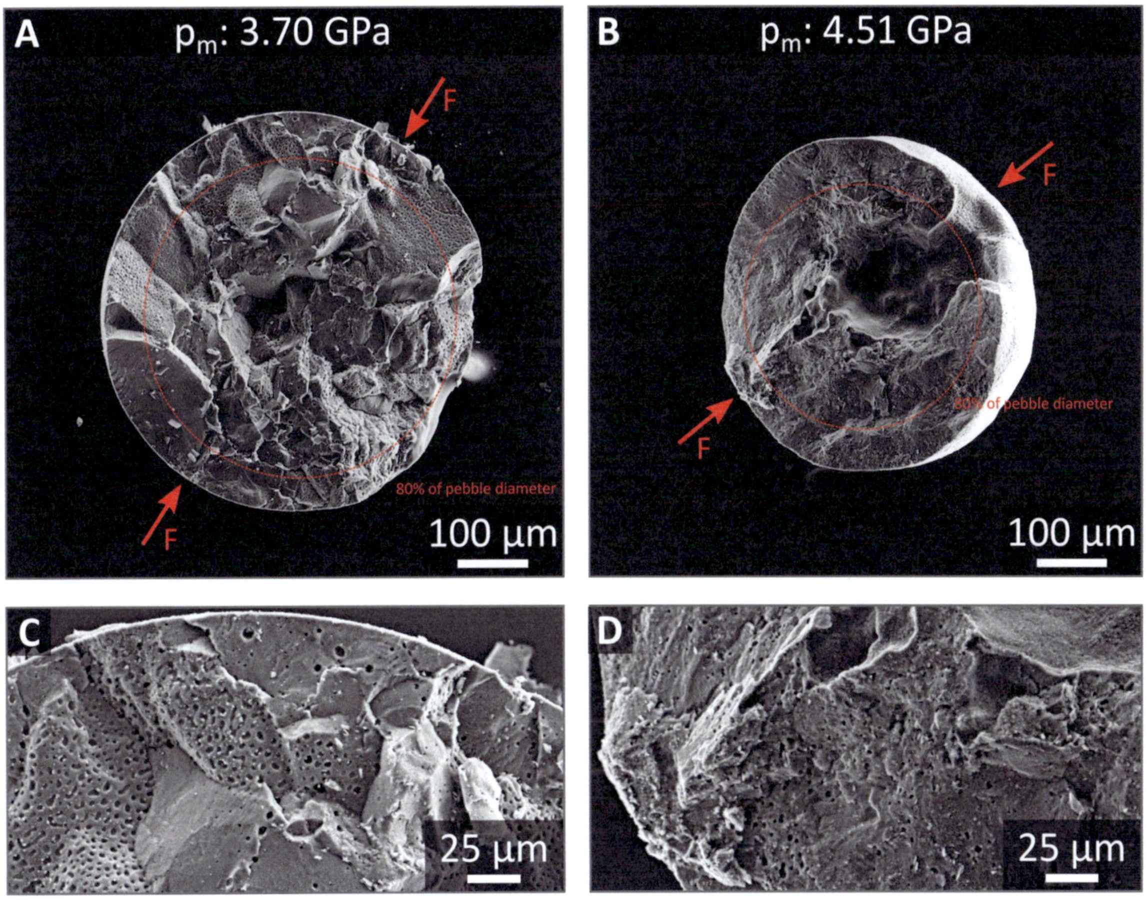

Figure 4.58: Fragments of 500 μm pebbles which nominally contain 20 mol% LMT and were fabricated by the KALOS process. The pebbles were annealed for 128 days. The mean contact pressure, at which the pebbles failed, is given in the headline. Also the direction, along which the load *F* was applied, is indicated in the images. The images C and D are magnified details of the images A and B, respectively.

For 1000 μm pebbles that were annealed identically, two fragments are shown in Figure 4.59. Compared to the 500 μm pebbles, the maximum mean contact pressure is significantly lower for these pebbles. Similar to the 500 μm pebbles of this sample set, one Weibull line satisfyingly describes the failure behavior of the samples. For the pebble that failed at just 1.22 GPa (see Figure 4.59A) it is difficult to locate the contact with the sapphire plates during loading. In contrast to that, the reason for the comparably low strength of the pebble is obvious. With the heterogeneously nucleated crystallization of the pebble, crack paths were established which start from the point of nucleation. Evidently, the failure of the pebble starts from a surface defect. That is also observed for the pebble that failed at 2.40 GPa (see Figure 4.59B). In this case, the pebble did not break into several fragments but remained largely intact. Judging from the opening of the largest crack that is observed on the pebble, the critical defect that led to the failure of the pebble was a surface defect, although its exact location is difficult to determine. It is also clear, that in principle a significant number of surface cracks are visible which could also have led to the failure of the pebble. On this particular pebble, the contact area with the sapphire plates broke off. In contrast to the smaller pebbles of 500 μm size, the strength determining defects, although they are also surface defects, are not located at the loading points.

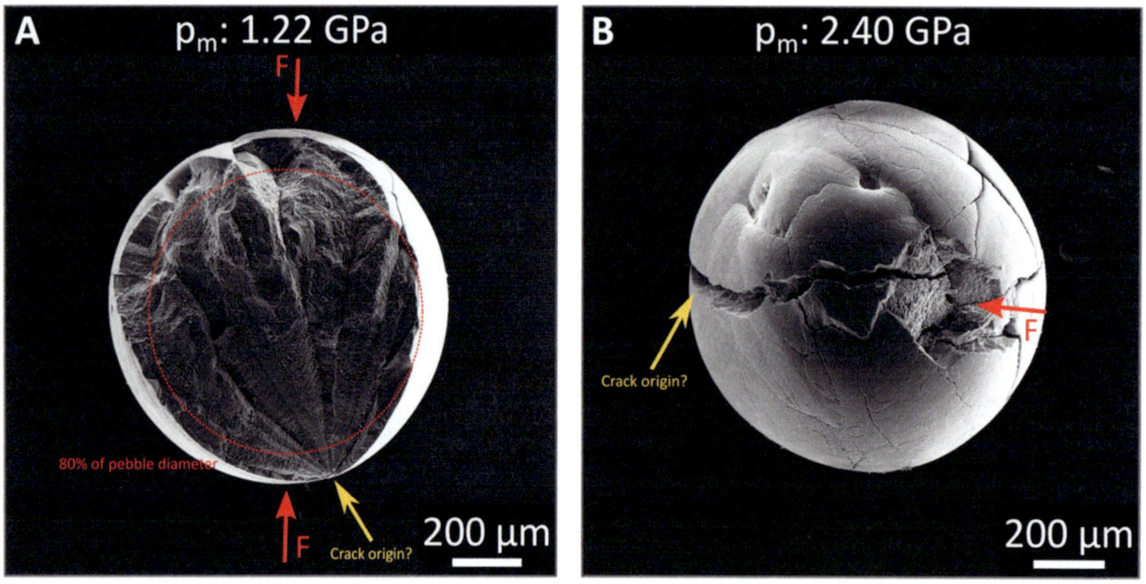

Figure 4.59: Fragments of 1000 μm pebbles which nominally contain 20 mol% LMT and were fabricated by the KALOS process. The pebbles were annealed for 128 days. The mean contact pressure, at which the pebbles failed, is given in the headline. Also the direction, along which the load *F* was applied, is indicated in the images.

In Figure 4.60 two fragments of an as-received 500 μm pebble that nominally contains 30 mol% lithium metatitanate and was fabricated by the KALOS process are shown. The respective pebbles failed at significantly different loads but both can be assigned to the same Weibull line. The mean contact pressures of the pebbles at failing amount to 3.88 GPa and 5.14 GPa. It is evidently clear, that both pebbles differ in their microstructure (see Figure 4.60C and E in comparison with Figure 4.60D and F). For all examined fragments of the in total 10 pebbles of this type, it was consistently observed that the pebbles that failed at high loads show a hypereutectic microstructure (see Figure 4.60F) which is dominated by primary lithium metatitanate, whereas the pebbles that failed at low loads show a hypoeutectic microstructure (see Figure 4.60E). For both pebbles that are displayed in Figure 4.60, the formation of a Hertzian cone crack near one point of loading is observed. At the opposite point of loading, these cracks are not observed. It is presumed that the pebble failed at these locations at a slightly later point in time and that consequently, the stress fields in the opposite region were much more chaotic and thus this contact region failed into multiple pieces. In general, only one contact region can reasonably be analyzed.

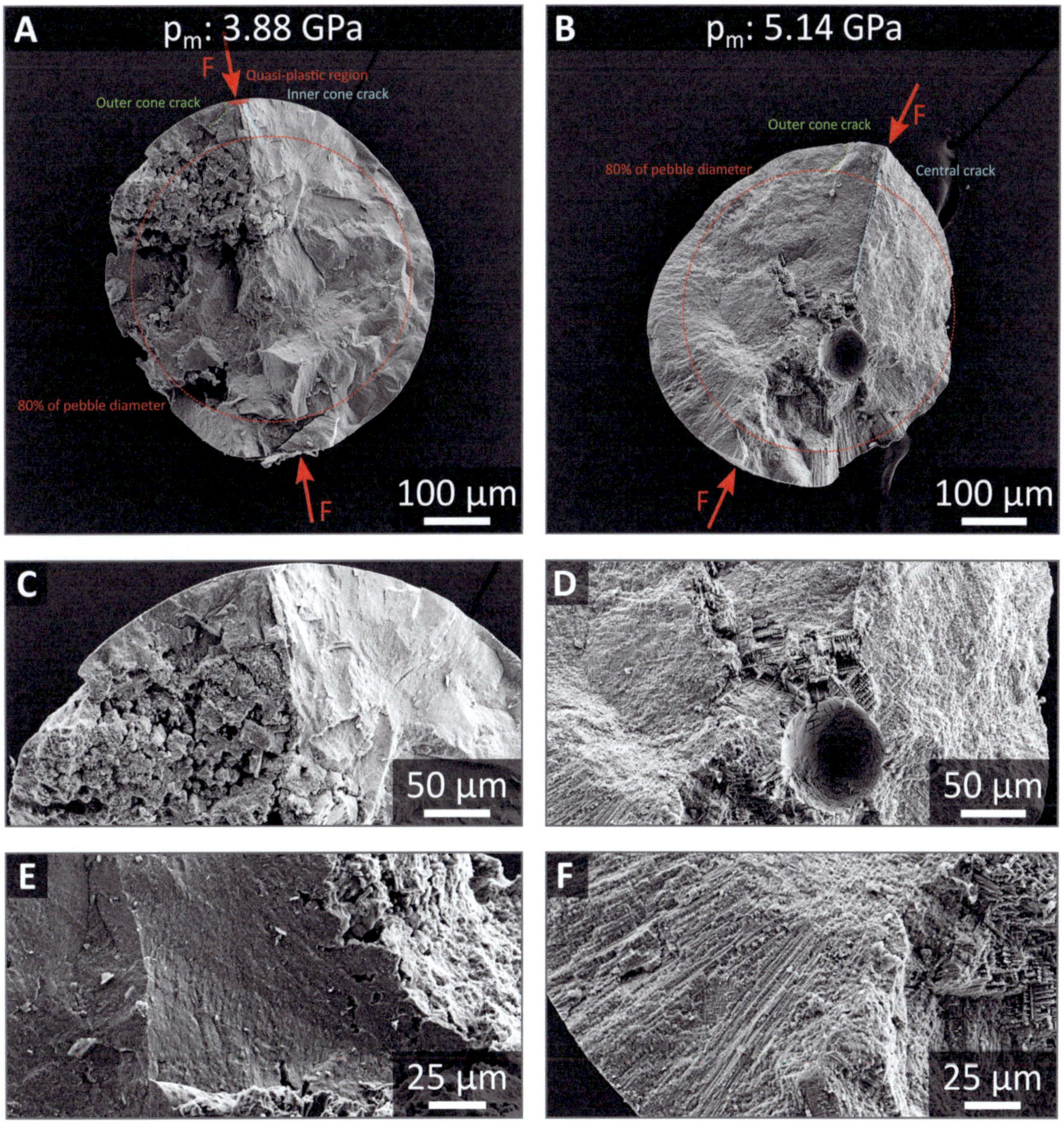

Figure 4.60: Fragments of as-received 500 µm pebbles which nominally contain 30 mol% LMT and were fabricated by the KALOS process. The mean contact pressure, at which the pebbles failed, is given in the headline. Also the direction, along which the load *F* was applied, is indicated in the images. The images C, E and D, F are magnified details of the images A and B, respectively.

For the pebble that failed at the lower load, also the formation of an inner cone crack is observed as well as a region that evidently showed quasi-plastic behavior (see Figure 4.60A and C). Both of these features are not visible for the pebble that failed at a higher load. Yet, a central crack underneath the loading point is observed for the hypereutectic pebble. From the micrograph of the pebble that failed at 3.88 GPa, it is clear, that an area of high porosity lies within the region of tensile stresses (see Figure 4.60A). The shape of the pores within that area is mostly angled (see Figure 4.60C). Although the critical defect that led to the failure could not be identified, it is expected that it is located within this porous area. In contrast, it is less clear where the critical defect is located for the pebble that failed at the higher load. The pebble fragment shows a large spherical cavity to which the central crack seems to extend, but which it apparently does not reach. Yet, the spherical cavity does not seem to be the critical defect, and neither do the pores within the pebble center. It rather seems that a surface defect may have been critical instead.

The annealing of pebbles of this kind for 32 days does not principally change the observations (see Figure 4.61). Again, the mechanically stronger pebbles show a hypereutectic microstructure (see Figure 4.61D and F) and the formation of Hertzian cone cracks is also apparent for both samples. However, for the pebble that failed at a lower load, the cone crack is only slightly visible. However, one fracture surface appears to have formed along the extension of the primary crack that led to failure. It is located within the area where tensile stresses should appear and also lies relatively close to the axis of compression along which these stresses should be maximal. A number of pores are located around and within the primary fracture surface (see Figure 4.61C). Most likely one of the pores that is located within this area led to failure, concluding from the pattern that is observable on the fracture surface. From Figure 4.61E it is also obvious that this pebble features the hypoeutectic microstructure. For the pebble that failed at a mean contact pressure of 5.44 GPa, a deeply extending inner cone crack is observed, while the characteristic outer cone crack of a Hertzian contact is not. Within the pebble volume, an area of high porosity is observed (see Figure 4.61D), which is similar to the one that was observed for the mechanically stronger pebble in the as-received state (see Figure 4.60). By coincidence, the porous area extends just until 80 % of the pebble diameter. As the tensile stresses are expected to be maximal here, failure probably started from this region. However, the surface of the pebble fragment does not give an indication where the critical failure is located. Therefore, the failure of the pebble might have as well started from the surface of the pebble.

The fragments of failed as-received 1000 µm pebbles that nominally contain 30 mol% LMT and were fabricated by the KALOS process are depicted in Figure 4.62 and Figure 4.63. All pebble fragments reveal a large irregularly shaped pore within the center of the pebbles whose appearance resembles an interlinked network of smaller pores in the case of the fragments in Figure 4.62C and D. The size of this pore or porous area decreases as the withstood mean contact pressure increases. In accordance with the observations that were made for the 500 µm pebble fragments, the fragments that failed at higher mean contact pressures show the characteristic hypereutectic microstructure (see Figure 4.63C, D and G, H), whereas the pebbles that fail at lower loads show the hypoeutectic one (see Figure 4.63A, B and E, F).

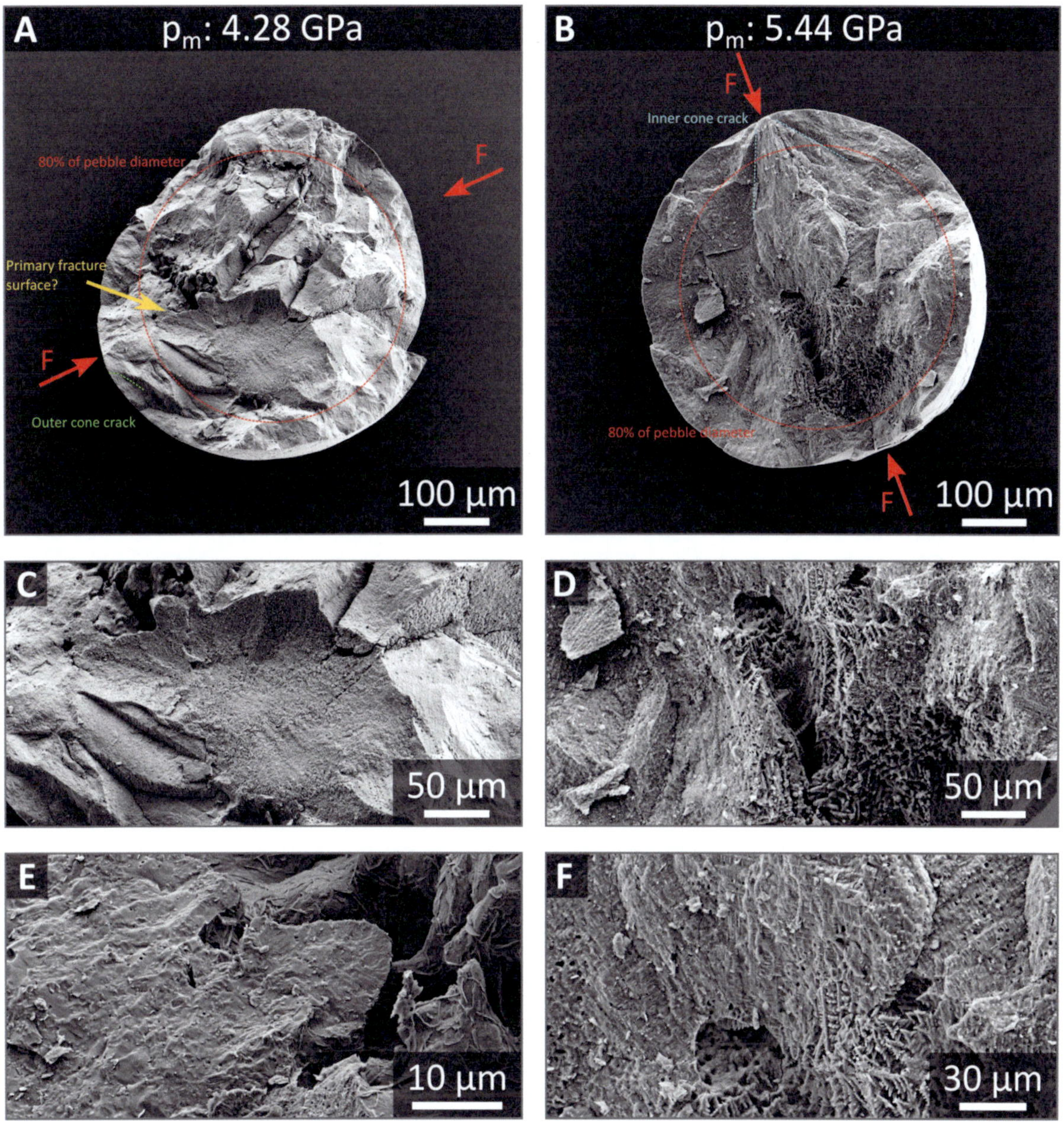

Figure 4.61: Fragments of 500 µm pebbles which nominally contain 30 mol% LMT and were fabricated by the KALOS process. The pebbles were annealed for 32 days. The mean contact pressure, at which the pebbles failed, is given in the headline. Also the direction, along which the load *F* was applied, is indicated in the images. The images C, E and D, F are magnified details of the images A and B, respectively.

The Weibull plot of this sample is more complicated than that of most other samples, and a single Weibull line can only moderately describe the mechanical test data. Judging from the Weibull plot of this sample (see Figure 4.50), the two samples that withstand the lower maximum mean contact pressure are part of the almost horizontal low-strength part, while the other two points belong to the steeper high-strength part.

In contrast to all other pebbles that are shown in Figure 4.62, the pebble that failed at the lowest load (see Figure 4.62A) shows a significantly large quasi-plastic region. The crack initiating defect, however, is difficult to locate. The most probable location is one of the many edged segments of the inner pore. For the pebble that failed at a mean contact pressure of 3.43 GPa, the critical defect is seemingly not part of the inner pore, but from some less porous area within the volume of the sample.

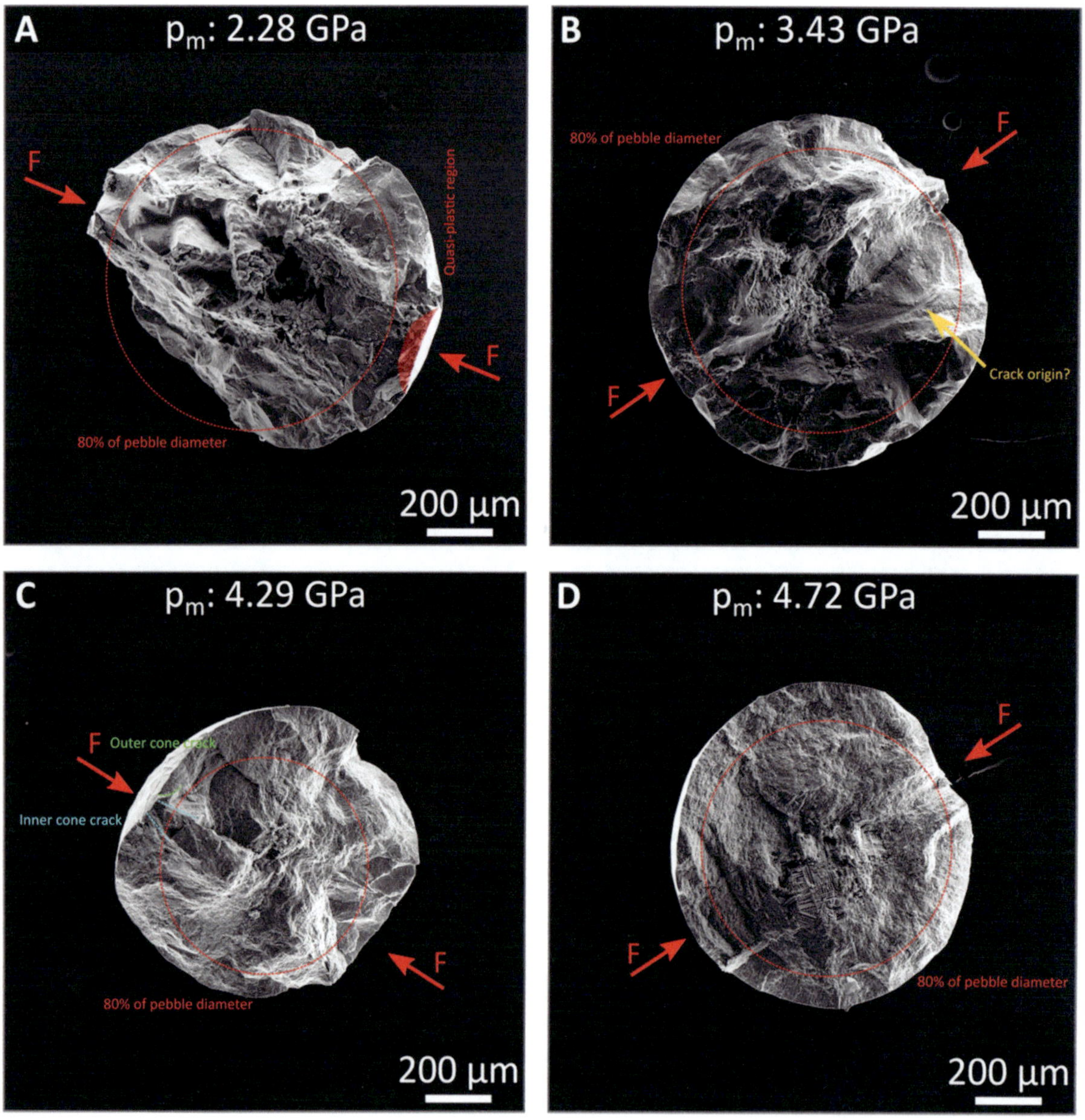

Figure 4.62: Fragments of as-received 1000 µm pebbles which nominally contain 30 mol% LMT and were fabricated by the KALOS process. The mean contact pressure, at which the pebbles failed, is given in the headline. Also the direction, along which the load *F* was applied, is indicated in the images. Magnified details of these images are shown in Figure 4.63.

Also a characteristic inner cone crack was formed during the loading as the top right loading point clearly shows. This feature is also present, although less prominent, for the third pebble fragment (critical mean contact pressure 4.29 GPa, Figure 4.62C), which in addition also features a small outer Hertzian cone crack.

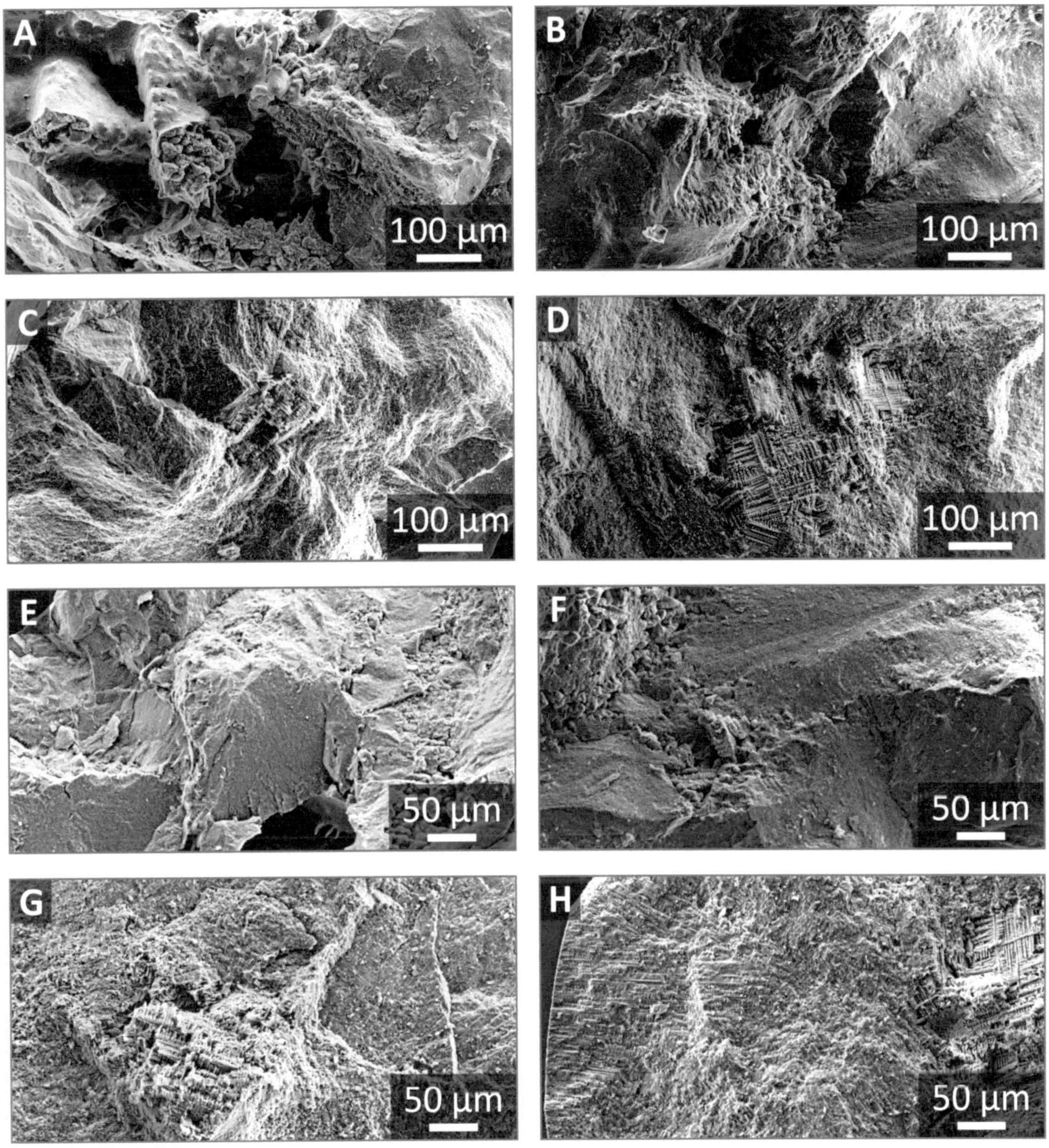

Figure 4.63: Magnified details of the fragments of as-received 1000 µm pebbles that are shown in Figure 4.62 as overview images. The A, B, C, D as well as E, F, G, H labeled images in this figure correspond to A, B, C, D of the overview images in Figure 4.62, respectively.

fracture. However, also some of the pebbles that show better adherence between the grains (see Figure 4.29) showed intergranular fracture. Apart from these differences, both pebbles show rather similar fragments, which also include a considerable quasi-plastic region at the contact points.

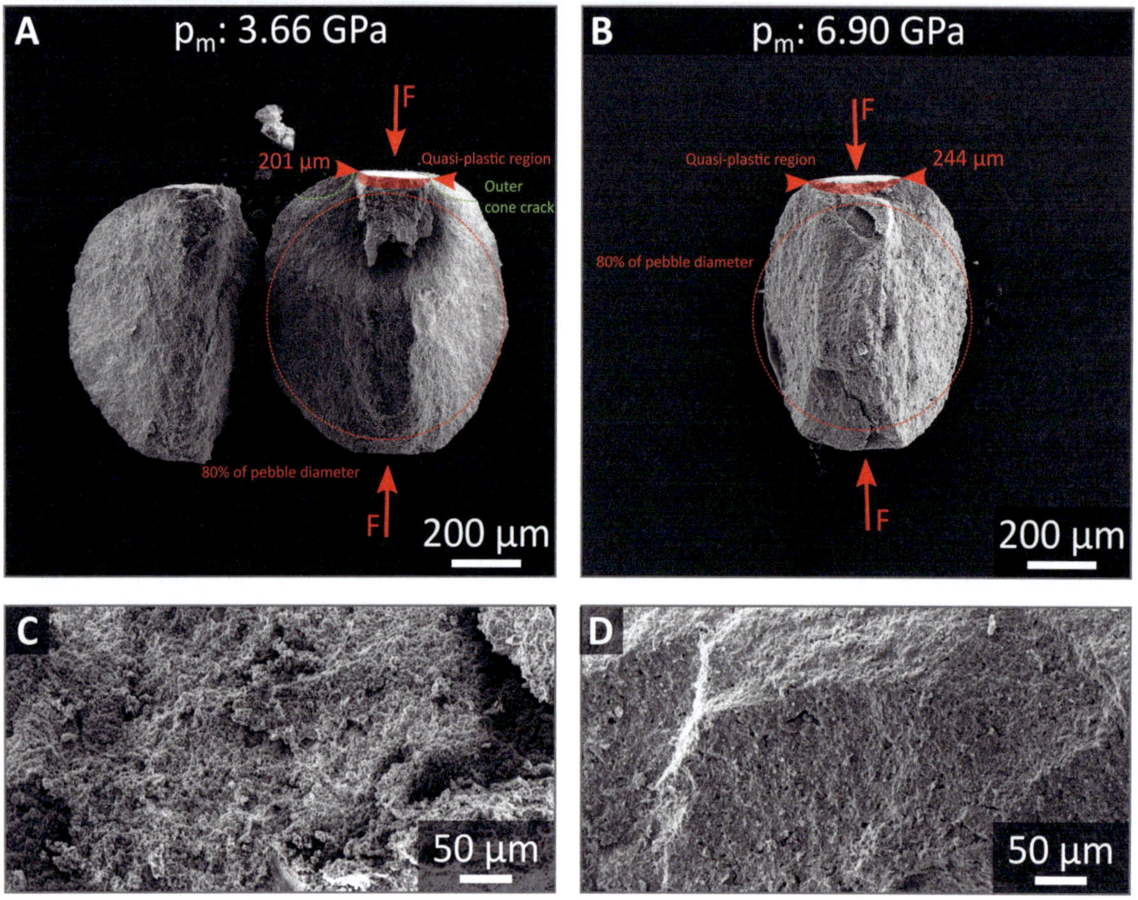

Figure 4.66: Fragments of as-received 1000 µm lithium excessive pebbles which were fabricated by the sol-gel method. The mean contact pressure, at which the pebbles failed, is given in the headline. Also the direction, along which the load *F* was applied, is indicated in the images. The two arrow heads indicate the flattened area and its diameter. The images C and D show magnified details of other fragments of the same pebbles, from which the largest fragments are shown in the images A and B, respectively.

Two representative fragments that are generated by mechanically testing the SG-LMT pebbles after annealing them for 128 days are depicted in Figure 4.67. The critical defect of both pebbles was a volume defects as the formation of a central ridge is observable, although not very clearly. For both samples, remains of the crust that forms on most of these pebbles are visible (compare with Figure 4.29). Both fragments also reveal that the crust does not influence the failure behavior of the samples in a way that only the failure of the crust at probably very low loads is tested. For the pebble that failed at 3.31 GPa, the crust adheres reasonably well to the core of the pebble at the loading points while it is detached from the core near the equator of the pebble. This may either be coincidence or a consequence of the compression. The second example of a fragment of these pebbles illustrates that the crust may also be blasted off from the core during the loading. Yet, at the contact points where the mechanical load is applied the crust stays in place. Also the characteristic inward pointing cones are formed which suggest the failure from a volume defect. From both samples also a notable difference in grain size is perceivable (compare Figure 4.67C and D).

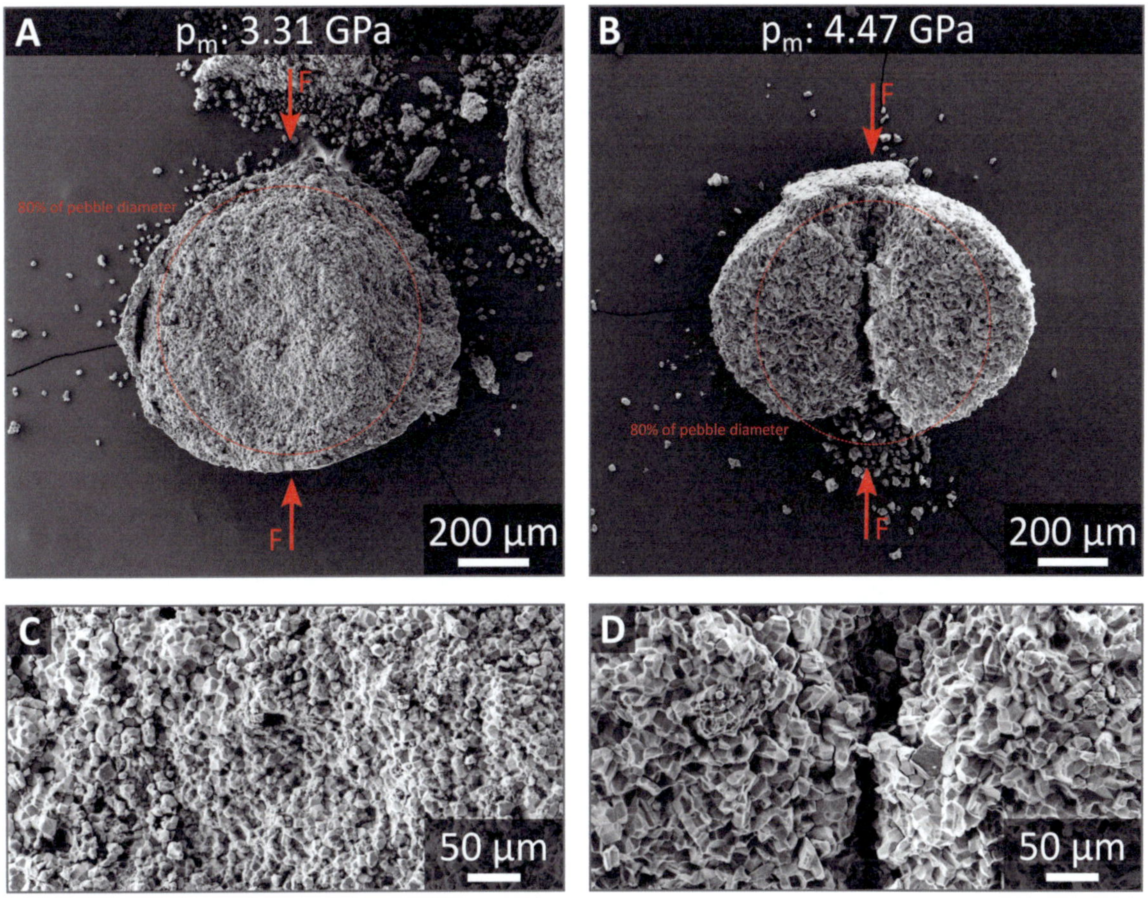

Figure 4.67: Fragments of 1000 µm lithium excessive pebbles which were fabricated by the sol-gel method that were annealed for 128 days. The mean contact pressure, at which the pebbles failed, is given in the headline. Also the direction, along which the load *F* was applied, is indicated in the images. The images C and D are magnified details of the images A and B, respectively.

For the biphasic pebbles that consist of lithium orthosilicate and nominally 50 mol% of lithium metatitanate, two fragments are shown in Figure 4.68. As for most samples, one Weibull line reasonably describes the failure behavior of these pebbles. In both cases, a significant flattening of the contact area with the sapphire plates can be observed. Where this part did not break off, a region of quasi-plastic behavior is visible beneath the contact area. Although the pebble that failed at 4.37 GPa shows a fabrication related defect which runs through the volume of the pebble, the failure does not seem to be triggered by a volume defect, but rather by a surface located defect near the edge of the loading.

Whether the fabrication related defect led to the failure or whether it did not play a significant role is not possible to judge from the available information. For the pebble that failed at 5.05 GPa the origin of the critical crack also seem to be located at the edge of the loaded area. Yet, there is some ambiguity about the location of the crack origin, but volume defects certainly do not play a role for the failure of this pebble. Underneath the indicated quasi-plastic region, this pebble also shows a thin crack which formed perpendicular to the contact area from its center and extends into the volume of the pebble.

Two fragments of biphasic pebbles with a nominal content of 70 mol% LMT that were fabricated by the emulsion method are detailed in Figure 4.69. While a single Weibull line is also sufficient to describe the failure behavior of the majority of these pebbles, all pebbles that fail significantly below a mean contact pressure of 4 GPa are not represented by this Weibull line. Similar to the biphasic pebbles that feature a nominal LMT content of 50 mol%, also the flattening of the contact points is observed for the pebbles with nominally 70 mol% LMT. For the pebble that seemingly failed at 1.68 GPa, this flattened area is comparatively large with respect to the pebble that failed at 5.26 GPa. Also, the visibly quasi-plastic region below the flattened area is considerably larger for the pebble that failed at lower load.

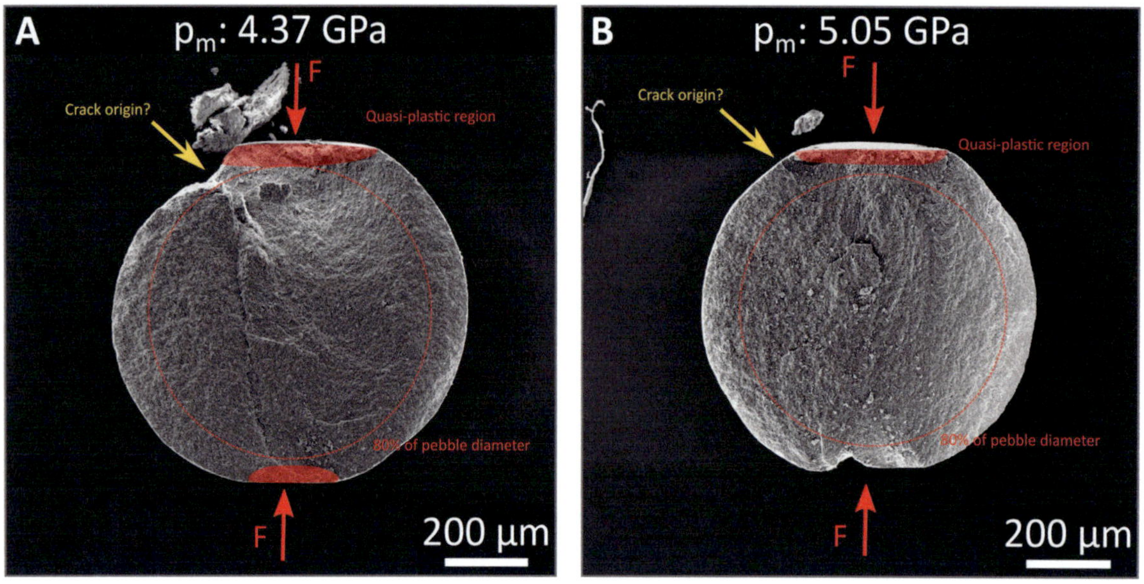

Figure 4.68: Fragments of about 1000 µm pebbles which nominally contain 50 mol% LMT and were fabricated by the emulsion method. The mean contact pressure, at which the pebbles failed, is given in the headline. Also the direction, along which the load *F* was applied, is indicated in the images.

For both of these fragments, a sharp ridge is formed within the pebble volume along the axis of compression, which extends within the inner 80 % of the pebble radius where significant tensile stresses are expected. In both pebbles, also a second ridge of similar character is observed parallel to the axis of compression but off-center. The most likely point at which failure was initiated for these pebbles is found on the central bar. Both pebbles seemingly show inward pointing cones at the loading points. In the case of the pebble that failed at 1.68 GPa, the initiation point of failure lies relatively close to the estimated 80 % of the pebble diameter, which coincides with the tip of the cone. For the other sample that failed at high load, the point of fracture initiation lies probably much more central as the two fracture lines indicated in Figure 4.69 suggest.

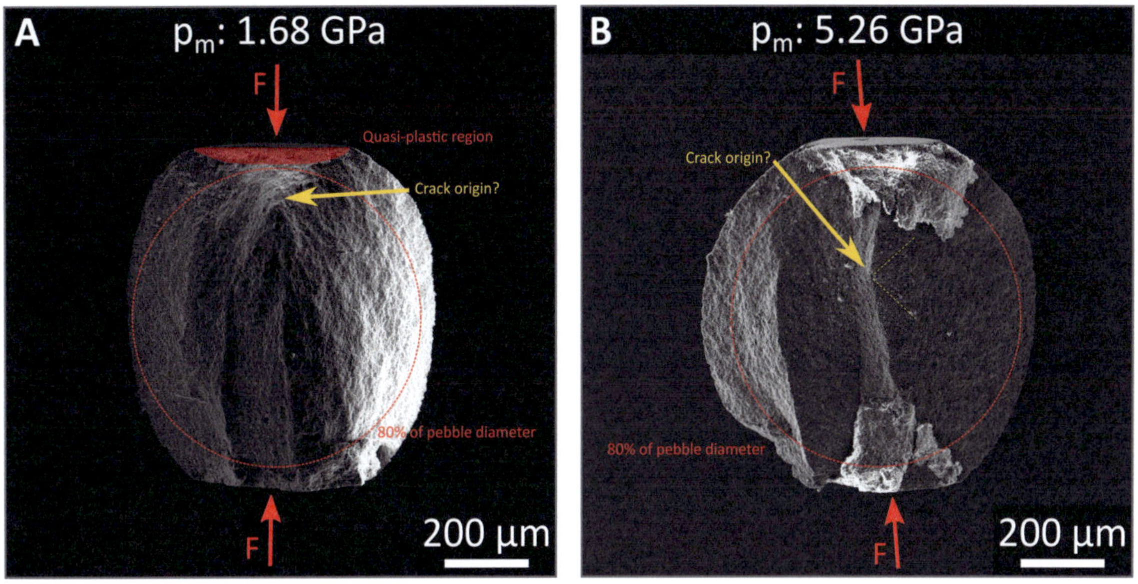

Figure 4.69: Fragments of about 1000 µm pebbles which nominally contain 70 mol% LMT and were fabricated by the emulsion method. The mean contact pressure, at which the pebbles failed, is given in the headline. Also the direction, along which the load *F* was applied, is indicated in the images.

For a content of 90 mol% LMT in the biphasic pebbles that were fabricated by the emulsion process, two typical fragments are displayed in Figure 4.70. The failure behavior of both pebbles can be described by the same Weibull line which very well describes the failure behavior of all pebbles of this kind. The pebble that failed at 4.73 GPa shows a relatively large area which seemingly displayed quasi-plastic behavior.

At the edges of this area small outer cone cracks are visible which reach the pebble within a relatively short distance. From the area that seems to have deformed severely during the mechanical loading, a ridge extends through the complete volume of the pebble, which is in contrast to the observations for the 70 mol% containing biphasic pebbles. Also the observed ridge shows a significant curvature. From the observable fracture lines, the location of the critical defect rather seems to be located at the center of the loading contact. This is similarly observed for the pebble that failed at higher load. Yet, for this pebble no pronounced ridge is observable on the fragment. In summary the fracture of both pebbles seems to be related to surface defects rather than volume defects as observed for the 70 mol% containing pebbles.

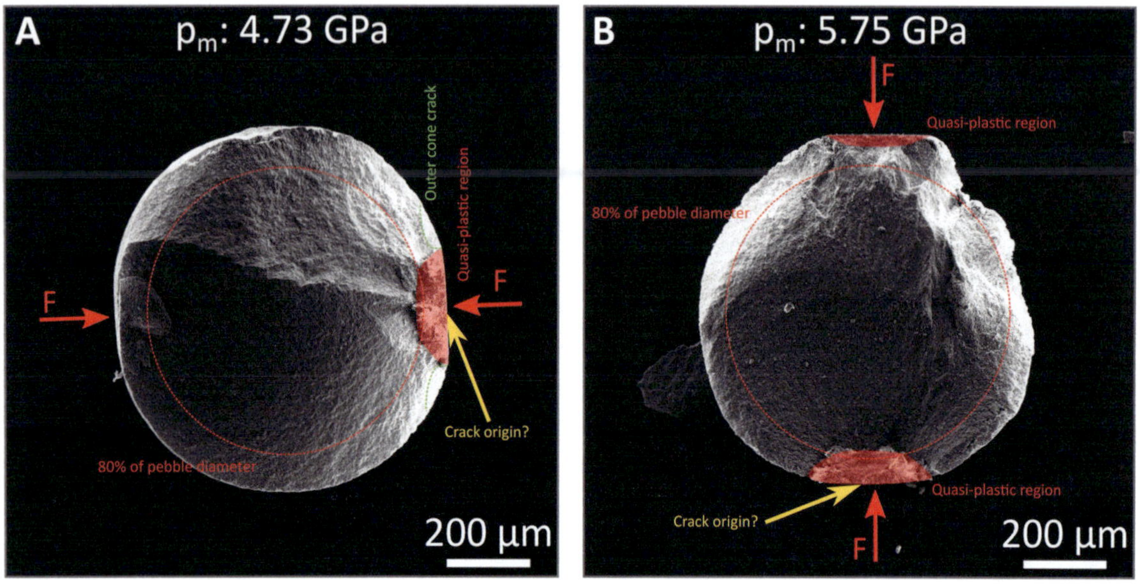

Figure 4.70: Fragments of about 1000 µm pebbles which nominally contain 90 mol% LMT and were fabricated by the emulsion method. The mean contact pressure, at which the pebbles failed, is given in the headline. Also the direction, along which the load F was applied, is indicated in the images.

5 Discussion

The effects that determine the mechanical strength after fabrication and after annealing for different durations are necessary to understand in order to adapt the fabrication processes as well as the operating conditions or to advance the material accordingly. The available results are therefore discussed in this regard in this section.

5.1 The influence of the fabrication method on the mechanical strength

From the elemental analysis (see section 4.1) it is clear, that with each of the three fabrication methods the anticipated composition of the pebbles could reasonably well be achieved. Yet, it is equally evident, that the fabrication method has significant implications on the mechanical strength of the fabricated pebbles (see 4.5). In the following the reasons for these differences will be discussed on the basis of pebbles of comparable composition, i.e. pebbles with a nominal composition of 20 mol% and 30 mol% lithium metatitanate in lithium orthosilicate that were fabricated by the KALOS process and the emulsion method (section 5.1.1) as well as the lithium excessive pebbles that were fabricated by the emulsion method and the sol-gel method (section 5.1.2).

5.1.1 Biphasic lithium orthosilicate based pebbles

The two fabrication methods, which were used to fabricate biphasic pebbles consisting of lithium orthosilicate and lithium metatitanate, are undoubtedly different. While the KALOS process is a melt-based fabrication technique, the wet-chemical emulsion method is powder-based and involves a sintering step. It is obvious from the explanations in sections 2.2.2.2 and 2.2.3.2, that these methods involve very different thermal treatments of the pebbles. This leads to a striking difference in the qualitative phase composition of both pebbles.

Fabrication related thermal stresses within the KALOS pebbles

The rapid quenching of the just crystallized KALOS pebbles during their flight through the cooling section indisputably leads to a suppression of the γ to β phase transformation in lithium metatitanate (see Figure 4.1). Therefore, the KALOS pebbles consist of lithium orthosilicate and metastable, cubic γ-LMT, whereas the pebbles that were fabricated by the emulsion method feature no visible amount of cubic γ-LMT but monoclinic β-LMT instead (see section 4.2.1). But not only leads the rapid quenching of the KALOS pebbles to different phases, it also leads to notably higher tensile strain within the lithium orthosilicate phase for the KALOS pebbles in comparison. Equally significant is the strain of the cubic γ-LMT phase in the KALOS pebbles which is strongly compressive. In principle, also a significant deficit in lithium could lead to a considerable contraction of the unit cell and thus possibly explains this observed compressive strain since cubic γ-LMT shows a considerable homogeneity range (see Figure 2.10). Yet, the ICP-OES measurements of the KALOS samples show a perceivable lithium surplus for the nominally 20 mol% LMT containing pebbles, while they show a slight lithium deficit for the nominally 30 mol% LMT containing pebbles. Yet, the observed strain of the γ-LMT phase differs just marginally for both types. Furthermore, the annealing of the pebbles for 4 days leads to an almost entire annihilation of the strains. The absence of these strains in these samples also rules out that the observed strains in the as-received samples were introduced by the preparation, i.e. during the swing milling, and handling of the material. Consequently, thermal stresses are the only plausible explanation for the observed strains within the γ-LMT phase of as-received KALOS pebbles.

These thermal stresses are proportional to the temperature difference between the pebble volume and the surface of the pebble according to the equations (2.8), (2.9), (2.10) and (2.11), whether it is fully crystallized or has just developed a solid crust. The formation of a crust during the early stages of solidification can be regarded as evidenced, because the majority of pebbles shows crystallization by homogeneous nucleation and the presence of an enveloping surface layer with finer microstructure than the bulk of the pebbles (see section 4.3.3). The presence of the enveloping layer of finer microstructure (see Figure 4.24 and Figure 4.26) can be assigned to a higher supercooling of the melt in this area and a thus higher probability of homogeneous nucleation. Consequently, the crystallization of the melt droplets begins at the surface. From the surface pictures of the as-received pebbles (see Figure 4.12 and Figure 4.13), it is obvious that for the nominally 30 mol% LMT containing samples predominantly lithium metatitanate covers the surface instead of lithium orthosilicate which covers the surface of the nominally 20 mol% lithium metatitanate pebbles. These observations are easily explained when a eutectic system is assumed with the eutectic between both nominal compositions. In this case lithium orthosilicate and lithium metatitanate crystallize exclusively at the very beginning of the solidification for the nominally 20 mol% LMT and 30 mol% LMT containing pebbles, respectively, which correlates well with the observations of, for instance, large two dimensional lithium metatitanate structures at the surface. The heterogeneous nucleation of crystallization that a few pebbles show, is most certainly a result of an impacting smaller, already crystallized pebble. For these pebbles, the resulting stresses during crystallization are difficult to assess. Yet, as most pebbles form a crust in the early stages of solidification, the assessment of the thermal stresses during solidification of Timoshenko and Goodier [112] as well as Forgac et al. [128] are applicable for most of the pebbles. As explained in section 2.3.2, the situation for pebbles during their solidification depends very much on the cooling that the pebbles experience as well as the thermomechanical response of the liquid and solid phases. Yet, if tensile stresses are generated within the liquid core of a solidifying pebble, the crust may buckle in response or deform plastically [127], with the latter being unlikely for lithium orthosilicate as well as lithium metatitanate. Another response to the tensile stresses is the formation of internal cavities [127]. The pebbles that were fabricated by the KALOS process clearly show surface buckling as well as cavities within their volume, which implies a liquid core under significant tensile forces. The model of Forgac et al. [128] does not provide a principal answer to the tangential stress state at the surface, however, it is clear that the more substantial the tensile stresses of the liquid core are the more is the tensile nature of the tangential stresses at the surface of the crust reduced.

Yet, Forgac et al. [128] show in their work that even mild tensile stresses within the core are enough to solely form compressive stresses in the crust. Consequently the formation of cracks in the pebble crust during the stage of solidification is implausible. Because the pebbles show a considerable amount of buckling and also internal cavities (see Figure 4.12, Figure 4.13, Figure 4.24 and Figure 4.26), it is reasonable to assume that the stresses that are built up by the volume contraction during solidification were removed for the most part as a result. The observed buckling is much more severe for the nominally 30 mol% LMT containing pebbles, while the measured closed porosity is very low for both samples. Apparently, the tensile stress within the liquid core of these pebbles is higher than for the nominally 20 mol% LMT containing ones, which results in an increased buckling of the crust. The formation of internal cavities is only expected when the crust is rigid enough to withstand the tensile forces within the liquid core. According to Forgac et al. [128], just a few parameters are increasing the tensile stress when the cooling of the pebbles is considered to be the same for these two types of material. Either the bulk modulus of the liquid is increased with a higher LMT concentration or the density ratio of liquid to solid material. Both parameters may have a significant influence as neither is known. However, the more probable effect seems to be that of an increased density ratio, as the resultant effects are predicted to be larger than for a change of the bulk modulus, according to Forgac et al. [128]. Once the pebbles are completely solidified, they are further cooled to room temperature during the fabrication, and consequently thermal stresses according to Grünberg [129] are generated, which are tensile in tangential direction with the maximum stress at the surface (see Figure 2.16).

While the tangential stresses may therefore lead to the formation of cracks at the surface of a pebble, the radial stress is always compressive for a fully crystallized cooling pebble and therefore uncritical in this regard. These tangential thermal stresses can grow to substantially high values for high differences in temperature. Yet, for the present case, the actual temperature difference is not well defined. Nevertheless, the stresses are undoubtedly high enough to form cracks at the pebble surfaces. They preferably initiate from the dents that result from the buckling of the pebble crust in case of the nominally 20 mol% LMT containing pebbles, while the nominally 30 mol% LMT containing pebbles hardly show such cracks even though they feature much graver dents in comparison. Clearly all cracks that form as a consequence of these stresses are nucleated at the pebble surface and should in theory extend perpendicular to the pebble surface into the pebble volume. When it is assumed that the cooling of the crystallized pebbles happens in a similar fashion for both types of pebbles, it is difficult to explain why the different LMT content has such a strong impact on the formation of cooling cracks. Certainly, the most trivial explanation is that the increased LMT content leads to an increased mechanical strength. Yet, assuming entirely crystallized pebbles which purely consist of LOS and LMT, the elemental composition and the quantitative XRD results (see sections 4.1 and 4.2.2) suggest a less than double volumetric content of LMT for the LMT-richer pebbles at relatively low values (12.9 vol.% vs. 22.8 vol.%). It is therefore unsure, whether a 30 % increased strength of pure lithium metatitanate over pure lithi um orthosilicate, as reported in section 2.2.3.1, is sufficient to prevent the generation of surface cracks during cooling.

Generally, considering the published ternary phase diagram of Li_2O-SiO_2-TiO_2 (see Figure 2.11) as well as the studies of Hanaor et al. [108] and Knitter et al. [73], lithium orthosilicate and lithium metatitanate do not show a solid solution phase in which both compounds are quantitatively present. These results also suggest that the quasi-binary system Li_4SiO_4-Li_2TiO_3 is a eutectic system with the eutectic point at a composition of about 25 mol% LMT in LOS (see Leys et al. [74] as well as section 4.3.3). Hence, the nominally 20 mol% LMT and 30 mol% LMT containing pebbles should exclusively show a strongly hypoeutectic and a strongly hypereutectic microstructure, respectively. As the contrary is observed, it is evident that on average the pebbles of the KALOS process show the expected composition (see Table 4.2), but that the deviation from this average value is of considerable magnitude. Apparently, the deviations are large enough to reach the eutectic composition for a significant number of KA-30-LMT pebbles, as these pebbles almost purely consist of lamellar eutectic domains showing neither primary lithium orthosilicate grains nor primary lithium metatitanate. Consequently, the pebbles that are dominated by lithium metatitanate dendrites have to feature a significantly higher concentration of lithium metatitanate so that on average the determined composition is established. Yet, only few pebbles exist that feature a microstructure which matches the nominal composition of the pebbles and thus shows a limited amount of primary lithium metatitanate.

Assuming a eutectic composition of 25 mol%, the solidification of the pebbles in the very beginning will lead to the predominant crystallization of LOS for the hypoeutectic, nominally 20 mol% LMT containing pebbles in contrast to the predominant crystallization of LMT for the hypereutectic LMT-richer pebbles. The surface images of the as-received pebbles (see Figure 4.12 and Figure 4.13) accord with these expectations. Yet, this effect only applies to a very narrow surface region and should therefore have no considerable effect. In terms of the generated stresses according to Grünberg [129] (see section 2.3.2), also no significant differences between both compositions are expected as the stresses are proportional to the product of thermal expansion coefficient and Young's modulus. The thermal expansion coefficient of γ-lithium metatitanate is not known. However, if the thermal expansion coefficient is not too different from that of β-lithium metatitanate (i.e. about 50 % of the thermal expansion of LOS at high temperatures, see Figure 4.10), the thermal stress actually increases in the order of 2 % for the LMT-richer pebbles for the assumed volume contents of both phases. Eventually, this difference should be insignificant. As especially the pebbles of the LMT-richer sample with pearlite-like microstructure are virtually free of surface cracks, it has to be assumed that the composition nevertheless has an important impact on the pebble quality beyond the discussed points. For the nominally 30 mol% LMT containing pebbles that evidently feature a LMT content that is close to the eutectic composition, the temperature difference between the liquidus temperature and the eutectic temperature will be small, while it will be relatively large for the nominally 20 mol% LMT containing pebbles. During solidification, the regions of the droplet whose temperature drops below the eutectic temperature will be fully solid, i.e. form a crust on the pebble, when the possibility of supercooling of the melt is excluded for simplicity. Between the still liquid core of the pebbles and this crust, a two-phase solid-liquid region develops whose radial dimension increases with the increasing temperature difference between the liquidus and the eutectic temperature, if the temperature gradient in both droplets is comparable (see Figure 5.1). Since in both cases dendrites are growing into the two-phase region, the effective thickness of the crust during solidification is probably different, because the secondary dendrite arms may interfere with each other in a much larger range along the radial coordinate of the droplet, if the two-phase region is considerably larger. Also, the effective size of the crust does not change as much, if a large two-phase region exists.

The difference in thickness will change the distribution of compressive stress that acts on the crust because of the contracting core. An effectively thick crust will experience a relatively homogeneous distribution of stress which does not change too much with progressing solidification. In contrast to that, an effectively thin crust, which considerably grows in size with time, will experience a more in-homogeneous distribution of stress, as newly solidifying material 'freezes' the compressive stress in the outer layers of the crust. After solidification and upon cooling, this residual compressive stress near the pebble surface counteracts and to a degree compensates for the tensile stress that originates from the contraction of the cooling surface. Although this model plausibly explains the low amount of surface cracks in the nominally 30 mol% LMT containing that show a pearlite-like microstructure, the available results do not provide an experimental proof.

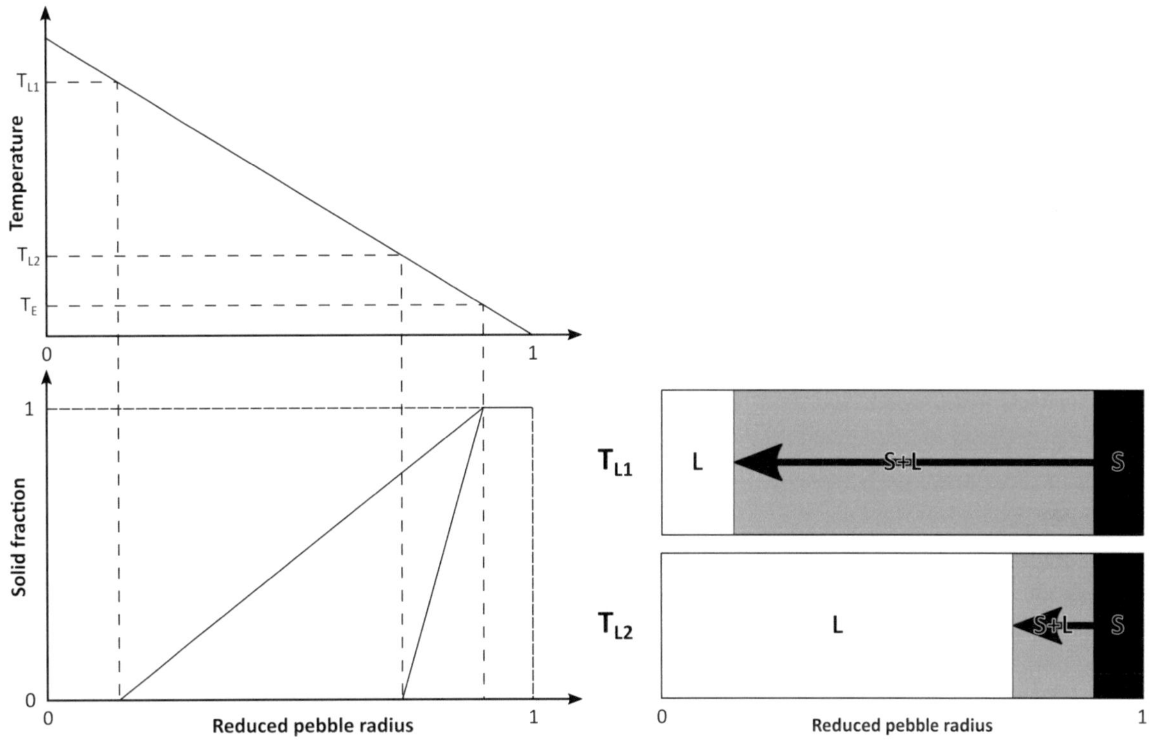

Figure 5.1: Schematic illustration of the size of the phase fields during the solidification of a cooling droplet for two different liquidus temperatures T_{L1} and T_{L2} in a eutectic system with T_E being the eutectic temperature. For simplicity a linear temperature gradient from the surface to the droplet center is assumed. The grey arrows indicate the direction of growth of the dendrites and their potential length.

The thermal stresses that are generated during cooling should also lead to a remaining elastic tensile strain in either phase of the pebble crust. The strain should also be noticeable by XRD measurements as the tensile stresses significantly exceed the compressive stresses that also form according to Grünberg [129]. However, the presented results in section 4.2.2 do not show tensile strains in both phases but rather mild compressive strain in the γ-LMT phase. The appearance of these strains can easily be explained when the thermal expansion of both phases is considered. As shown in Figure 4.10, the thermal expansion coefficient of lithium orthosilicate is comparably high. The thermal expansion coefficient of γ-LMT is not known, yet it can be approximated by the literature values reported in Table 2.3. Compared to the data for the thermal expansion coefficient of β-LMT, these data do not suggest an immoderately different, but larger value. Consequently lithium orthosilicate contracts much more during cooling than γ-LMT does, which leads to a tensile strain in the lithium orthosilicate phase while γ-LMT experiences a compressive strain. The generation of large thermal strains is mostly restricted to the lamellae within individual eutectic domains as within these domains both phases are ideally aligned. The overall thermal strain of a eutectic domain shows thus an intermediate value. Therefore, the primary grains of either lithium orthosilicate or lithium metatitanate should experience less thermal stresses. Yet, for the hypereutectic microstructure that is strongly dominated by primary lithium metatitanate dendrites, no eutectic domains are visible between the dendrite arms. In this special case also grains of primary lithium metatitanate should experience high thermal strains.

Theoretically, large differences in the generated stresses have to be expected as lithium metatitanate is clearly the minor phase with 12.9 vol.% and 22.8 vol.% for the nominally 20 mol% and 30 mol% LMT containing pebbles, respectively. If a lamellar assembly of γ-LMT and LOS is assumed that is cooled from 1000 °C to room temperature, a maximum of stresses of 980 MPa and 1552 MPa in lithium orthosilicate and -6637 MPa as well as 5249 MPa in γ-lithium metatitanate are calculated for nominally 20 mol% LMT as well as nominally 30 mol% LMT containing pebbles, respectively. For eutectic pebbles of (assumed) 25 mol% LMT the theoretical stresses within lithium orthosilicate grains amount to 1260 MPa and -5960 MPa within LMT grains. Any of these stresses in lithium orthosilicate clearly exceeds the reported tensile strength of lithium orthosilicate which amounts to about 430 MPa in the best case (see Table 2.2). Therefore the formation of cracks within the lithium orthosilicate lamellae or along the grain boundaries to lithium metatitanate may be a consequence. As hardly any cracks within the lithium orthosilicate lamellae are observable, the formation of cracks along the grain boundaries is obviously favored. If the description for the tensile strength that is given in Table 2.2 is extrapolated to lower grain sizes, a size of about 300 nm is necessary to arrive at a sufficiently large value to withstand a stress of 1500 MPa. This dimension is occasionally reached by the finest of the lamellae along their least extension, yet not in direction of the tensile forces.

Generally, as the estimation of the maximum thermal stresses shows, the absolute values of these thermal stress should be largest in the lithium metatitanate grains, which should translate into a strain of -2.49 % and -1.97 % compared to a strain of 0.89 % and 1.41 % for the lithium orthosilicate grains of the LMT-poorer and LMT-richer KALOS samples, respectively. Yet, rather the contrary is observed with notably larger absolute strains within the lithium orthosilicate grains. In section 5.3.1 it will be shown that a considerable portion of this strain originates from either impurity elements or a lithium surplus within the lithium orthosilicate lattice or both. For the LMT-richer pebbles, a reasonable value for tensile residual thermal stress is obtained (261 MPa, see section 5.3.1), whereas there seems to be no clearly identifiable residual thermal strain within the LMT-poorer pebbles although the LMT grains show compressive strain. It is thus reasonable to assume that the LOS grains that share grain boundaries with LMT grains also experience some tensile stress, although it is not clearly detectable. As a considerably amount of lithium orthosilicate crystallizes primarily for the LMT-poorer pebbles, these dendrites will experience considerably less thermal stress, because the pearlite-like eutectic domains that surround them certainly act as quasi-single phase grains. The thermal behavior of these pearlite-like domains resembles that of pure lithium orthosilicate to a much larger degree than that of lithium metatitanate given the volume fraction of lithium metatitanate of about 17.5 %. As the primary lithium orthosilicate grains show a much larger grain size than the LOS lamellae in the pearlite-like domains which can easily reach nanoscale, the primary lithium orthosilicate grains dominantly contribute to the diffractogram and as a result, the strain of the lithium orthosilicate lattice may be underestimated.

The magnitude of the measured contraction of the two phases is dependent on the temperature difference, which should only vary marginally. Therefore, the thermal strain of the lamellae should be about equal throughout the volume of a pebble. From a mechanical point of view, these residual thermal stresses should generally not play a significant role during the extension of an intergranular crack in parallel direction to the LOS/LMT lamellae. Yet, for any other angle of inclination, the lithium orthosilicate lamellae lose a small fraction of their apparent mechanical strength as a result of the tensile strain that they experience. In principle, the opposite is true for the lithium metatitanate lamellae, which should show an increased apparent mechanical strength. As the strain in the lithium orthosilicate lamellae is notably higher than for the lithium metatitanate ones, an overall negative effect on the mechanical strength should for all that prevail. Furthermore, the thermal stresses that originate from thermal expansion mismatch weaken the grain boundaries and therefore increase the probability of intergranular cracks [185]. Thermal anisotropy, which is significant especially for lithium orthosilicate (see section 4.2.3), may also weaken the grain boundaries as Duba and Heard [185] show. Yet, for the case of anisotropy, it very much depends on the grain orientation whether the generated stresses have a negative effect on the grain boundaries or not. Cai et al. [186] showed that the stress fields that are generated by residual thermal stresses can lead to crack deflection along the grain boundaries between the involved phases, which is also certainly facilitated by the weakening of the grain boundaries according to the explanation of Duba and Heard [185]. Cai et al. [186] furthermore show that when such a behavior is observed, the material usually also shows a notably increasing resistance to crack growth with increasing crack extension, i.e. classical R-curve behavior.

Fabrication related stresses and properties of the biphasic emulsion method pebbles

For the biphasic pebbles that were fabricated by the emulsion method and that contain a nominally equal amount of lithium metatitanate, the situation after fabrication is principally different. In contrast to the KALOS pebbles, these pebbles are fabricated by sintering after wet shaping (see section 2.2.3.2). As the sintering temperature of 1000 °C of these pebbles lies below the temperature of the β to γ phase transformation (see Figure 2.10), the pebbles consist of lithium orthosilicate and β-lithium metatitanate. The sintering step of the emulsion method does not involve the (deliberate) establishing and crystallization of a melt, therefore the buckling of the pebble surface and the consequential formation of dents should be no issue for these pebbles. Yet, also the LOS/LMT pebbles that were fabricated by the emulsion method are not perfectly spherical. The non-sphericity of the pebbles can originate from the incomplete coalescence of two or more slurry droplets which have not yet gelled completely during the shaping process (see for instance Figure 4.17B). Yet, the perfect spherical shape of these pebbles is also deteriorated by dents which show some similarity to the ones that are observed for the KALOS pebbles. Apparently, also the formation of considerable tensile forces in the pebble core that act on a rigid crust, which leads to the buckling of the KALOS pebbles, takes place for the emulsion method pebbles. Therefore, the theoretical framework of the thermoelastic model of Forgac et al. [128] is also applicable in this case. As all of these defects are already visible in a precursor state on the surface of the green pebbles (see Figure 4.19), the model of Forgac et al. [128] rather applies to the drying of the pebbles, which of course also involves an increase in shrinkage with decreasing water content of the (swelled) agarose gel. Consequently, it is a gradient in the water content instead of a gradient in temperature that defines the behavior of the system.

The drying of the gel will commence at the surface of the gel pebbles, while an insignificant amount of water is removed from the core. This leads to a contraction of the surface/crust and when this contraction is large enough, drying crack-like structures are formed. These crack-like structures, however, are not sharp-edged as one might expect but rather rounded. As the gel is actively washed during the first stages of water removal (see section 2.2.3.2) and as the gel is still relatively plastically flexible when it contains some water, the surface will smooth to a considerable degree. Once the drying of the pebbles' crust reaches a state at which the majority of water is removed while the core of the pebbles still holds a considerable amount of water, the crust will experience a contraction towards the core of the pebbles. These contractionary forces will also close the initially formed drying cracks, since the crust experiences compressive tangential and radial stress in this case. Yet, with the smoothed edges of the cracks, they are still visible on the surface and most probably cannot be considered as healed. From the contraction of the core, the drying pebbles develop the 'seams' and dents as Forgac et al. [128] predict. The drying also leads to an obvious inhomogeneous green density, with a higher green density near the surface of the pebbles (see Figure 4.31). These regions of higher green density are very probably the result of the washing and drying procedures described in section 2.2.3.2. Apparently, as Figure 4.31 shows, these steps do not lead to a significant migration of the solid content towards the surface. But during the drying at 55 °C, the water that the gel contains, might form the pores that are visible within the core of the pebbles due to partial evaporation and as a result the green density within the core is reduced. It is probably also due to these pores, in comparison to the KALOS pebbles, that the closed porosity of the biphasic emulsion method pebbles is higher, as with decreasing green density the achievable density after sintering is usually also reduced despite a higher shrinkage. The cause for the formation of the observed seams of pores is not clearly identifiable. A very probable cause is the low adherence of the already dried crust surface to itself, which is already evident from the non-gaping but non-healed drying cracks. As the dents and especially the 'seams' form from the contraction of the core, the surface of the crust is pulled towards itself, which is especially perspicuously visible for the 50 mol% LMT and 90 mol% LMT containing pebbles in Figure 4.19B and D.

During the sintering, no significant contractionary forces that originate from the cooling of the pebble surface have to be assumed, as there are no steep temperature gradients within the heated volume of the furnace and the sintering program does also not involve steep heating or cooling rates. To some degree, tensile forces within the core of the emulsion method pebbles may, however, originate from the inhomogeneous green density with a higher green density near the surface of the gelled droplets. This would clearly lead to a higher shrinkage in the core during sintering, which in return has a similar but not as strong an effect as a crystallizing liquid. Yet, the pebbles do not show any defects that have formed after the densification by sintering. As the pebbles are cooled to room temperature after sintering at 1000 °C, also thermal stresses according to Grünberg [129] could, at least in principle, form. Yet, as the cooling of the pebbles is performed much slower than during the KALOS process, the so-generated temperature gradients are considerably less steep and therefore the resulting thermal stresses at the surface of the pebbles are much lower. Apparently, as no cracks are observable which extend from the surface into the pebble volume, these stresses are uncritically low. Regarding the thermal stresses that form as a consequence of the thermal expansion mismatch between lithium orthosilicate and lithium metatitanate, the XRD measurements (see Figure 4.6) do not show any sign. The determined unit cell volume of β-lithium metatitanate accords perfectly with the reference values with a deviation below 0.1 %. This is almost equally true for the lithium orthosilicate grains for which just barely larger unit cell strains are observed which can be regarded as slight tensile strain, as an apparent correlation with the nominal LMT content is perceivable (see Figure 4.6). Yet, none of these observations can be related to the mismatch of the thermal expansion, despite it being substantial (see Figure 4.10). Apparently the slow cooling of the pebbles after sintering leads to the thermal annihilation of these stresses.

Effects of the fabrication related stresses on the mechanical strength of the nominally 30 mol% LMT containing pebbles

The described defects and residual thermal stresses determine the mechanical behavior of the pebbles to a certain degree. This is evident when the determined Weibull parameters are compared. For the pebbles of approximately 1000 μm diameter both Weibull moduli and characteristic mean contact pressures differ considerably and statistically significantly. As will be shown in section 5.2.2, the mechanical strength of the pebbles with nominally 20 mol% and 30 mol% lithium metatitanate that were fabricated by the emulsion method is related to their open porosity and therefore by surface defects. For the pebbles that were fabricated by the KALOS process such a simple classification is not possible. In Figure 5.2 the likelihood contour plot of the relevant samples is shown, and it is evident, that for the nominally 30 mol% LMT containing sample either two Weibull lines are necessary to describe the failure behavior of the pebbles (after 32 days of annealing) or a single Weibull line describes the failure data just moderately well (for the as-received pebbles, see Figure 4.50). Furthermore, it is also clear, that both Weibull parameters of the as-received sample are notably lower than those that were determined for the emulsion method samples.

However, there is are two examples of subsets of 30 mol% LMT containing KALOS pebbles (subset I in each case) that show an almost equally high Weibull modulus as the samples that were fabricated by the emulsion method, but an annealing for 4 and 32 days is necessary to observe them (see Figure 4.49 for the sample that was annealed for 4 days). As already indicated in Figure 4.50, a few data points of the as-received samples are in very good agreement with the respective subsets of the annealed samples and are an almost perfect extension of the Weibull line of subset I of the for 32 days annealed KALOS pebbles. It is therefore justified to consider this extension as another true subset with basically identical Weibull parameters as subset I of the 32 days annealed pebbles. Figure 4.62 and Figure 4.63 show that the failure of pebbles that can be assigned to this group is attributed to volume defects rather than being caused by surface defects.

As already stated, the nominally 30 mol% LMT containing pebbles, especially those that show a micro-structure which predominantly consists of pearlite-like eutectic domains, show virtually no surface cracks because of the way they crystallize. However, as Figure 4.26A shows, these pebbles feature cracks within their volume. For these cracks to form, the core of the pebble has to experience tensile forces. There are two moments during fabrication when such stresses may be generated. The first one is during crystallization according to Forgac et al. [128] when the crust is not cooled fast enough to follow the contraction of the core as it solidifies. These stresses may lead to the formation of cracks, yet, as long as a considerable amount of liquid is present within the core, the formation of pores seems to be more likely. The second moment during fabrication that such stresses may arise is at the instant of time when the core of the pebble is (virtually) solid but still has a significantly higher temperature than the surface. The cooling of the core, with respect to the surface also leads to tensile forces, which cannot be relieved by the formation of pores. On top of that, the frozen compressive stresses in the crust/surface of these pebbles also require the formation of balancing tensile forces within the core. The combination of both of these stresses can eventually explain the formation of the volume cracks that are observed for these pebbles.

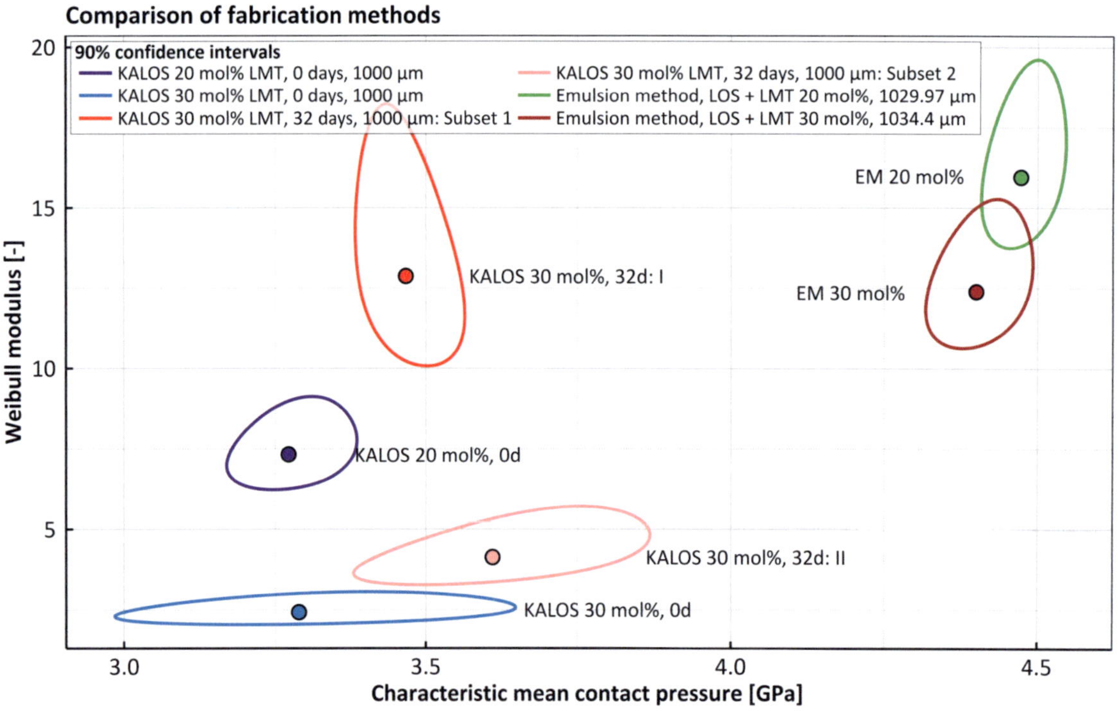

Figure 5.2: Likelihood contour plot of the comparable KALOS and biphasic emulsion method pebbles that nominally consist of 20 mol% LMT and 30 mol% LMT in LOS. Additionally, also the KALOS sample that was annealed for 32 days and nominally contains 30 mol% LMT is shown for comparison.

In Figure 5.3 the Weibull plot of three of the samples for which the likelihood contour plot is shown in Figure 5.2 are depicted. From this Weibull plot, it is evident that the strongest nominally 30 mol%

LMT containing KALOS pebbles also reach similar strengths as the strongest pebbles that were fabricated by the emulsion method for this composition. From the slight upward curvature of segment II of the 32 days KALOS sample in Figure 5.3, it seems possible that, despite being modelled reasonably well by a Weibull line, the data points asymptotically approach the Weibull line of the emulsion method fabricated pebbles. As the data points of the as-received sample show a very similar behavior for these mean contact pressures, this hypothesis equally applies for that sample. If one assumes that the Weibull lines of segment I of the 32 days sample as well as the Weibull line of the emulsion method pebbles represent two non-competing failure mechanisms, i.e. each failure mechanism exclusively exist in one of two sub-populations of the pebble sample while they do not exist in the other one, the resultant compound Weibull plot would closely resemble the data of the samples that were annealed for 32 days (see Figure 5.17 for example). Yet, this is only true when the Weibull moduli of both generating Weibull distributions are relatively similar as in this case.

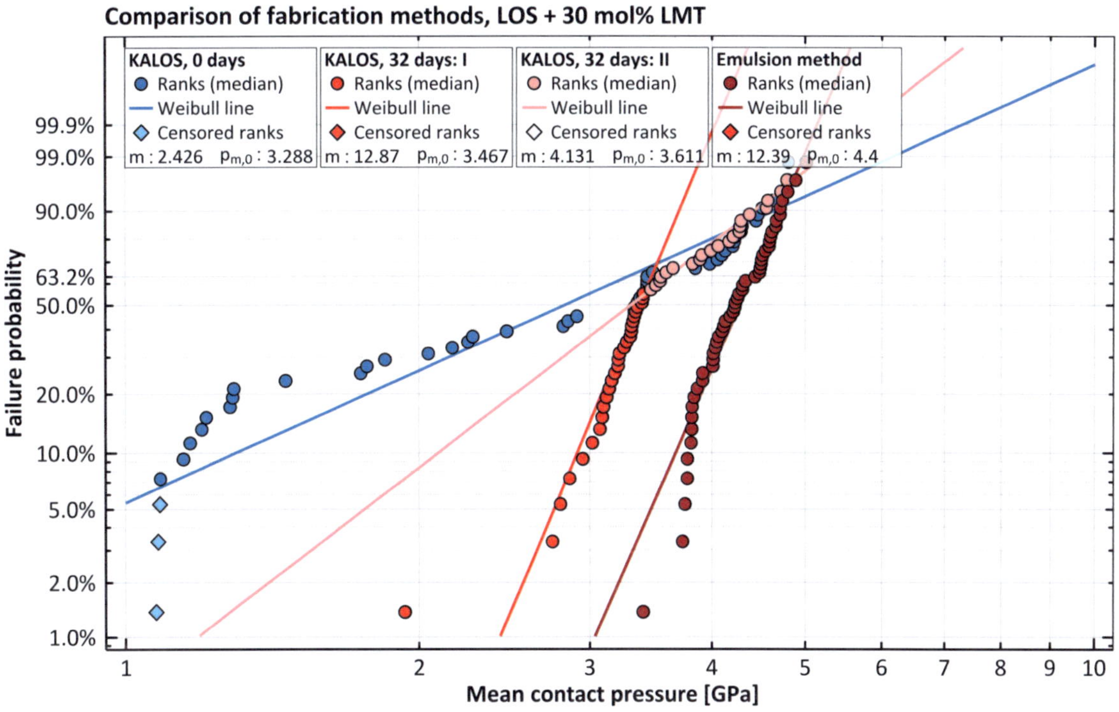

Figure 5.3: Weibull plot of the comparable KALOS and biphasic emulsion method pebbles that nominally consist of 30 mol% LMT in LOS. Additionally, also the KALOS sample that was annealed for 32 days is shown for comparison.

The relevant compound distribution of two non-competing failure modes is given by Abernethy [137] and features, in addition to the Weibull parameters of both Weibull distributions (F_1 and F_2), the probability p of testing a pebble of either sub-population. With these distributions, the high strength part of the Weibull plot can also be modelled for the as-received sample. If another Weibull distribution (F_3) is introduced into the model to represent the low strength part of the data of the as-received sample and it is assumed that this distribution is also not competing for failure with the other Weibull distributions, the compound distribution reads as

$$F_{compound} = p_1 F_1 + p_2 F_2 + (1 - p_1 - p_2) F_3 \tag{5.1}$$

with two probabilities p_1 and p_2 to test a pebble which fails according to the respective Weibull distribution. Equation (5.1) is simply an extension of the case for two non-competitive failure modes as detailed by Abernethy [137]. In principle, other compound distribution functions that also consist of three individual Weibull distributions are viable as a model. Since the determination of one of these compound distributions is not unambiguously possible, the simplest one, i.e. equation (5.1), is chosen.

If the said two experimentally determined Weibull distributions are used in this model as an educated guess, the obtained data is reproduced moderately well (see Figure 5.4, see Appendix D for the computational approach), especially for the high strength data. The presence of an R-curve behavior within the low strength subset of the three subsets because of crack deflection due to residual thermal stresses, as suggested by Cai et al. [186], cannot be verified with sufficient certainty. Yet, it is also a plausible explanation for the downward curvature of the Weibull plot. The parameters for the probabilities were guessed on the basis of the experimental values which can be assigned to one of the three segments. Yet, the unusually pronounced central segment that resembles segment I of the pebbles that were annealed for 32 days leads to a considerable ambiguity for these values. For instance the number of 17 failures (i.e. 34 %) that show the highest strength, and are thus assigned to the Weibull distribution with the parameters of the biphasic emulsion method pebbles, is probably underestimated because of the large gap between the two segments in x-direction, which hints at 'missing' experimental values. Consequently the same applies for the central segment. For the low strength portion of the Weibull plot a Weibull distribution with a Weibull modulus of 4 and a characteristic strength of 1.5 GPa was guessed. However, there is considerable ambiguity of the nature of this part as a range of parameters are equally fitting. Therefore, the confidence of these Weibull parameters for this failure mode is low.

Two effects may contribute significantly to this ambiguity. Firstly, approaching the lower testing limit of the measurement equipment can result in a strong downward curvature of the data at very low contact pressures, when the condition of random drawing of samples is not fulfilled anymore (see also section 4.5.2). In this particular case, a downward curvature in a Weibull plot can be modelled by a three parameter Weibull distribution (see equation (A.6) and Abernethy [137] or as a varied proof test [132], which also essentially uses a three parameter Weibull distribution). Secondly, if the samples show an R-curve behavior, then the representation of the respective subset by an ordinary straight Weibull line can only be a rough approximation of the behavior.

Although the chosen model function (5.1) allows for relatively steep segments in a Weibull plot, the central segment cannot be represented very well. Yet, the transition between two dominant failure modes is always relatively smooth in an idealized mathematical model as both failures have a similar probability of appearing at the transition. The missing of the smooth parts may either indicate that by chance pebbles of the corresponding strength were not tested or, in case such pebbles essentially do not exist in the sample, that the model is inadequate to reproduce this part of the sample data.

The interpretation of the Weibull plot on the basis of the proposed model (as illustrated in Figure 5.4) is supported by the fractographic analysis that is presented in Figure 4.62. Both pebbles that fractured at mean contact pressures which have been assigned to high strength segment in the Weibull plot (failures at 4.29 GPa and 4.72 GPa), show a hypereutectic microstructure that is fully dominated by lithium metatitanate dendrites, while the other two fragments clearly show the microstructure that mainly consists of eutectic pearlite-like domains (see Figure 4.63). Of course, also the emulsion method samples show a considerably different microstructure in comparison to the pebbles that were fabricated by a melt-based process, but for the KALOS pebbles, the change in microstructure is unambiguously correlated with a change in composition. Consequently, the mechanical behavior of the pebbles whose microstructure is dominated by lithium metatitanate dendrites may be, as limiting case, reasonably similar to the nominally 30 mol% LMT containing pebbles that were fabricated by a shaping and sintering process, as the LMT content is presumably more similar. Yet, from the experimental data only about 17 of 50 pebbles can be assigned to the presumed failure mechanism that also the emulsion method pebbles show, for which failure is initiated by surface defects.

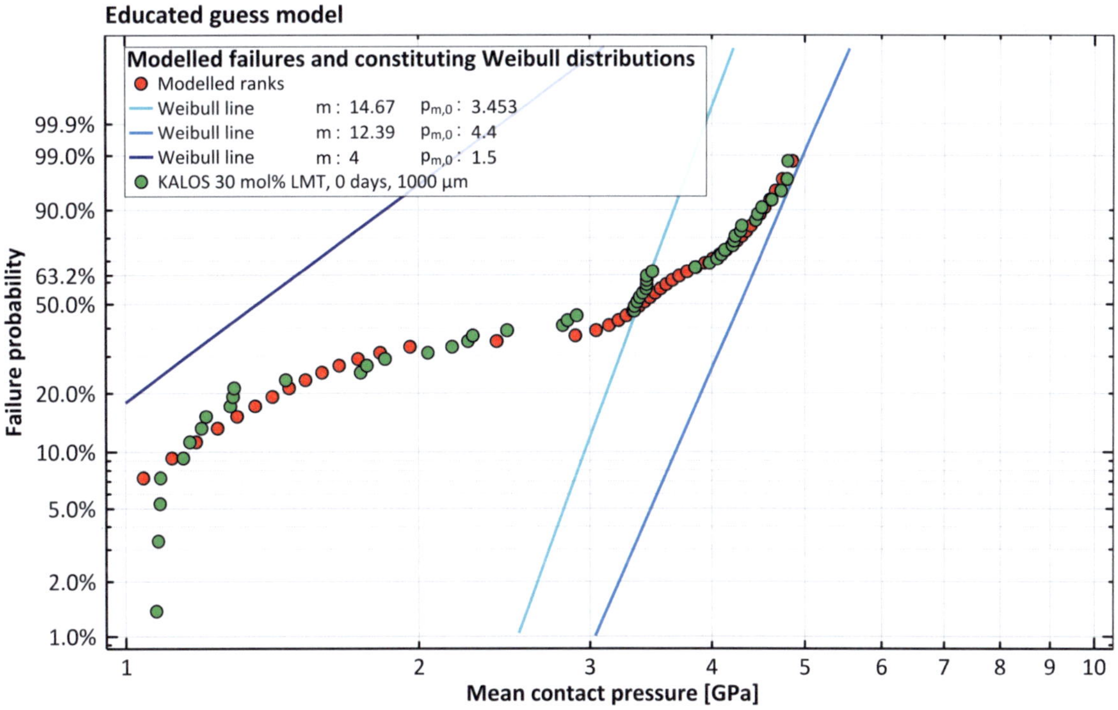

Figure 5.4: Modelled compound Weibull plot from three constituting Weibull distributions according to equation (5.1). The distributions with Weibull moduli of 14.67, 12.39 and 4 are appearing with a probability of 25 %, 40 % and 35 % respectively.

However, from the modeling of the failure data in Figure 5.4 a higher fraction of these pebbles is suggested, which is in better agreement with the observation that pebbles of either of the two microstructures appear approximately equally often. Ultimately, a definite attribution of these KALOS pebbles to the failure mechanism of the emulsion pebbles is impossible as any different failure mechanism which by chance shows similar Weibull parameters cannot be distinguished. The SEM analysis of the fragments of the KALOS sample (given in Figure 4.62), however, gives some indication. As explained in section 4.5.3, these pebbles seemingly failed from surface defects. Therefore both types of pebbles might effectively share the same mechanism of failure. For the failure of the nominally 30 mol% lithium metatitanate containing pebbles that fail below mean contact pressures of 3 GPa, the predominant portion of observed failures are 'apparent failures' for which a complete failure of the pebble happens at considerably higher loads. In Figure 4.62 an example of a pebble fragment (p_m: 2.28 GPa) that showed an apparent failure before being brought to complete failure is depicted. This fragment shows, in contrast to the other pebble fragments, a considerable quasi-plastic region at the point of mechanical loading. As Lawn et al. [114] describe, a quasi-plastic behavior of the contact zone in Hertzian contacts is commonly observable for coarse-grained ceramics. The authors furthermore developed a model for this quasi-plastic behavior which relies on the formation of shear faults (see section 2.3.1). These cracks may coalesce above a certain load [123] which in return may lead to a momentary drop in strength. As Cai et al. [123] point out, these cracks generally form at grain boundaries which are considerably weak and this weakness may be even increased by residual thermal stresses. For the pebbles that were fabricated by the KALOS process such stresses were determined, as explained earlier. During the uniaxial loading of the pebbles, such a drop in strength leads to a drop in the detected mechanical load, which is, as a result, interpreted as apparent failure of the pebble.

Within a Weibull plot like in Figure 5.3, these events of quasi-plastic behavior appear as a separate failure mode of low Weibull modulus. As soon as the relevant stresses within the sphere are high enough, the failure by other mechanisms is more probable and thus the observation of an apparent failure is rare, which is in very good accordance with the experimental observations. Consequentially three distinct failure mechanisms are identified for the nominally 30 mol% LMT containing KALOS pebbles. At least two of these failure mechanisms are caused by the two distinct microstructures, one being dominated by lithium metatitanate dendrites with a high characteristic strength while the other is dominated by pearlite-like domains with a medium characteristic strength, in the same set of pebbles.

Effects of the fabrication related stresses on the mechanical strength of the nominally 20 mol% LMT containing pebbles

As Figure 5.2 indicates, the 1000 µm, nominally 20 mol% LMT containing KALOS pebble sample behaves significantly different than the KALOS samples that nominally contain 30 mol% LMT. For the pebbles with lower nominal LMT content a single Weibull line adequately describes the failure behavior. It is in consequence also not possible to unambiguously verify a notable R-curve behavior for these samples with regard to the analysis of Cai et al. [186], which correlates residual stresses with R-curve behavior. Figure 4.45 shows that the Weibull modulus for all different pebble samples of this material does not vary considerably and therefore all samples should share the same dominant failure mechanism. Consequently Figure 4.59, which depicts fragments of two 20 mol% LMT containing KALOS pebbles that were annealed for 128 days, can also be used to conclude on the dominant failure mechanism of the as-received pebbles, which is the failure from surface defects.

In comparison with the failure from volume defects of the KA-30-LMT pebbles which almost purely consist of pearlite-like domains, the determined Weibull modulus is expectedly and considerably lower than the one of subset I of the nominally 30 mol% LMT containing pebbles that were annealed for 32 days. Yet, compared to the pebbles that were fabricated by the emulsion method and which also fail from surface defects, the Weibull modulus is much lower (see Figure 5.2). This observation can be regarded as clear indication that the surface defects of both samples were generated by two different processes, which is self-evident when the surface morphologies are compared that are established by the two fabrication methods (see Figure 4.12 and Figure 4.17). The cracks and defects that the thermal stresses generate in the surface of the KALOS pebbles upon cooling unambiguously lead to a significantly broader defect size distribution than one can expect from the comparably undisturbed surface of the emulsion method pebbles. As a result, a lower Weibull modulus is observed for the KALOS pebbles compared to the biphasic emulsion method pebbles. The results furthermore clearly show that the Weibull modulus for bulk failure of the almost eutectic 1000 µm KALOS pebbles, which is approximately equal to that of surface defect induced failure of the biphasic emulsion method pebbles, is significantly higher than the Weibull modulus for surface induced failure, which complies well with the expectations (see Munz and Fett [132]).

The differences in the characteristic strength of emulsion method pebbles and KALOS pebbles cannot be related to a certain comprehensive property, which is in principal due to the more complicated failure behavior of the KALOS pebbles. For instance, in section 5.2.2 it will be shown that the characteristic strength of the biphasic emulsion method pebbles is closely correlated to the open porosity. Yet, the value of the open porosity does not correlate with the mechanical strength of the KALOS pebbles in this way. According to equation (2.14) and (2.20) the crack length determines the characteristic strength for the most part. The cross section images (see Figure 4.25 and Figure 4.33) clearly illustrate, that the surface cracks of the KALOS pebbles run much deeper than those of the respective emulsion method pebbles. Considering the surface morphology, it can be concluded that not only a broader crack size distribution is found for the KALOS pebbles, but also the crack density is generally higher. Equation (2.20) furthermore shows that the characteristic strength is also reduced when the defect density increases. Therefore the increased defect density very probably reduces the characteristic strength of the KALOS pebbles additionally.

5.1.2 Lithium excessive pebbles

The two powder-based fabrication routes for the lithium excessive pebbles (see section 2.2.3.2) are in principal quite similar. Both, the emulsion method as well as the sol-gel method, rely on a pre-synthesized powder and a sintering step. The major difference for both methods is the method for shaping. In addition to that, the sintering of the shaped green pebbles is also slightly different (see Table 3.4). As already explained in section 5.1.1 for the emulsion method fabrication of biphasic pebbles, also the lithium excessive pebbles that were fabricated by either method should only experience tensile thermal stresses at the surface because of a temperature gradient between the warmer center of the pebble and the cooler surface of the pebble during the cooling after sintering unless significant density gradients within the green pebbles exist.

Both types of pebbles lack any perceivable amount of buckling, therefore the green density distribution within the green pebbles can be considered to be homogeneous. Also similar to the observations for the biphasic emulsion method pebbles that contain 20 mol% and 30 mol% LMT, the XRD analysis does not reveal a considerable residual strain within the sintered pebbles. The surface morphology of both as-received lithium excessive LMT pebble batches (emulsion method and sol-gel method) is, however, different. While the surface of the emulsion method pebbles shows a considerable amount of open porosity, the pebbles that were fabricated by the sol-gel method show a relatively dense surface with a number of cracks running through it. With high certainty, these cracks originate from thermal stresses during the cooling of the pebbles after sintering and thus large thermal stresses are dissipated. As Figure 4.15G shows, only small cracks in comparison are formed in the emulsion method pebbles. Most probably the high porosity of the pebble surface helps to reduce the thermal stresses for the emulsion method samples as the effective Young's modulus of the surface is significantly reduced by the presence of the pores. As a result, the generated stresses at the surface are reduced as well, as equation (2.11) shows. Also the different lithium excess as determined by ICP-OES, which is more than double for the pebbles that were fabricated by the sol-gel method (see Table 4.3 and Table 4.4), does not seem to have a visible impact on the unit cell of β-lithium metatitanate, of which both pebbles exclusively consist. As the pebbles overwhelmingly consist of β-lithium metatitanate (see Figure 4.2), stresses originating from thermal mismatch are neither expected nor observable.

A large difference between both fabrication methods can also not be observed for the determined porosity of the pebbles. While the open as well as the closed porosity of the pebbles that were fabricated by the emulsion method are notably lower than that of the pebbles that were fabricated by the sol-gel method, the differences are rather mild. As explained in section 4.4, the obtained value for the sum of the open and imaginary porosity of the sol-gel method pebbles is almost certainly significantly increased by a high amount of imaginary porosity, because of the ellipsoidal pebble shape of the sol-gel method pebbles while a spherical pebble shape is assumed for the calculations. This point is supported by the surface morphology of these pebbles (see Figure 4.15 and Figure 4.16) as it does not suggest a high open porosity for the sol-gel method pebbles especially compared to the emulsion method pebbles. Basically, the mercury intrusion porosimetry measurements suggest a close to insignificant deviation of the open porosity values, although a slightly higher porosity for the emulsion method pebbles was obtained (see Table 4.8).

These results are in better accordance with the images of the pebble surfaces. However, the surface morphology of both pebble types suggests a notably higher open porosity for the emulsion method pebbles, and thus the mercury intrusion measurements obviously underestimate the open porosity of the emulsion method pebbles. Although the green density of both types of pebbles is not known, the obviously necessary 'calcination' step for burning the PVA binder from the pebbles suggests that a larger amount of organic material has to be burnt before sintering. This is in contrast to the emulsion method pebbles, for which such a step is not necessary. Therefore, the green density of the sol-gel pebbles is very probably lower than that of the emulsion method pebbles. Furthermore, the sintering duration is significantly longer for the sol-gel pebbles, even though the sintering temperature is slightly lower by 20 K.

In principle, the increased sintering duration should visibly increase the density of the sol-gel pebbles over that of the emulsion method pebbles, all other things equal. Yet, the effect of the different sintering temperature cannot be evaluated without a detailed study. Eventually, the grain size of the pebbles that is observed at the surface and also within the bulk in the as-received state is relatively similar, despite the significant differences during sintering.

Differences between the mechanical strength of emulsion method samples and the sol-gel method samples

From all investigated properties of the pebbles, the mechanical strength differs the most, with the emulsion method pebbles showing a significantly higher strength than the sol-gel method pebbles. Clearly, both pebbles are not of the same size, with the emulsion method pebbles being considerably larger (see Table 3.4). The calculation of the mean contact pressure that the pebbles experience during the uniaxial compression should compensate for the larger pebble size. Yet, the fragments of both types of pebbles reveal considerable plastic deformation at the contact points during the loading as large flattened areas are formed. The calculation of the mean contact pressure, however, only considers elastic deformation and is therefore not unrestrictedly correct in these cases. However, if the observed flattened regions of some of the pebbles are used for calculating a contact pressure by relating the applied force to the flattened area, the results do not qualitatively differ (see Table D.1 in Appendix D). This also complies with the results of Kogut and Etsion [187], who note that the area of plastic deformation is quite constrained and that the vast majority of the pebble volume is only elastically loaded. Therefore, the calculated mean contact pressure can be regarded to be valid as a measure of the strength of both types of pebbles for direct comparison.

The differences in the mechanical strengths of the two types of pebbles are most visible in a common Weibull plot (see Figure 5.5). It is obvious, that the failure behavior of the sol-gel method pebbles has to be approximated by two Weibull lines, whereas a single Weibull line is sufficient for the emulsion method pebbles. Clearly, all three Weibull lines are statistically significantly different from each other. The cross section images of the sol-gel pebbles reveal that two different microstructures can be found in these samples (see Figure 4.29). While one microstructure shows a monomodal distribution of grains which do not seem to adhere too much to each other, the other microstructure shows a bimodal grain size distribution with reasonably adhering grains. The latter microstructure shows a considerable amount of dispersions of a second phase, which are also larger than the grains of the major phase. The microstructure of the emulsion method pebbles shows, as in the first case of the sol-gel method pebbles, a small grain size and scanty intergranular adherence. However, the small grains form larger blocks, which is different from the microstructure of the sol-gel method pebbles. From the analysis of pebble fragments (see Figure 4.64 and Figure 4.66) it is clear that the perceived adherence between the grains is not indicative for a failure by intergranular fracture, although too little adherence certainly favors intergranular fracture.

Furthermore, the analysis of the fragments shows that intergranular fracture is mostly found for pebbles that fail at low mean contact pressures. Technically, the fragments of the sol-gel method pebble that failed at 3.66 GPa, i.e. the fragment that showed the lower strength of the two presented ones, is assigned to the high strength subset II in the Weibull plot. Yet, at the point of transition from one failure mechanism to another one, both mechanisms have a similar probability of causing failure. As no fragments were found that show intergranular fracture at high mean contact pressures, it is reasonable to conclude, that the intergranular fracture is a unique feature of the low strength subset I. Also, the low strength subset I of the sol-gel method pebbles shows a rather similar Weibull modulus as the emulsion method pebbles as a whole. As intergranular fracture is appendant to both pebble types, it is a further indication that the same failure mechanism determines their strength. The collocation of the sol-gel pebbles' Weibull lines with a low strength, high modulus segment that is followed by a high strength, low modulus segment, shows that two competitive failure mechanisms cannot be active in one common type of pebbles. Therefore, either two non-competitive failure modes, i.e. both failure modes exclusively appear in separate type of pebbles, determine the failure behavior of the sol-gel method pebbles. Or, a single failure mechanism determines the failure behavior, for which the fracture toughness is a function of the crack length and increases with increasing crack length, i.e. the so-called R-curve behavior.

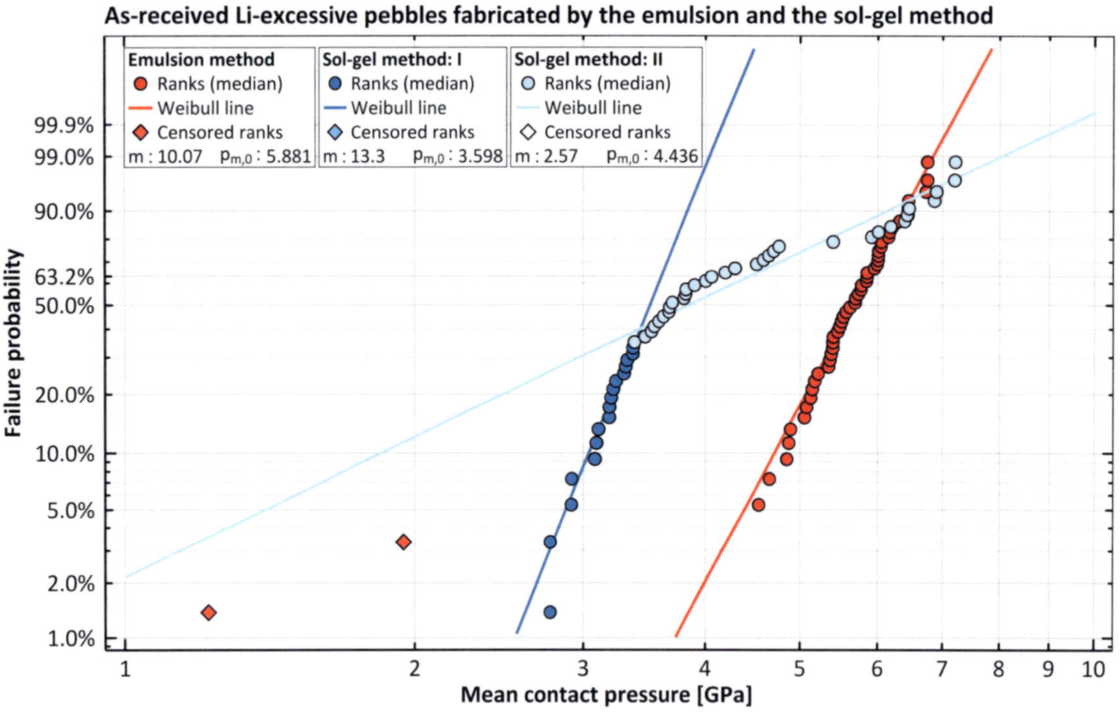

Figure 5.5: Weibull plot of the as-received lithium excessive pebbles that were fabricated by the emulsion method and the sol-gel method.

The observation of inter- as well as intragranular fracture, especially with the intragranular fracture appearing at high mean contact pressures, therefore may also hint at two non-competing failure modes, which is justified by the presence of two distinct microstructures (see Figure 4.29). Yet, the failure data within the Weibull plot cannot be represented satisfyingly by this model. Danzer et al. [136] detail the effect of the R-curve behavior on the Weibull plot, which can resemble the Weibull plot of the sol-gel method pebbles. The data points for observed failure transition between two asymptotic cases which feature the same Weibull modulus. Such behavior is observed if the material is intrinsically or extrinsically strengthened by for instance crack-bridging, local phase transformations, crack deflection or significant crack-branching at the crack tip [132,188]. Yet, the crack bridging is reported to occur in perceivable amounts only for coarse grained materials. In case of alumina, which is regularly chosen as model material, this effect seems to be operative for grain sizes of more than 10 µm. This is a significantly larger grain size than in the as-received sol-gel pebbles, however the shape of the R-curve is certainly material dependent [189] and it is thus reasonable to conclude that the pebbles nevertheless show an R-curve behavior.

Of course, a toughening of the pebbles that are composed of β-lithium metatitanate by local phase transformation at the crack tip can be ruled out as the β-phase already represents the equilibrium phase at room temperature. Principally, the observation of an R-curve behavior as a result of considerable crack-branching in the region around the crack tip is possible but very difficult to prove. The existence of the two modes of crack propagation may also hint at crack deflection as dominant mechanism on the pebble strength. For the present case of the sol-gel method pebbles, the Weibull line for subset II approximately represents the low strength asymptotic case [136] where intragranular fracture dominates. The deflection of the growing crack, i.e. the intergranular fracture, leads to an increase in toughness but the too low number of experimental values prevents the observation of the high strength asymptotic segment in the Weibull plot. It is furthermore clear, that the failure of the pebbles is initiated by volume defects for all sol-gel method pebbles, as well as for the emulsion method pebbles. Gorham and Salman [190] describe the observed failures as meridian plane fracture, which initiate at or close to the tip of a conical region of deformation and extend from there through the pebble to the mirrored failure.

If, as in this description, two major fragments are generated a relatively even fracture surface will be formed. But when multiple fragments are formed along this central crack, a ridge will be established as it was observed for several samples in section 4.5.3. In case of the emulsion method pebbles, for which only intergranular failure was observed, also the failure by intragranular crack propagation should be the limiting case for crack propagation because the differences between both types of pebbles are very little. Consequently, the observed failures of these pebbles are still in the transitional part and have not reached the high strength asymptotic limit. As Danzer et al. [136] also show, the apparent Weibull modulus in the transitional zone increases until it reaches its maximum at the kink when reaching the asymptotic high strength limit. It is therefore fitting that the observed Weibull modulus of the emulsion method pebbles is slightly lower than that of segment I of the sol-gel method pebbles. As a result, the observation of blocks of more adhering grains for the emulsion method pebbles does not seem to have a significant effect on the failure behavior of the as-received pebbles.

With regard to the characteristic strengths of the two types of pebbles, the observed characteristic strength of the larger pebbles should be reduced compared to the smaller pebbles if there are no other effects on the samples that are a function of the pebble size. For volume defects, following equation

(2.18) and as more clearly described by Munz and Fett [132], the volumes V_1 and V_2 of two samples of different size relate inversely to their characteristic strengths as follows:

$$\frac{\sigma_1}{\sigma_2} = \left(\frac{V_2}{V_1}\right)^{\frac{1}{m}}. \tag{5.2}$$

Yet, using the pebble volumes in this equation is in most cases an oversimplification of the problem as the effective volume is the correct choice. The effective volume correctly takes the stress distribution within the sample into account with regard to its potential to extend cracks and thus lead to failure [132]. Yet, the complicated stress distribution in uniaxially compressed spheres made it difficult to calculate an exercisable mathematical description of the effective volume. Therefore the pebble volumes, although being imperfect, are used to estimate the size effect nonetheless. Certainly, relation (5.2) will lead to a lower characteristic strength for a larger pebble, which is also clearly illustrated by Danzer et al. [136]. Using the Weibull parameters of segment II of the sol-gel method pebbles and assuming perfect spheres, the characteristic strength for the emulsion method pebbles would amount to 3.78 GPa. Such a characteristic strength is well within the observable limits. Yet, instead of a decrease of this parameter, an increase is observed, which is not to be confused with the determined characteristic strength of the as-received emulsion method pebbles. A decrease in the characteristic strength of the asymptotic limit at high strengths would move the whole curve to lower strengths as well as to higher values of unreliability (again see Danzer et al. [136]).

Consequently, the characteristic strength of the emulsion method pebbles has to have a higher value compared to the sol-gel method pebbles, which is not accessible on the basis of the available data. The determined values of closed porosity, which certainly represent one form of volume defects, may give some indication for a higher strength of one type of pebbles with respect to the other one, as it is significantly lower for the emulsion method pebbles than for the sol-gel pebbles. If the determined value of closed porosity is a measure of the pore size, very much like in the case of the biphasic emulsion method pebbles (see section 5.2.2), it can as well be determinant for the mechanical strength in this case. Probably, the bimodal grain size distribution and the thus larger potential critical defects around the larger grains is causative for the lower characteristic strength of the sol-gel method pebbles, as they in return lead to a lower critical stress (see equation (2.13)) and consequently to a lower characteristic strength (see section 2.4). In contrast the emulsion method pebbles do not show grains of the large mode. Consequently the pore size does not reach these high values, although also these pebbles show large pores at the boundaries of the blocks that have formed. Yet, the effective size of these approximately two dimensional pores is significantly reduced when they are not well-aligned to the occurring stresses and therefore these pores do not appear to be as critical.

5.2 The influence of the composition on the mechanical strength of the pebbles

The composition of the samples, i.e. the differences in the nominal composition, clearly has a significant influence on the mechanical strength of the pebbles. This is obvious for the KALOS pebbles, although the average contact pressures that these pebbles withstand do not vary as considerably as the ones that were determined for the biphasic pebbles that were fabricated by the emulsion method.

5.2.1 Pebbles fabricated by the KALOS process

The principal effect of the KALOS fabrication process on the pebble properties is already discussed in section 5.1.1. In principle, the effect of the crystallization behavior on the formation of surface cracks, which is discussed in section 5.1.1, can also be considered as a composition related effect. The most apparent effect of the pebble composition is the change of the microstructure of a portion of the pebbles from a microstructure that is dominated by pearlite-like domains to one that is dominated by lithium metatitanate dendrites, when the nominal composition is changed from 20 mol% LMT to 30 mol% LMT. As shown in Figure 5.3 this circumstance has significant implications on the failure behavior and therefore on the Weibull plot for the 1000 μm KALOS pebbles that nominally contain 30 mol% LMT. It is furthermore clear, that the pebbles that nominally contain 20 mol% LMT do not suffer from this effect, although it cannot be excluded that the lithium metatitanate content also varies considerably for individual pebbles as for the KA-30-LMT pebbles. Yet, their failure behavior is easily described by a single Weibull line. If, in addition to the examination in section 5.1.1, the smaller 500 μm pebbles are included in the analysis, the obtained Weibull parameters can be illustrated as in Figure 5.6. It is obvious, that the smaller pebbles show a significantly higher characteristic strength for both compositions. Other than the difference in characteristic strength, the failure behavior of the 20 mol% LMT containing pebbles is unaffected as the Weibull modulus shows essentially the same value for large and small pebbles. Emanating from the Weibull parameters of the 1000 μm pebbles sample and by applying equation (5.2), the characteristic strength of a 500 μm pebble sample is expected to be 3.94 GPa for surface defects, which is very close to the obtained value of 4.08 GPa and clearly within the confidence boundary as Figure 5.6 shows. As a result, there is no doubt that the classical size effect of the Weibull theory explains the observations for these samples.

From Figure 5.6 it is also clear that the obtained Weibull parameters for the two pebble sizes of the nominally 30 mol% LMT containing pebbles are significantly different from each other. Like the nominally 20 mol% LMT containing pebbles, the smaller pebbles show a considerably higher characteristic strength than the larger ones, while the Weibull modulus shows a comparable value. However, the description of both data sets by a single Weibull line is only a moderately satisfying approximation as already stated and as the detailed analysis in section 5.1.1 shows for the 1000 μm pebbles. If the Weibull plots are compared directly, as in Figure 5.7, the high similarity of the failure behavior of both pebble sizes is very obvious.

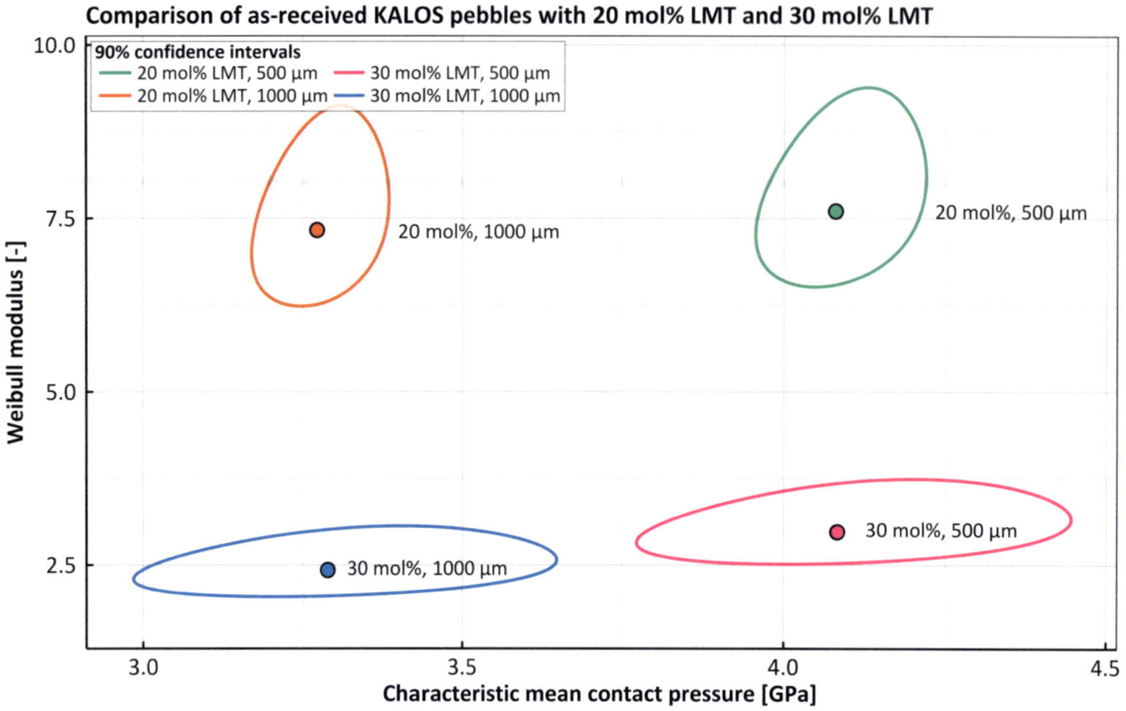

Figure 5.6: Likelihood contour plot of the as-received KALOS pebbles that nominally consist of 20 mol% LMT and 30 mol% LMT in LOS.

Especially for the low strength part of the data, the graph seems to be moved to higher strengths, whereas for medium and higher strengths the similarities moderately decrease. Yet, the pronounced steep central subsegment of the larger pebbles which coincides with segment I of the 32 days annealed sample (see also Figure 5.3) is just observable as a 'bump' in the Weibull curve of the 500 µm pebbles. Still, the SEM analysis of the pebble fragments that is shown in Figure 4.60 generally supports the similarity of the two Weibull plots, with the fragment that failed at 3.88 GPa showing pearlite-like microstructure and failure from a volume defect, which is equivalent to the observations for the larger pebbles. Also the pebble that failed at 5.14 GPa, which shows a lithium metatitanate dendrite dominated microstructure and a presumable failure from a surface defect, is equally observed for the larger pebbles.

A comparison of the failure data from the smaller pebbles with a model according to equation (5.1) is, in contrast to that of the larger pebbles, not only in good agreement with the experimental values for the high strength pebbles, it is also able to model the bump in the Weibull plot much better (see Figure 5.8). With this result, the presence of an R-curve behavior of these pebbles is still plausible, but the chosen model with three failure modes excellently describes the experimental data too. The comparison of the results of both pebble sizes does therefore lead to no common conclusion for this issue and further investigations are necessary to fully understand the failure behavior of these pebbles. However, both explanations lead to reasonable descriptions of each set of data. For the application of the model, the same Weibull moduli as for the pebbles of 1000 µm diameter were used and the characteristic strengths were adapted to reasonable values, except for the ambiguous Weibull distribution to model the low strength part. Equation (5.2) is applied to derive the characteristic strength of the steep sub-segment for a 500 µm pebble. A value of 3.98 GPa is obtained in this way.

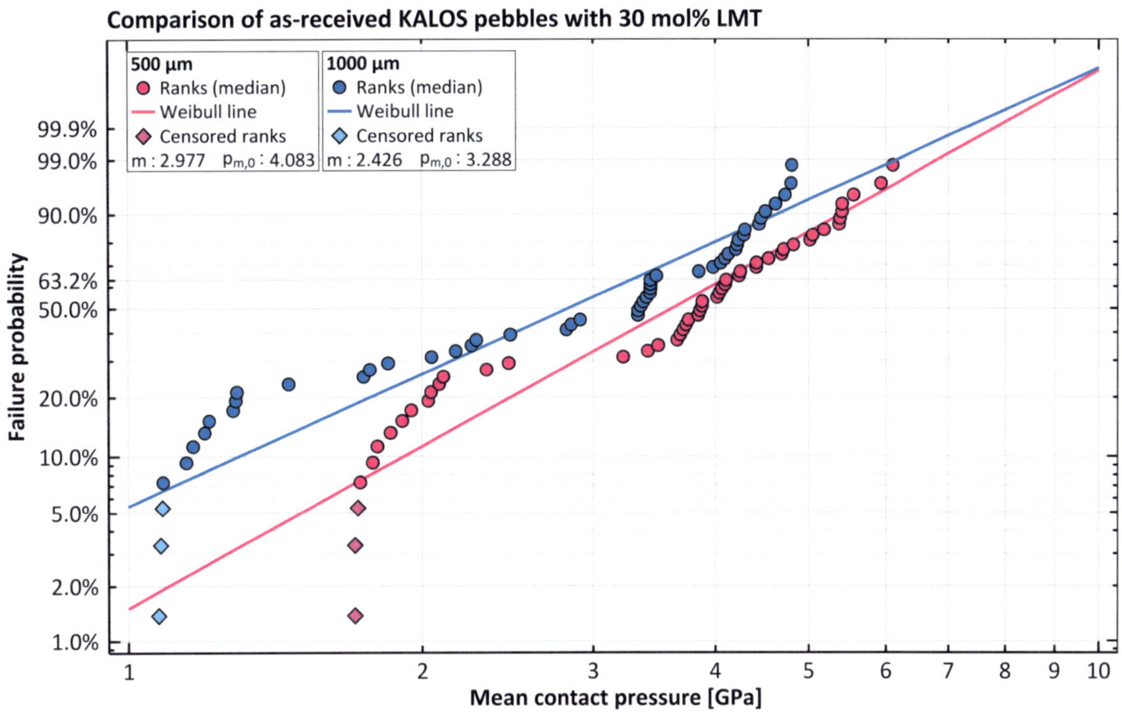

Figure 5.7: Weibull plots of the as-received KALOS pebbles that nominally consist of 30 mol% LMT in LOS.

Yet, if the same equation is used to calculate the expectable characteristic strength for small pebbles from the Weibull distribution which represents the data of the biphasic emulsion method pebbles (pebble diameter: 1034.4 µm), a value of 4.95 GPa results, which is significantly lower than what the experimental data suggests. Also, if volume defects were assumed instead of the identified critical surface defects, a too low value of 5.25 GPa would be obtained. For the applied model, a characteristic strength of 5.5 GPa seems to be a reasonable value. Also the probability parameters were adjusted slightly according to the experimental data. Generally, these observations undeniably support the conclusion that the failure behavior of nominally 30 mol% LMT containing KALOS pebbles is governed by the same mechanisms, independent of their size. Yet, a fraction of the smaller pebbles seems to reach higher strengths than predicted by classical Weibull theory. Because of the limited data basis, it cannot be defined whether this observation is significant or not.

The Weibull analysis clearly highlights, that, compared to the strictly hypoeutectic pebbles that nominally contain 20 mol% LMT, the mean crush load of the nominally 30 mol% containing pebbles is only slightly higher (see Figure 4.40 and Figure 4.41) because these pebbles show a significant number of 'apparent' failures at low loads. If this failure mechanism would not be present and if the strength were determined by the two remaining failure mechanisms, the mean crush load would be significantly higher. The reason for not observing these apparent failures in such a quantity in the nominally 20 mol% LMT containing samples, while they appear as a non-competitive failure mode in the Weibull plots of the LMT richer pebbles, is not clearly identifiable from the present experiments.

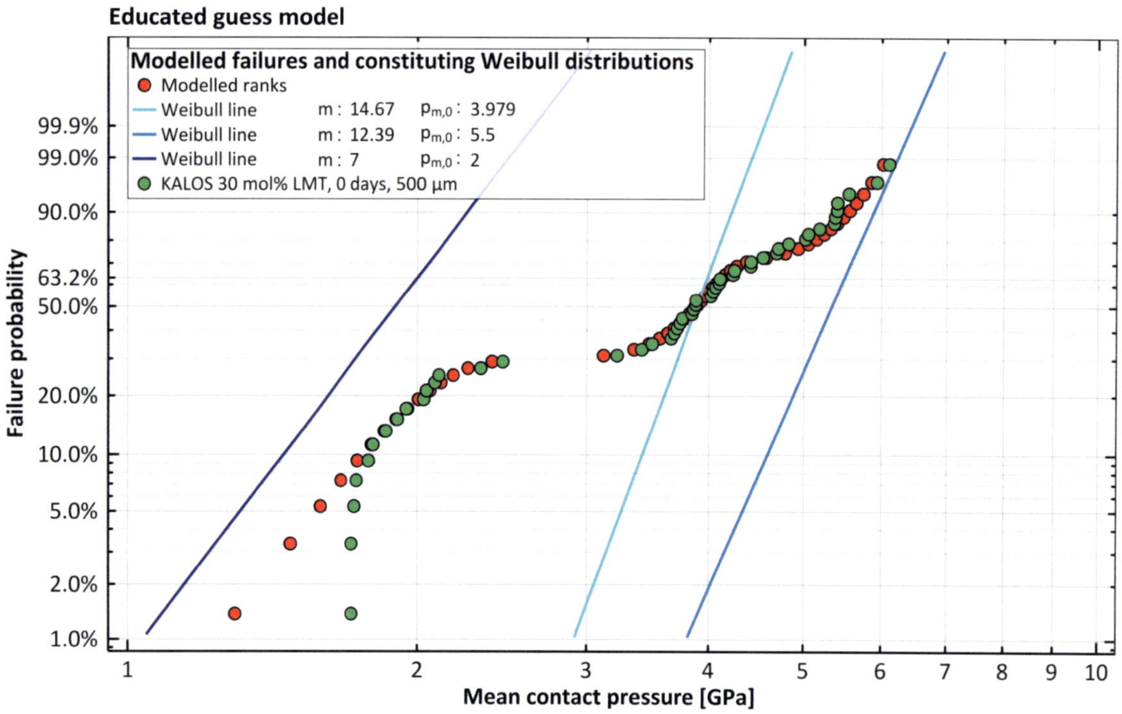

Figure 5.8: Modelled compound Weibull plot from three constituting Weibull distributions according to equation (5.1) compared to the experimental values of the nominally 30 mol% LMT containing KALOS pebbles of a diameter of 500 µm. The distributions with Weibull moduli of 14.67, 12.39 and 7 are appearing with a probability of 40 %, 30 % and 30 % respectively.

Yet, with some consideration a potential mechanism can be proposed. In general the effect of the thermal expansion mismatch on the adhesion between the grain boundaries should be comparable between both nominally different KALOS samples as the differences in the determined strains (see section 4.2.2) are low. In principal, the thermal stresses within the nominally 20 mol% LMT containing pebbles are higher than for the LMT-richer pebbles. This is especially explicit for the strain of the LOS phase, for which a 1.31 % and 0.65 % strain is observed for the nominally 20 mol% LMT containing pebbles and the nominally 30 mol% containing ones, respectively. Yet, the differences in these stresses are attributed to the differences in microstructure. As the microstructure of the nominally 30 mol% LMT containing pebbles varies considerably and the crystallization in eutectic domains is overall reduced in comparison with the nominally 20 mol% LMT containing pebbles, slightly reduced thermal stresses should be observed by XRD. Within the eutectic domains, however, the weakening effect of the thermal stresses should be identical. Consequently, the thermal stresses do not play a role in enabling significant microplastic behavior for one particular type of pebbles. In section 2.3.1 the necessity of a large enough grain size for enabling microplasticity is discussed. When the common definition of grain size is applied, it is very low for the KALOS pebbles in general.

However, if the eutectic domains are treated as quasi-grains, some differences are visible between the samples. Especially for the nominally 30 mol% containing pebbles, a significant variation of this quasi-grain size is visible (see Figure 5.9). Of course, as the cooling rate at the surface of a droplet is much higher than at any other location of it, the (quasi-)grain size is significantly reduced at the surface. For the pebbles which feature larger eutectic domains in their core (see Figure 5.9C), also the region of very low (quasi-)grain size at the surface appears to be far less extensive.

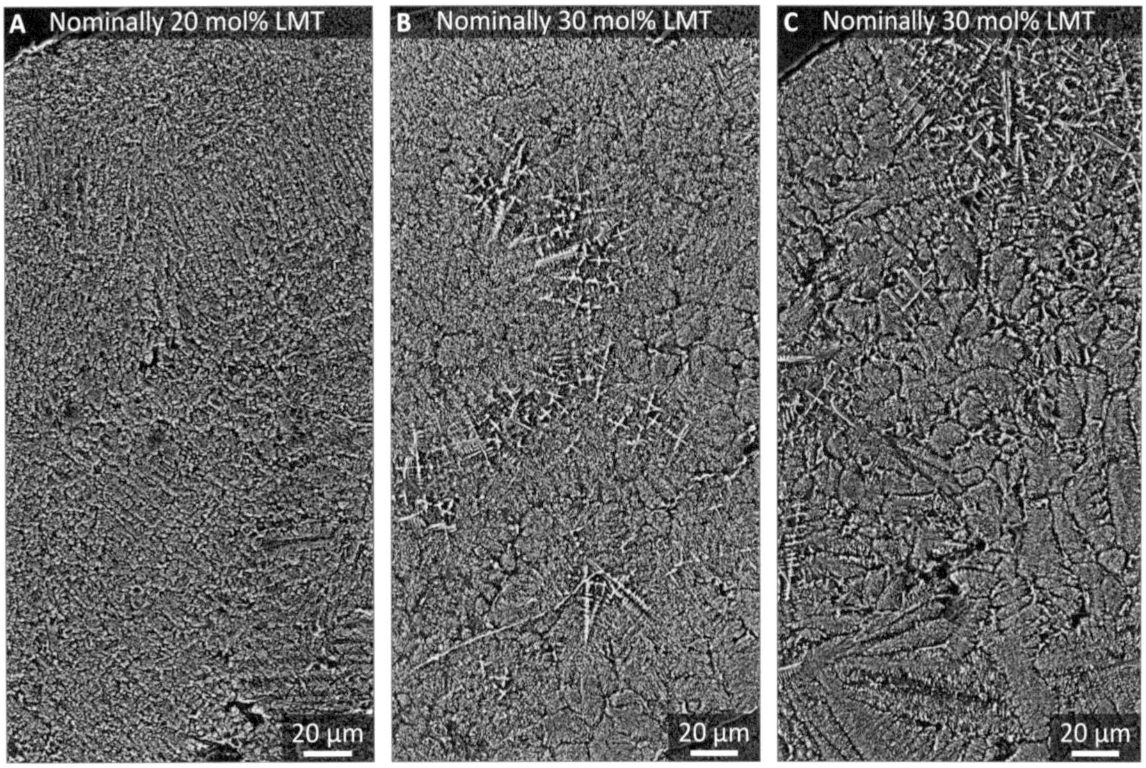

Figure 5.9: Comparison of the (mainly) eutectic microstructure of three different pebbles that were fabricated by the KALOS process with their nominal composition given. For all images, the surface of the pebble is at the top of the image.

For the pebble depicted in Figure 5.9C, the eutectic domains already reach considerable sizes only a few micrometers distant from the pebble surface. Compared to the nominally 20 mol% LMT containing pebbles, the (quasi-)grain size seems to be generally larger for the nominally 30 mol% LMT containing pebbles and the region of reduced (quasi-)grain size at the surface seems to be accordingly larger for the nominally 20 mol% LMT containing pebbles. The shear faults that lead to microplastic behavior develop within the volume below the Hertzian contact. From the fragment that is shown in Figure 4.62, it becomes clear, that the region of microplasticity extends up to 50 µm into the volume of a 1000 µm pebble. Consequently, if the eutectic domains behave like usual grains in this context, the increased domain size near the surface of a number of nominally 30 mol% LMT containing pebbles is potentially enough to enable microplastic behavior for these pebbles.

As a result another failure mechanism is introduced into the respective pebbles, which should be competing with the failure from volume defects for the pebbles that feature a pearlite-like microstructure. This conflicts somewhat with the applied model. However, as the apparent failures happen at rather low contact pressures, the differences between the applied model and a model that takes the competition into account may be insignificant after all. The effect of grain boundary weakening thermal stresses may still play a role in enabling microplasticity, even if the domains are treated as quasi-grains. Guiberteau et al. [124] explain the formation of microplasticity in alumina as being caused by intragrain defects, for instance by crystal twins. In the case of the KALOS pebbles the interfaces between LOS and LMT lamellae are certainly plausible intragranular defects within the quasi-grains, i.e. eutectic domains, if their adhesion is weak. It is thinkable that these interfaces qualitatively act in a similar way as twin boundaries in single phase ceramics, and therefore are a potentially significant factor.

In comparison with the failure mechanism that is assigned to 500 µm pebbles that feature a pearlite-like microstructure in the nominally 30 mol% LMT containing sample, the Weibull modulus of the nominally 20 mol% LMT containing pebbles is considerably lower which is explainable by failure from surface defects in contrast to volume defects, which is also observed for the 1000 µm pebbles and in accord with the SEM observations.

5.2.2 Biphasic pebbles fabricated by the emulsion method

The results of the XRD investigations as well as the microstructure examinations clearly show that pebble samples that predominantly consist of lithium orthosilicate and lithium metatitanate can be fabricated by the emulsion method for any given composition. The observed variation of the diameter of the fabricated dried pebbles (see Figure 4.21) can only be attributed to insufficient control of the fabrication process as the variations are clearly independent of the slurry composition. Coagulation of non-gelled droplets within the cooled oil bath, because of too little separation is a possible reason for these modest variations. The various maxima in the observed multimodal size distributions of the gelled pebbles appear at much too similar pebble diameters for coagulation of two droplets of roughly similar size to be a possible explanation for these maxima. That suggests that a different effect is also considerably influencing the droplet size. However, the pebble size of the 80 mol% LMT containing samples, and to a lesser extent that of the 70 mol% LMT containing pebbles, is significantly larger than that of the other samples, which is not sufficiently explained by random variations of the fabrication process. The larger pebble size does not coincide with a distinctly reduced green density, what implies that the slurry composition was close to the expectations and that the drying of the pebbles was similarly effective as for all other samples. Presumably more severe operational inaccuracies during the shaping of the gel pebbles are causative for the observed larger green pebble sizes, while the multimodality of the size distributions of the green pebbles are presumably an intrinsic effect of the experimental equipment.

Effects on the pebble densification and the resultant properties

As the shrinkage (see Figure 4.22) does not largely vary as a function of the LMT concentration, the variations in the shaping are basically directly translated into variations of the sintered pebble size. This explains the observed variations in the pebble size of the sintered pebbles as shown in Figure 4.21. However, this is not true for the pebbles that feature a content of 90 mol% LMT. Still, Figure 4.22 suggests that the green density of the pebbles is inversely related to the observed shrinkage. To estimate the effect of the nominal LMT content on the shrinkage, a linear regression is performed for pebble samples with rather similar green density (i.e. samples that contain 20 mol% LMT to 50 mol% LMT) to exclude effects of the varying green density to a reasonable degree. The result shows, equally to the general observation that is given in section 4.3.2, that the shrinkage slightly decreases with increasing LMT content. Ideally, the LOS forming reaction according to equation (3.1) involves shrinking of 29 % for the given crystallographic data (see Table 2.1 and Table D.2 in Appendix D). It goes without saying that the measured shrinkage is larger than that, because of the burning of the binder and the sintering of the pebble. Yet, with increasing LMT concentration, the influence of the LOS forming reaction on the overall shrinkage is reduced to insignificant levels. This should, in principle, lead to a decreasing tendency of the shrinkage with increasing LMT concentrations as the lithium metatitanate phase does not undergo a chemical reaction during sintering.

The mean value of the green density of these pebbles is referred to as 'reference green density' in the following. For all but one composition (namely 90 mol% LMT), the observed shrinkage can indeed be explained by a variation in green density as Figure 5.10 shows. In this plot, the deviations of the measured shrinkage from the estimated, LMT-dependent shrinkage are given as a function of the deviation of the determined green density from the reference green density. All data points, with the exception of the one for 90 mol% LMT, correlate very well. Furthermore, this shows that the implicit assumption that the shrinkage linearly depends on the LMT content is also valid for all but one tested composition. The observation that higher green densities lead to lower shrinkages is evidently logical. However, if the green density is too low, a coherent sintered pebble will not be obtained. Therefore, the achieved green densities are obviously high enough to not obstruct the sintering into a single pebble.

The increased shrinkage 'compensates' for lower green densities to some degree (i.e. to about a third of every percent of deviation in green density according to the linear regression in Figure 5.10), yet the finally reached pebble density after sintering is still influenced by the initial green density as a frail tendency to lower open porosity with higher green density is observed. This trend is most obvious for the samples that nominally contain 70 mol% LMT and 80 mol% LMT as these prominently stand out (compare Figure 4.22 and Figure 4.39). An unambiguous explanation for the substantial deviation of the 90 mol% LMT pebbles cannot be given, but it is assumed that the established estimation of the LMT-dependent shrinkage is not valid for very high contents of lithium metatitanate.

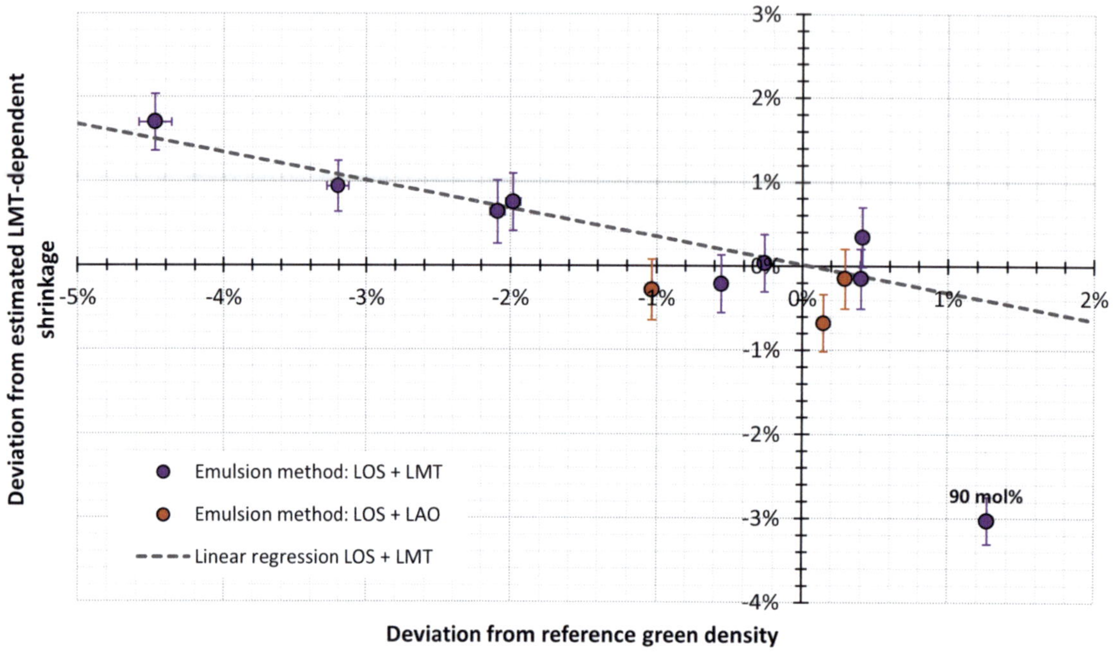

Figure 5.10: The deviation from the estimated LMT-content dependent shrinkage is given as the deviation from the reference green density. The linear regression disregards the data point for 90 mol% LMT. The coefficient of determination of the linear regression is 0.84. For the LAO containing pebbles the respective values are obtained identically. In x-direction an uncertainty of ±2.5 % is assumed, while in y-direction a measurement uncertainty of ±11 µm (see section 3.6) is applied.

The reason might be that the presence of lithium orthosilicate does not have a significant effect on the shrinkage at these levels of LMT content and thus the sintering behavior changes. In contrast, no detectable effect on the closed porosity of the pebbles is observed. Presumably the long sintering time together with not too different green densities makes it difficult to clearly observe such effects.

The fabricated pebbles show the anticipated phase composition, however, with increasing LMT concentration, lithium metasilicate as impurity phase is observable, although in very small concentrations. As lithium metatitanate can take up some excess lithium (see Figure 2.10) the used lithium surplus might eventually be inadequate for compensating this consumption of lithium, as well as for losses of lithium during fabrication and sintering. In section 5.3.2 indications are presented that within lithium excessive β-lithium metatitanate an increase of the lithium excess leads to an increased a-parameter of the crystal lattice, while the parameters b and c are reduced. Probably because of the less laborious sample preparation, the values of a/c-ratio appear to be more stable than the as-determined parameters since some experimental deficiencies are neutralized by the formation of a quotient.

For the deliberately lithium excessive lithium metatitanate samples, there is no significant difference in the development between both values (compare Figure 5.19 and Figure D.5). Consequently, this ratio is discussed in the following instead of the a-parameter. All biphasic lithium metatitanate containing pebbles show a significantly increased a/c-ratio (from 0.5202 to 0.5207) in comparison with the literature value of 0.5190. The a/c-ratio almost monotonically decreases with increasing nominal LMT concentration, as presented in Figure 5.11. With increasing LMT concentration, the internal loss of lithium is of course expected to increase, but as more LMT and less lithium surplus are introduced into the green pebbles, the relative amount of lithium that the LMT can take up is reduced (see Appendix D) which explains the decreasing a/c-ratios. Yet, although the values of the a/c-ratios are notably larger than those that were determined for the deliberately lithium excessive samples (see Figure D.5), the determined ratios of a/c do not saturate at low nominal LMT concentrations, which suggests that the LMT phase is also not saturated with lithium. However, it has to be noted that impurity elements, whose concentrations were not determined in this study, may also lead to a stretched crystal lattice.

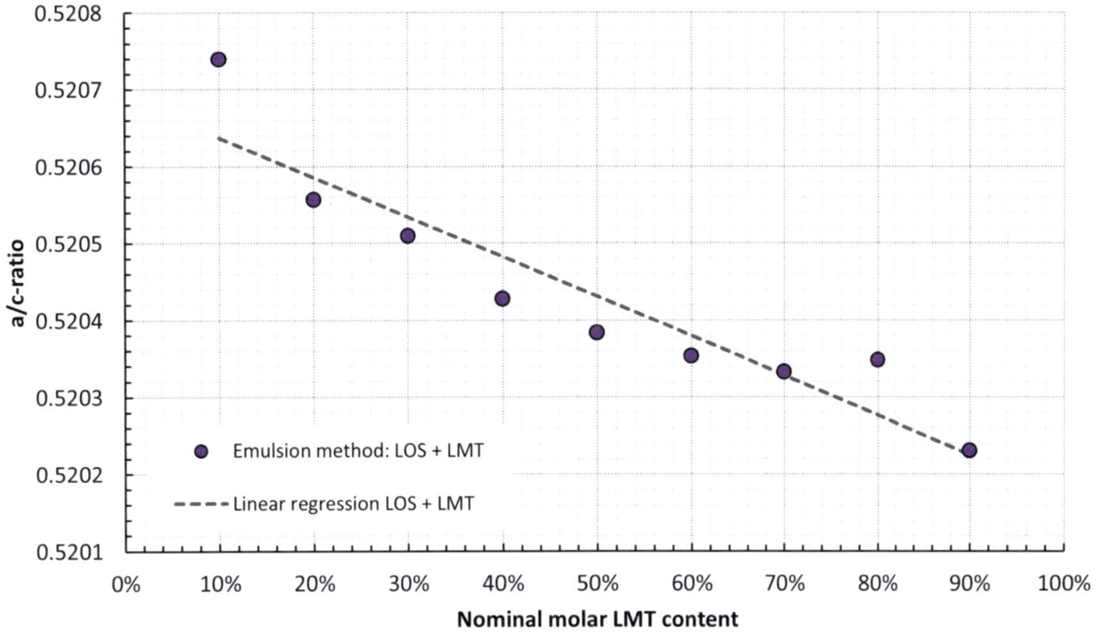

Figure 5.11: The determined a/c-ratio of the β-lithium metatitanate unit cell as a function of the nominal LMT concentration of the biphasic emulsion method pebbles. The estimated error of the Rietveld refinement is too low to be displayed in this graph. The coefficient of determination of the linear regression is 0.874.

It is consequently not sensible to postulate a higher overall lithium excess within the lithium metatitanate phase compared to the deliberately lithium excessive samples. Of course, when the crystal is already close to saturation, the incorporation of additional lithium is increasingly less favorable. Consequently all samples have probably formed some lithium metasilicate, even at concentrations below 50 mol% LMT. Presumably, lithium metasilicate is simply not observable for pebbles with LMT contents below 50 mol% by XRD (see Figure 4.3) as the detection limit for lithium metasilicate is undercut. An inclusion of a lithium surplus within the lithium orthosilicate phase does not seem likely as the unit cell strain of LOS (see Figure 4.6) is comparably low, even though the results of the KALOS pebbles suggest this possibility (see section 5.3.1) in opposition to the published phase diagram (see Figure 2.9).

Overall, the elemental analysis confirms, that the expected concentrations are met with a very slight tendency towards a surplus of lithium. These observations fit in the explanation of the XRD analysis, as there is no significant loss of lithium within the pebbles to form the less lithium rich lithium metasilicate phase to a significant degree. Therefore, the formed lithium metasilicate is also not observable in the cross sections of the biphasic lithium metatitanate containing pebbles.

The qualitative XRD results (see Figure 4.3) prove that not only biphasic samples which consist of lithium orthosilicate and lithium metatitanate can be fabricated by the emulsion method, but by this method also pebbles can be produced that consist of lithium orthosilicate and lithium aluminate. As the produced dried pebbles are of great uniformity, the sintered pebbles also show narrow pebble size distributions with no significant influence of the slurry composition. Compared with the samples consisting of LOS and LMT, the fabrication is obviously better controlled as only insignificant deviations occur. The shrinkage of the LOS+LAO pebbles during sintering is almost identical to that of the LOS+LMT pebbles with comparable composition. Yet, actually a slightly higher shrinkage is to be expected as two phases are formed by solid state reaction sintering. According to equation (3.2) and the theoretical densities that can be derived from the crystallographic data in Table 2.5 (and Table D.2), a theoretical shrinkage of 17 % is calculated for pure LAO. However, only 2-5 % shrinkage may result from this reaction for the tested compositions in addition to the shrinkage that results from LOS as the LAO concentration is relatively small. The green density of the LAO containing pebbles is almost identical to the values of the LOS+LMT with the respective second phase concentration. As the observed shrinkage is also comparable (see Figure 4.22), these results suggest that the densification behavior of the LOS+LAO pebbles is dominated by the solid state reaction sintering of the lithium orthosilicate phase. This observation is also supported by the plot in Figure 5.10 in which the shrinkage of these pebbles can be reasonably explained by their green density, although the correlation for the second phase content dependent shrinkage as well as the reference green density are derived from the LOS+LMT samples. However, this plot also suggests, that in tendency, the LAO containing pebbles are shrinking less than expected. The cross section images (see Figure 4.34) furthermore underpin the assumption that the densification of lithium orthosilicate dominates the overall densification as they show that the LAO particles are solely in contact with LOS particles or grains and thus the densification of pure LAO should not play a considerable role.

For the sintered LOS+LAO pebbles, the closed porosity clearly increases with the LAO concentration (see Figure 4.38). The cross sections of the LOS+LAO samples accordingly display an increase of large pores that formed during sintering as a function of the LAO concentration. The mostly spherical or elliptical pore shape hints at gas bubbles (most probably CO_2, see equation (3.2)) which formed because of too rapid sintering of the entire pebble. Also an effect of the zirconium impurities on the formation of porosity cannot be fully excluded, as the lithium zirconate, Li_2ZrO_3, forming reaction generates additional CO_2. Yet, the amount of lithium zirconate within the pebbles is low and consequently an effect of the impurities on the formed porosity should be clearly subordinate. As the cross sections also clearly show, the pebbles have developed a dense crust during sintering while a porous core is formed. The tendency to forming a porous core seems to be increased with the amount of Li_2CO_3 that is introduced into the green pebbles. The highest amount with 50.1 wt.% is introduced into the pebbles with a nominal content of 30 mol% LAO. The amount of lithium carbonate in the green 10 mol% LAO samples was 47.7 wt.% of Li_2CO_3 compared to 43.5 wt.% of Li_2CO_3 in the green pebbles that feature a nominal content of 10 mol% LMT. For the LMT containing samples this constitutes the highest amount of lithium carbonate since the amount is decreased with increasing nominal LMT content (all values are given in Table D.3 in Appendix D). Obviously, the crust sintered to a gastight structure before all gaseous species could leave the pebble. If CO_2 is encased during the sintering, as the results suggest, it will form rounded, ideally spherical pores, which corresponds to the cross section images. This considerable closed porosity surely reduces the observed shrinkage and thus explains that it is slightly lower than expected. As already mentioned in section 5.1.1, the formation of a denser crust is already structured in the green pebbles (see Figure 4.31 for the LMT containing green pebbles as an example, cross section images of the green LAO pebbles are given in Appendix D) as the green density near the surface is considerably higher.

The helium density measurements in conjunction with the cross section images (see Figure 4.33 and Figure 4.38) also clearly show that there is an increase in closed porosity within the 10 mol% LOS+LMT sample. These findings suggest that the observed pores are also generated by the same mechanism that leads to the closed pores within the LAO containing samples. Obviously, for all samples which contain a higher amount of LMT, such pores are not formed as markedly as for the 10 mol% LMT sample. Yet, the encasing of gaseous species might very well lead to the formation of pores in the other samples, even if it is not as apparent. Although a higher amount of closed pores was formed for the 10 mol% LMT sample, it appears that the amount of open pores is comparable to that of the 20 to 40 mol% LMT samples. For the samples from 10 to 60 mol% LMT, the majority of pores that are visible in the cross sections are highly likely closed pores, as the pore size is relatively small and the amount of pores is also small so that a connection to the surrounding atmosphere is not evident. In reverse, the majority of open pores are logically located near the surface of the pebbles. For the 70 to 90 mol% LMT pebbles, the pore shape changes from rounded to polygonal and also the visible amount in the cross sections increases significantly. This indicates a transition toward a LMT dominated microstructure where the pores form between platelet-like LMT grains.

Yet, as Figure 5.10 shows, a significantly different densification behavior is only observed for the pebbles with a nominal LMT content of 90 mol%. Why the pores form much more readily within these pebbles, as opposed to the LOS richer samples, can only be explained by a higher 'mobility' of the LOS phase during sintering as the sintering temperature is much closer to the melting point of lithium orthosilicate (1258 °C, see Figure 2.9) than to lithium metatitanate (1533 °C, see Figure 2.10). For the LMT-richer samples, most of the visible pores are presumably open pores. The amount of open porosity for the 70 and 80 mol% LMT samples is considerably increased over all other LOS+LMT samples. The only significant differences between these pebbles and the other pebbles, beside the difference in LMT content, are the significantly larger mean pebble size as detailed in Figure 4.21 and the lower green density as shown in Figure 4.22. Because of the larger volume of the pebbles, a higher amount of organic material has to be burnt during sintering, which may lead to larger channels at the surface from which the gases can pass off. The lower green density should also lead to a lower overall density, as the shrinkage of the pebbles only fractionally compensates for that. In principal the lower green density should translate into a higher open as well as closed porosity of the samples, but an increased closed porosity is not observed for both types of pebbles. Considering the magnitude of the differences in green density, the effect on the porosity should be relatively low and may therefore be not observable.

Similarly, the comparable shrinkage of the LAO containing pebbles does not seem to translate into pebbles of equal relative density. The measured total porosity (which contains a certain amount of imaginary porosity, see Table C.16) of these pebbles is lower than that of the LOS+LMT pebbles. Especially the 30 mol% LAO pebbles show the lowest sum of total and imaginary porosity of all bi-phasic emulsion method samples, either with LMT or LAO as second phase. Yet, the amount of imaginary porosity in both series of samples is difficult to evaluate. Principally, given the already low values of porosity of the 30 mol% LAO containing sample, the amount of imaginary porosity has to be lower than about 4 % (see Figure 4.39). If the LMT containing samples and the LAO containing ones had the same true relative density, the amount of imaginary porosity for the LMT containing samples would amount to about 9 %, when the values of the 90 mol% LMT containing pebbles and 30 mol% LAO containing pebbles are used. Yet, such a stark difference in sphericity in favor of the LAO containing pebbles is not supported by the surface images of the samples (see Figure 4.17 for the LOS+LMT samples and Figure 4.20 for the LOS+LAO ones).

As a result, it is justified to assume a higher relative density for the LAO containing pebbles with regard to the LMT containing ones. The observed increase in density goes along with an increased amount of LAO for this reason, while the high melting point of LAO (1785 °C, see Figure 2.12) with respect to LOS (1258 °C, see Figure 2.9) does not suggest a higher sintering activity of LAO. Potentially, the increased amount of Li_2CO_3 leads to a higher densification, as a significant surplus of lithium may lead to the temporary formation of either lithium oxosilicate, Li_8SiO_6 or Li_5AlO_4, which both form liquid phases at eutectic temperatures of 990 °C and 1050 °C, respectively (see section 2.2.5). Yet, at the sintering temperature only lithium oxosilicate would be liquid judging from these data. The actual phase equilibria for high lithium concentrations in the system Li_2O-SiO_2-Al_2O_3 are, however, unknown and therefore no terminal assessment is possible. If liquid phases are formed they may promote material transport, which is certainly increased if the concentration of liquid is increased, which is most likely a function of the Li_2CO_3 content.

With the overall Li_2CO_3 content being largest in the LOS+LAO samples, this would explain the comparably low open porosity of these samples with respect to the LOS+LMT samples (see Figure 4.39). An increased sintering rate also leads to a higher tendency of encasing forming gaseous species. Overall, the cross sections reveal that LOS and LAO are relatively homogeneously distributed within the pebbles and that LAO is mainly found at the grain boundaries of the LOS grains. The EDX analysis shows that in some areas aluminum and silicon may be part of a common phase, which would be in accordance with the observed stretched lattice of the LAO phase in the XRD analysis (see Figure 4.6). Yet it is difficult to draw definite conclusions from the EDX analysis in this regard.

It has to be noted, that the shape of the 10 mol% LAO containing pebbles is occasionally significantly non-spherical. According to the model of Forgac et al. [128], which can sensibly be applied to the pebbles that were fabricated by the emulsion method as explained in section 5.1.1, a significant contraction in the core of these pebbles must have occurred. Yet, this contraction does not seem to be considerably large for the LAO-richer pebbles. Of course, the formed closed porosity is counteracting the contractionary tendencies, which is in principal in accordance to the model of Forgac et al. [128] although the authors do not consider the possibility of the formation of gaseous species. Yet, they consider the formation of pores in the core of the pebbles as one possibility of the system to compensate for the increased density within the core [127]. With regard to the non-sphericity of the pebbles, the formed porosity cannot compensate for the immense contraction that obviously took place within one of the pebbles detailed in Figure 4.20. As the concentration of lithium carbonate in the slurry is increased with increasing LAO content and the deformation of the pebbles is only visible for the 10 mol% LAO samples, a too high lithium carbonate concentration is unlikely the cause for this effect. In consequence, the reasons for the formation of this amount of contraction remain unclear, whereas the model of Forgac et al. [128], in conjunction with the formation of a dense crust, explains the moderate dents and defects in the biphasic emulsion method pebbles sufficiently.

The mechanical strength of the lithium metatitanate containing pebbles

Regarding the mechanical strength of the biphasic emulsion method pebbles, an increased LMT concentration also increases the determined mean contact pressure (see Figure 4.43). Yet, it is a rather volatile trend, especially for pebbles with nominal contents of 50 mol% LMT to 80 mol% LMT. The determined Weibull parameters of each of the test series are in principal showing the same result (see Figure 4.55). But, specifically for the samples that contain 70 mol% LMT and 80 mol% LMT, a considerably higher Weibull modulus is observable, which is a clear indication for a different failure mechanism than for the other samples, which seem to share a relatively equal Weibull modulus, maybe with the exception of the 20 mol% LMT sample. The calculated closed porosity of the samples, as derived by the helium pycnometry measurements, does not correlate well with the observed mechanical strength or in other words, the closed porosity is low enough to not deteriorate the mechanical strength of the pebbles significantly. Yet, the determined characteristic strength of the pebbles correlates extraordinary well with the determined open porosity (assuming a constant amount of imaginary porosity), if the 70 and 80 mol% LMT samples (with the sum of open and imaginary porosity larger than 30 %) are disregarded (see Figure 5.12).

A similarly high correlation cannot be found if the total porosity of the samples is used instead. As open porosity reaches the surface by definition, open pores can also be regarded as one form of surface cracks which obviously lower the overall strength of the pebbles. It must be noted that usually solely the largest flaw within a pebble determines the reduction in strength, as at its tip the stress concentration factor is usually the highest (see equation (2.14)), and not the overall fraction. But, as equation (2.20) shows, the flaw density also determines the characteristic strength to some degree. In the present case, the flaw density has to decrease by a factor of roughly 200 to span the observed range of characteristic strength from a nominal content of 10 mol% LMT to 90 mol% LMT if the pebbles are otherwise identical. The surface images, however, in tendency suggest the opposite (see Figure 4.17 and Figure 4.18), i.e. an increasing porosity and pore size with increasing LMT concentration. Also the grain size of the major phase is not changed largely as a function of the LMT content (see Figure 4.33). It is therefore clear that neither of these features has a significant effect on the strength of the pebbles.

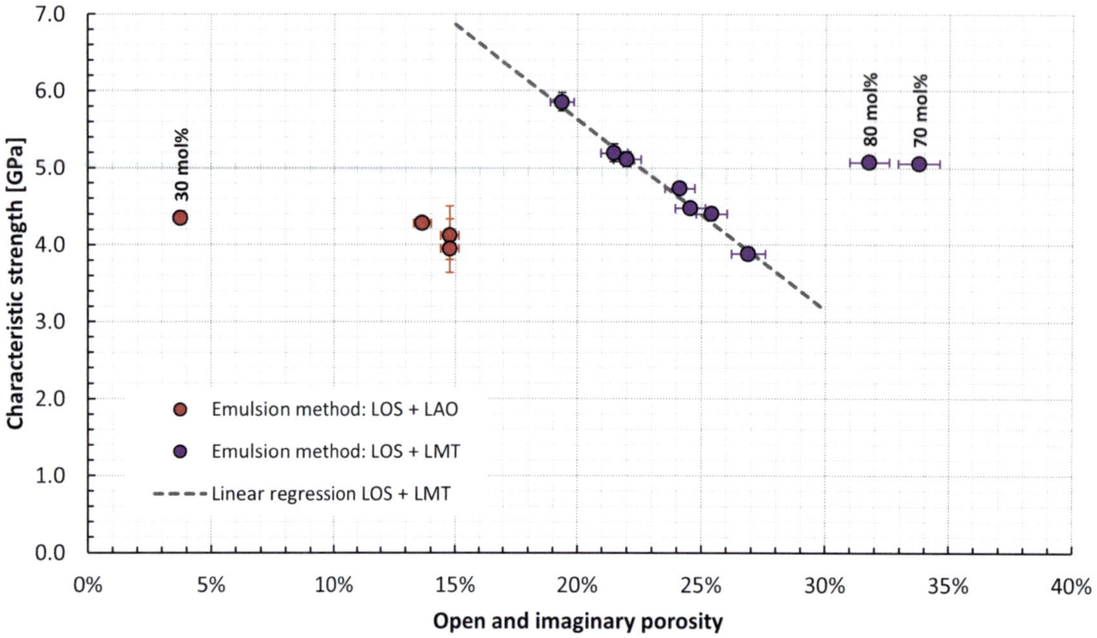

Figure 5.12: The characteristic strength of the biphasic LMT or LAO containing pebbles that were produced by the emulsion method is presented as a function of the sum of the open and imaginary porosity. In x-direction, an uncertainty of ±2.5 % of the determined porosity is assumed. The error bars in y-direction represent the one dimensional projections of the 90 % likelihood ratio confidence boundaries. The linear regression for all but two LOS+LMT data points (70 mol% and 80 mol% LMT) shows a coefficient of determination of 0.99. Since two Weibull distributions are necessary to represent the experimental data of the 10 mol% LAO containing sample, two data points are given.

However, the determined open porosity also comprises the sphericity of the pebbles and therefore also large flaws like dents and 'seams' are subsumed in this number. The dents of the pebbles are presumably one major contributor to the imaginary porosity as they are macroscopic deviations from ideal sphericity. In consequence, the seams of the pebbles have to be primarily considered as cause for the observed correlation in Figure 5.12.

In Figure 4.68 the failure from such a 'seam' of pores is relatively clearly observed for the pebble that failed at the lower load. The cross section images (see Figure 4.32), however, show that the 'seams' to not reach deeply into the pebbles but are rather a surface effect. Presumably, the extension of these seams in tangential direction varies with the LMT content and therefore a higher or lower open porosity is observed. The existence of the seams is one consequence of the densification during the drying of the pebbles as already explained (see section 5.1.1). This is also supported by a reasonably high correlation between green density and the open porosity of the pebbles as Figure 5.13 shows, i.e. the structure of the green pebbles is largely translated into that of the sintered pebbles. Even though the coefficient of determination of the linear regression function is notably lower than that of the other linear regression functions that are presented in this section, the fact that the determined y-axis intercept is almost identical to the ideally expected value of 100 % shows, that it very reasonably describes the relationship of the two quantities. Pebbles that do not show any open porosity would require a green density of about 48 % according to this linear function. As the density of the green pebbles is determined by the same method as the porosity of the sintered pebbles, the effects of macroscopic non-sphericity, i.e. the imaginary porosity, may actually be compensated for in Figure 5.13, as the almost ideal y-axis intercept suggests.

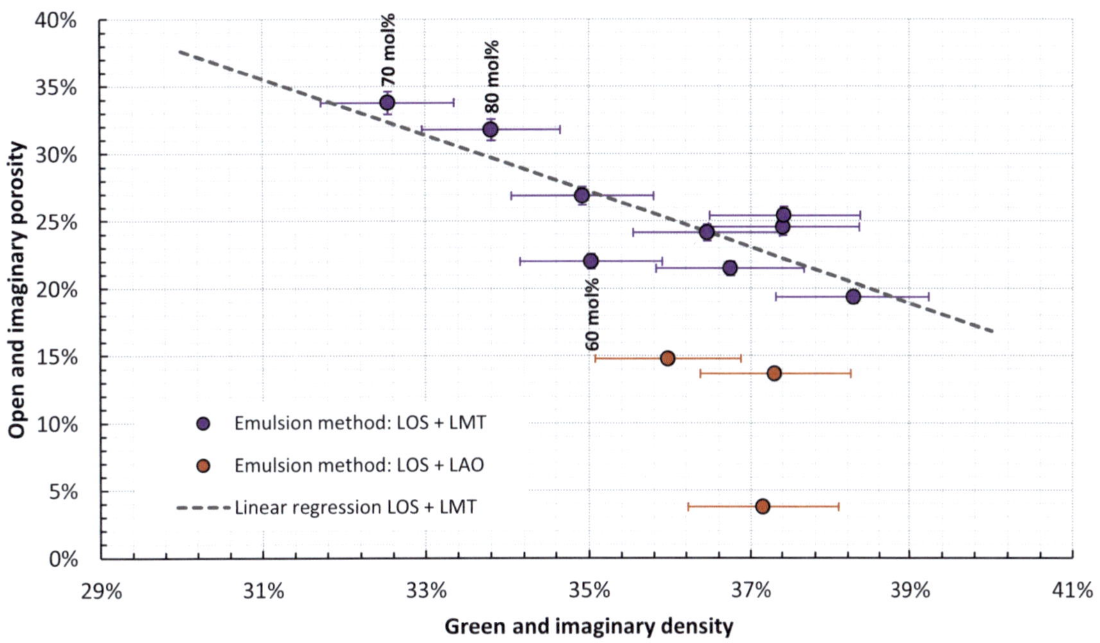

Figure 5.13: The sum of the open and imaginary porosity of the biphasic emulsion method pebbles as a function of the sum of the green and imaginary density is shown. An uncertainty of ±2.5 % for the determined porosity value and the density is assumed. For selected data points, the respective nominal LMT content is indicated. From the linear regression no data points are excluded and a coefficient of determination of 0.695 is obtained. The y-axis intercept of the regression function is at 100.04 %.

Therefore the established linear function can also be considered to be quantitatively correct and it is also concluded that no considerable differences between the macroscopic shape of the green and the sintered LOS+LMT pebbles are visible. The increase of the characteristic strength of the pebbles over the complete compositional range implies a reduction of the crack length in radial direction to roughly 44 % (see equation (2.14)) if no other effects contribute to the increased strength. Yet, the surface images do not suggest such a substantial reduction (see Figure 4.17). Consequently, a significant portion of the increase in strength of the LOS+LMT pebbles has to stem from an increase in fracture toughness with increasing LMT content. All other things equal, the fracture toughness has to increase by roughly 150 % over the tested range. Yet, whether this is quantitatively true, i.e. whether this increase is counteracted by the surface morphology and thus larger or vice versa, cannot be judged from the available results. Of course, this increase should also include the samples that nominally contain

70 mol% LMT or 80 mol% LMT. The failure from surface defects is also verified for all samples, except for the 70 mol% LMT and 80 mol% LMT containing pebbles, by the analysis of fragments, as Figure 4.68 and Figure 4.70 show.

Interestingly, for the two samples that show the largest open porosity, the samples that nominally contain 70 mol% LMT and 80 mol% LMT, the characteristic strength does not correlate with the open porosity. However, the surface morphology of these pebbles is similar to that of the other samples of this series (see Figure 4.17), except for the absence of large seams. Apparently, the present open porosity does not deteriorate the mechanical strength in the same way as for the other pebbles. In the Weibull analysis, these samples show an increased Weibull modulus and comparably low characteristic strengths with regard to their LMT content. In many cases, a higher Weibull modulus can be attributed to volume defects, while a lower Weibull modulus is attributed to surface defects, i.e. open porosity. Figure 4.69 depicts fragments of the nominally 70 mol% LMT containing pebbles. These fragments show a different fracture behavior as the pebbles whose strength is correlated with the determined open porosity value. With reasonable likeliness these pebbles fail from volume defects, which explains the higher Weibull modulus and that there is no apparent effect of the open porosity on the mechanical strength. From the cross section images (see Figure 4.33), however, it is evident that, as the transition towards a LMT dominated sintering progresses, the closed pores increase in size and show more angular shape than the pores that can be observed otherwise within this series of samples.

Principally, the pores within the nominally 90 mol% LMT containing pebbles are also of angular shape and their size is comparable, yet, the presence of the seams at the pebble surface are obviously the critical defects for these pebbles. The LMT containing pebble fragments also show that for the LMT-richer pebbles a considerably amount of quasi-plastic yielding is observable at the contacts with the compressing sapphire plates (see Figure 4.68, Figure 4.69 and Figure 4.70). The observed plasticity can be unambiguously attributed to the presence of β-lithium metatitanate as the LMT-poorer pebbles do not show any plastic behavior. Several results clearly indicate that β-lithium metatitanate shows a notable amount of plastic behavior by slip along at least one crystal plane (see also section 5.3.2). The XRD analysis of the lithium excessive pebbles revealed that these pebbles show a strong preferred orientation when placed on the sample holder only after the milling in a swing mill (see section 3.5.1 and 4.2.2). Such platelet-shaped grains are also predominately present in the biphasic LOS+LMT pebbles (see Figure 4.33). Also in this case, a significant milling step is part of the treatment. For the milling, also cleavage along a certain crystal plane instead of slip may be the operative mechanism. Yet, owing to the high similarity of the crystal structure of β-lithium metatitanate with graphite (see Figure 2.7), at least during low loading speeds, a considerable amount of plastic behavior along the (001)-planes, for which the preferred orientation is observed, is a reasonable assumption.

The plastic behavior of pebbles that feature high amounts of LMT is therefore completely different from the quasi-plastic behavior that is observable for a number of KALOS pebbles (see section 5.2.1). For the pebbles that are compressed, the plastic deformation leads to a considerably larger contact area than the Hertzian theory (see equation (2.4)) predicts as it assumes fully elastic behavior. For instance for the fragment of the nominally 50 mol% LMT containing pebble that failed at a load of 5.05 GPa (see Figure 4.68), a contact radius of 59 μm is expected, whereas the SEM image suggests a contact radius of 174 µm which is distinctly larger. However, for a lot of pebbles, especially the 70 mol% LMT containing ones (see Figure 4.69), the measured contact radius does not seem to vary significantly even for significantly different failure loads. It is therefore concluded that the plastic behavior at the mechanical contacts is not a significant factor for the evaluation of the strength of the pebbles. Yet, when the increase of strength as a function of nominal LMT content is considered, the LMT-richer pebbles may withstand larger loads but not necessarily larger contact pressures.

But, because of the large concentrated stresses at the mechanical contacts, a defined onset of plastic yielding may not be noticeable during the experiments. For the obtained data, these effects may not be of significant importance as no dramatic changes of the material behavior are noticed when significant plasticity sets in. This may be due to the fact that the region of plastic behavior is quite constrained as Kogut and Etsion [187] describe. In their analysis, for the comparably low amount of plastic deformation that is relevant in this case, they show that the region of plastic deformation is located below the mechanical contact, while all other regions of the pebble still behave elastic. Therefore, it is reasonable to assume that the tangential stresses at the surface of the pebble that were obtained on the basis of fully elastic behavior are still approximately valid, i.e. minor plastic behavior at the mechanical contacts may be negligible.

The mechanical strength of the lithium aluminate containing samples

With respect to the achieved mean contact pressures of the LOS+LAO pebbles, there is little difference to comparable LOS+LMT samples (see Figure 4.43), so apparently, the strengthening effect of LAO is comparable to the effect of LMT. However, it is not possible to correlate the mechanical strength to the open porosity of the pebbles (see Figure 5.12), also because of the few samples at hand, but mostly because the open porosity value of the nominally 30 mol% LAO containing pebble is significantly different. In principal, such a correlation is also not expected for the given results of the Weibull analysis. The nominally 10 mol% LAO containing pebbles have to be described by two Weibull lines (see Figure 4.57), in contrast to all LMT containing emulsion method pebbles, which reveals that two failure mechanisms determine the strength of these pebbles. As Figure 5.14 shows, the two Weibull distributions are competing for failure, i.e. the involved types of defects are both present in each pebble. This is a classic case of failure from surface defects at low strengths and failure from volume defects at high strengths. A similar case as for the nominally 30 mol% LMT containing KALOS pebbles (see section 5.2.1) cannot be made for these samples as apparent failures were not detected during the measurements. The samples that contain higher amounts of LAO show Weibull moduli which are highly comparable to that of the high strength segment of the nominally 10 mol% LAO containing pebbles (see Figure 5.14). This clearly illustrates that the LAO-richer pebbles also fail from volume defects. The surface morphology of all three pebbles that is shown in Figure 4.20 is in principal relatively similar with respect to smaller defects as they virtually do not exist. Yet, the nominally 10 mol% LAO containing pebbles show some pebbles which feature significant macroscopic surface defects in contrast to the other two sets of samples. It is most reasonable to assume that the surface defects of these pebbles lead to failure in addition to the usual failure by volume defects that the LAO-richer samples exclusively show. With the majority of pebbles failing from volume defects rather than from surface defects, it is clear that the strength of the LAO containing pebbles does not correlate with the open porosity values like the LMT containing pebbles. The significant increase of the characteristic strength of the pebbles that fail from volume defects that is observed for 10 mol% LAO and 20 mol% LAO containing pebbles cannot be explained by the grain size of the samples as it increases with increasing LAO content.

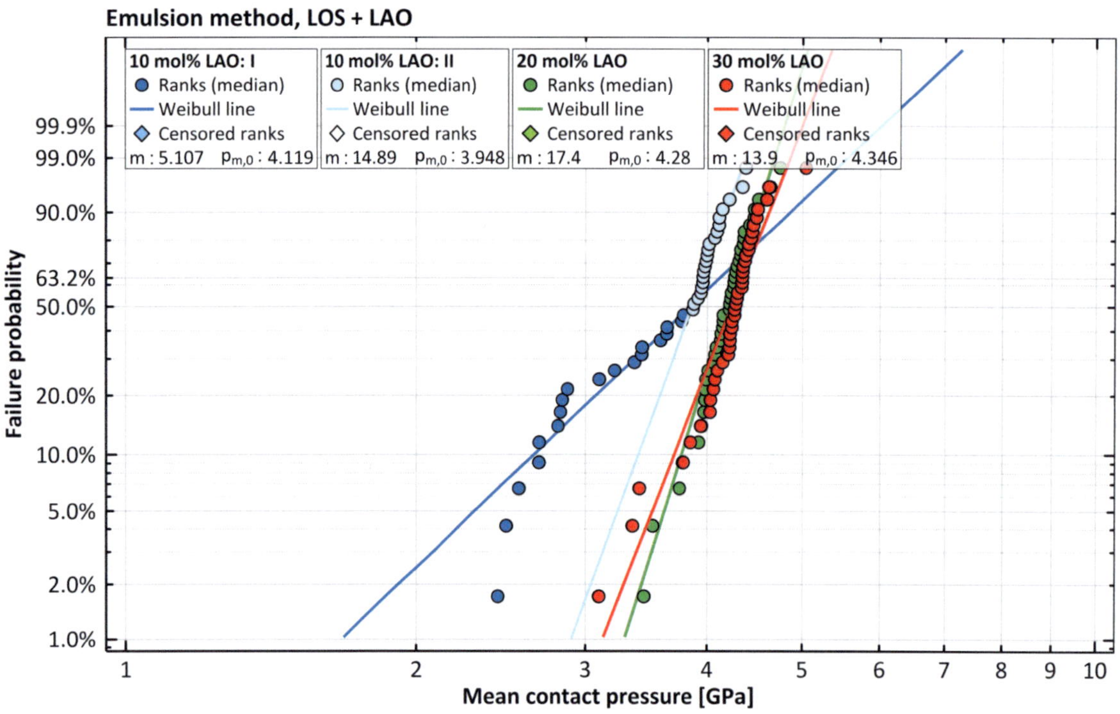

Figure 5.14: Weibull plots of the biphasic LAO containing pebbles that were fabricated by the emulsion method.

The same is true for the closed porosity which is considerably increasing with increasing LAO content and leads to very large pores. These large pores should in principal lead to a significantly reduced mechanical strength. Yet, the characteristic strength of the 20 mol% LAO and 30 mol% LAO containing samples is almost identical. Unless the increase in the pore size is not compensated by an equipollent increase in fracture toughness (according to equation (2.20)), another mechanism must be operative. If the pores within the pebble volume are assumed as crack initiating defects, the stress intensity factor must depend on their size. Although equation (2.14) is probably oversimplified for the present problem, the principal dependences on stress and defect size still apply.

If the approximate pore size which is reported in section 4.3.3 and the characteristic mean contact pressure that is obtained for pebbles that fail from volume defects (see Figure 5.14) are applied in equation (2.14), while neglecting any contribution of Y, the resultant stress intensity factor or in other terms the fracture toughness results. In Figure 5.15 the so-derived stress intensity factor is shown as a function of the nominal LAO concentration and a formidable correlation with a linear function is obtained. It can thus be assumed, that the fracture toughness of the LAO containing pebbles is linearly dependent on their LAO concentration within the tested composition range. The application of the characteristic mean contact pressure instead of the pressure referring to the pebble size leads to a significantly better coefficient of determination as when equation (2.7) is used to calculate the stress like Luscher at al. [121] recommend in their analysis. Yet, as Shipway and Hutchings [116] show, the factor of 1.4 (which is somewhat lower in their work) in equation (2.7) depends on the ratio of contact radius to pebble radius, especially for ratios of about 0.1 the theory seems rather sensitive to small changes. For the three LAO containing samples, these ratios vary from 0.110 to 0.116 to 0.115 for the 10 mol%, 20 mol% and 30 mol% LAO containing pebbles, respectively. It would therefore be necessary to assess this factor individually for all three samples, which was however not tried as the contact pressure seems so be an excellent estimator of the relevant stress.

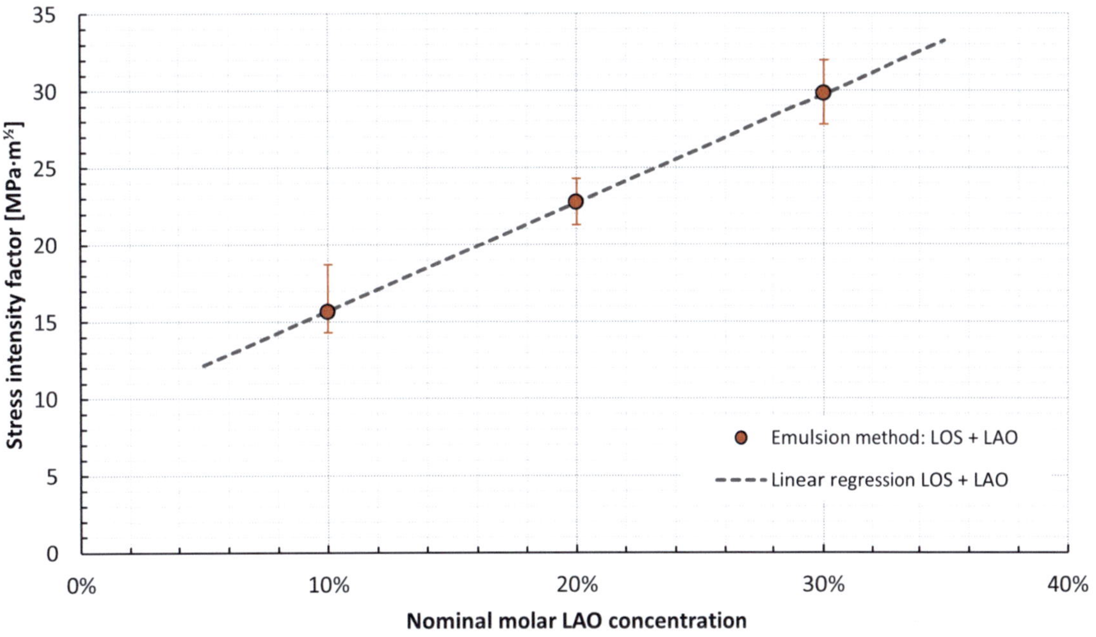

Figure 5.15: The stress intensity factor calculated from the characteristic mean contact pressure for pebbles that fail from volume defects and the approximate pore size (see equation (2.14)) as a function of the nominal molar LAO content of the EM-X-LAO pebbles. A coefficient of determination of 0.9999 is obtained for the linear regression. For the calculation of the error bars, an uncertainty of ±10 % of the approximate pore size and the one dimensional projection of the one dimensional 90 % likelihood ratio confidence boundaries have been used.

5.3 The influence of the composition on the long-term stability of the pebbles

The principal differences between the four pebble types that were annealed for up to 128 days and their implications on the mechanical strength are already detailed in the sections 5.1.1, 5.1.2 and 5.2.1. In the following the influence of the exposition to He/0.1 % H_2 atmosphere at 900 °C for this time span on the mechanical strength is discussed. The details of the annealing procedure are given in section 3.3. In Figure 4.40, Figure 4.41 and Figure 4.42, the evolution of the mechanical strength as a function of the annealing duration is shown. It is clear that the thermal treatments have a considerable influence on the strength of the pebbles, which is in most cases negative.

5.3.1 Biphasic lithium orthosilicate pebbles

For the pebbles that were fabricated by the KALOS process, one of the most apparent changes that the annealing causes is the phase transformation of cubic γ-lithium metatitanate to monoclinic β-lithium metatitanate. As Figure 4.1 shows, the phase transformation is already complete after 4 days of annealing for either type of pebbles. Going along with this transformation, the residual stresses are almost entirely relieved in lithium metatitanate after 4 days. Within the lithium orthosilicate phase, the strain of the lattice is also considerably lowered after this short period of annealing. However, the lattice remains significantly stretched compared to the literature values. Additionally, the lattice volume does not reach a constant value after 4 days of annealing as the volume is still, although less, reduced with prolonged annealing, if the value that was obtained for the LMT-richer pebbles after 4 days of annealing is neglected. After 128 days of annealing, both types of samples reach a similar decrease in strain, but just the initial drop in lattice strain can be attributed to the relieve of residual thermal stresses since the lithium metatitanate phase is then unstrained. These observations hint at (impurity) atoms which are dissolved within the lithium orthosilicate lattice and consequently stretch the lattice. The most probable element that may lead to a permanently stretched lattice of lithium orthosilicate may be aluminium as it is the most abundant impurity element in the KALOS pebbles as determined by the elemental analysis (see section 4.1). As Leys et al. [191] state, aluminium is a common impurity of the SiO_2 raw material. Nevertheless, also impurities that stem from the platinum crucible that is used during fabrication can be causative, but a quantification of the effects of impurities on the lattice volume does not exist. Compared to the data of Leys et al. [191], the pebbles that are used in this study show a significantly lower level of impurities.

In principle, also a surplus or deficit of lithium can act in the same way if a surplus of lithium can be integrated into the lithium orthosilicate lattice contrasting the published information on lithium orthosilicate (see section 2.2.2.1). As Figure 5.16 shows, there seems to be a reasonably high correlation between the lithium concentration within the pebbles and the lithium orthosilicate unit cell strain. As noted in section 5.2.2, the integration of a significant lithium surplus into the lithium orthosilicate lattice is indubitably not observed after a conventional sintering process. The melt-based fabrication of the KALOS samples, however, is significantly more prone to achieving metastable states and thus the apparent incorporation of additional lithium into the LOS lattice may be realized.

The linear regression of the KA-20-LMT samples also explains the data of the KA-30-LMT samples reasonably well, although these values by themselves would not suggest a correlation. This is also due to the fact that residual thermal strain within the lithium orthosilicate grains adds to the stretched lattice that is caused by incorporating additional atoms. This effect is clearly visible for the as-received KALOS pebbles that nominally contain 30 mol% LMT for which the remaining strain in the LOS unit cell after subtracting the respective value from the linear regression amounts to 0.24 %.

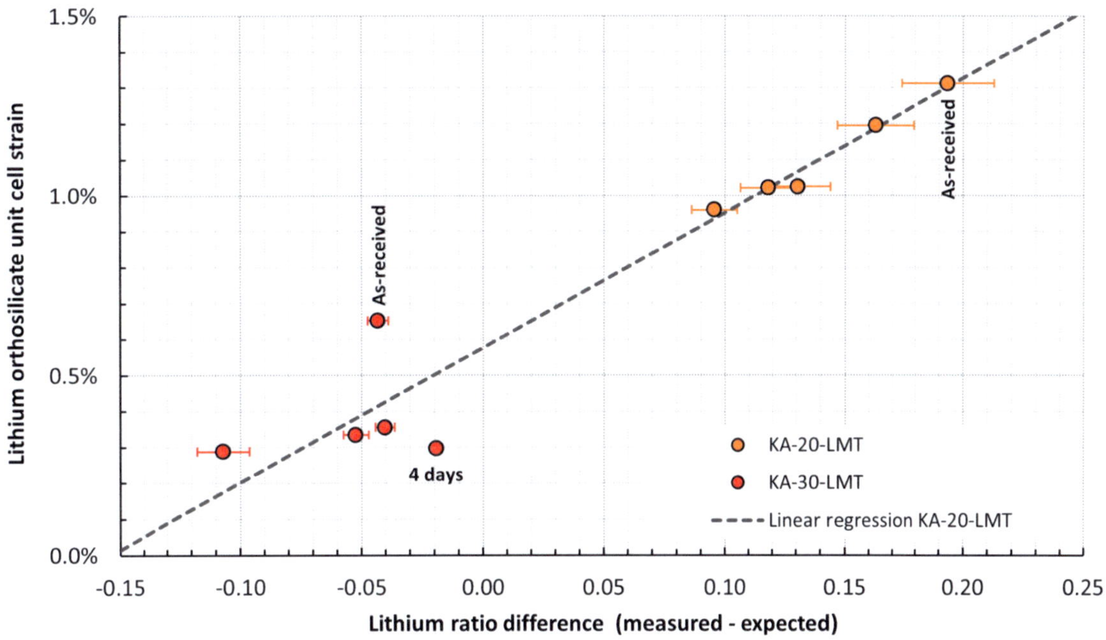

Figure 5.16: The measured strain of the lithium orthosilicate unit cell (from Figure 4.4) as a function of the difference between the measured and expected lithium content of the pebbles (from Table 4.1 and Table 4.2). The error bars in x-direction show ±10 % on the obtained value, whereas the error, which the Rietveld refinement suggests in y-direction, is too low to be displayed in this graph. The linear regression is only performed on the nominally 20 mol% LMT containing samples and shows a coefficient of determination of 0.970. The annealing duration is shown for selected data points.

This value corresponds to a stress of 261 MPa which is notably, although not excessively, higher than usually observed residual thermal stresses in alumina (about 200 MPa according to Lawn et al. [122]). In return, this also means, that the lithium orthosilicate phase of the LMT-poorer as-received KALOS sample does not show considerable thermal strain in the as-received state. An explanation for this finding is given in section 5.1.1 with the probably dominant contribution of primary lithium orthosilicate crystallites to the diffractogram over the pearlite-like domains. The data point of the KA-30-LMT pebbles that were annealed for 4 days seems to be an outlier as the lithium content of this sample is seemingly increased over the as-received state, which is implausible. From the graph it is also clear, that the y-axis intercept is not zero, which one would expect if the lithium surplus/deficit were the only effect that leads to an expanded unit cell. Presumably, the impurity atoms, i.e. aluminum, platinum and rhodium that are dissolved in lithium orthosilicate in the main are responsible for a strain of roughly 0.6 % (i.e. approximately the y-axis intercept) of the lattice. Generally, the observation that the pebbles lose some amount of lithium during annealing by vaporization is well known in literature for lithium metatitanate but also to some extend for lithium orthosilicate [192–198]. However, the work of other authors apparently disagrees with that, as they did not find significant changes of the lithium content of lithium orthosilicate pebbles during annealing [199].

The XRD analysis (see section 4.2.2 and Figure 4.4) also shows that the (in this work so-called) amorphous content, which is attributed to both phases present in the pebbles, is relatively sharply reduced after 4 days of annealing for both types of samples and then slowly continues to decrease over the annealing duration. Yet, the cross section images of the KALOS pebbles do not show any signs of a third, amorphous phase, which should be clearly visible given the expected amount of 10 to 20 mol%. Although a number of pebbles show the obviously metastable solution of titanium within the LOS lattice. It is therefore concluded that either a notably high amount of structures within the pebbles have to be small enough so that they rather contribute to the background signal of the measurement instead of generating detectable peaks or a high density of crystal defects has to be present. While the latter point cannot be verified directly with the applied methods, it is obvious from the microstructure of the pebbles (see Figure 4.24 and Figure 4.26) that there is a high amount of small structures within the pebbles in the pearlite-like eutectic domains. Consequently, the decreasing of the 'amorphous' content is, to some extent, a measure of the coarsening of the microstructure. The observed sharp drop in the overall amount of 'amorphous' material after 4 days of annealing coincides with the most notable change of the microstructure of the pebbles (see Figure 4.24 and Figure 4.26) towards coarser grains. Once a sufficient size of the grains is reached, they will not contribute to the signal of the 'amorphous' material but instead to the ordinary peaks in the diffractogram.

Although the observable coarsening is not excessively large, it may still be sufficient in many cases to pass over this boundary. Therefore, a substantial fraction of the observed drop is certainly due to the coarsening of the microstructure. Considering this, the assignment of the amorphous content to either phase is sensible, since both phases contribute to the amorphous signal and the composition of the pearlite-like domains reasonably reflects the overall composition of the pebble. Furthermore, the present amount of 'amorphous' material in the pebbles is consequently also a measure of the amount of pearlite-like structures as they indubitably show the finest structures within the samples and are in most of the cases the only structures that come into consideration for being too fine to generate detectable peaks. It is therefore reasonable that the KA-30-LMT samples show a higher 'amorphous' content of lithium orthosilicate, since, in contrast to the observed primary lithium orthosilicate in the KA-20-LMT samples, virtually all present lithium orthosilicate crystallizes in a lamellar fashion. The opposite behavior should be true for lithium metatitanate, yet no significant change of the determined amount is visible. For the nominally 20 mol% LMT containing samples, a number of pebbles most probably show the metastable incorporation of titanium into the lithium orthosilicate lattice. Yet, whether the lithium orthosilicate lattice actually prevails after the incorporation or whether it eventually crystallizes amorphously cannot be fully addressed by the XRD measurements. However, as the fraction of these pebbles is notable, the amorphous content should be also considerably increased with regard to the nominally 30 mol% containing samples, if these pebbles were fully amorphous. As this does not seem to be the case, it is certainly more likely that the lithium orthosilicate lattice remains intact.

In the as-received state, all crystal defects that were introduced by the rapid cooling during fabrication have not been healed yet. For instance thermal stresses may generate two-dimensional defects like crystal twins or low angle grain boundaries which reduce the effective grain size. Also the generation of one-dimensional defects like dislocations is highly plausible given the rapid cooling of the pebbles during fabrication. Any of these defects will reduce the amount of diffracted intensity at the actual Bragg angles and rather contribute to a diffuse background, which is interpreted as amorphous material. As mentioned in section 5.1.1, the primary crystallization of either lithium orthosilicate or lithium metatitanate has significant implications on the formation of thermal stresses within the pebbles. Therefore the amount of introduced crystal defects is different and consequentially the total amount of amorphous material also differs between the nominally 20 mol% LMT containing pebbles and the one that nominally contains 30 mol%. Since the amount of residual strain of the lithium metatitanate lattice is similar for both grades of pebbles in the as-received state, no significant effect on the determined amount of the apparently amorphous lithium metatitanate content is expected. Yet, the amount of residual thermal strain is significantly different in lithium orthosilicate and reflects similar differences in the amorphous content. Consequently, lithium orthosilicate that crystallized in the LMT-richer pebbles experienced higher thermal stress averaged over its volume, than in the case of the LMT-poorer pebbles and thus the amount of defects is very likely to be considerably higher. Which of the effects, the crystal defects or the structure size, contributes most to the observed amorphous content cannot be estimated as both contribute to the amorphous content in the same way.

Yet, these defects can annihilate during the initial 4 days of annealing, which may contribute to the observed sharp reduction in 'amorphous' lithium orthosilicate. This is similarly observed for the 'amorphous' lithium metatitanate, which additionally undergoes a phase transformation into the β-structure. A phase transformation is certainly very effective in annihilating crystal defects but also in removing residual thermal stresses, which the measurements clearly show. The effect of the structure size, especially of the LOS/LMT lamellae, is then dominant when the crystal defects are healed for the most part, i.e. most probably after 4 days of annealing. In this state, the fraction of 'amorphous' lithium metatitanate is significantly lower for the nominally 30 mol% LMT containing pebbles than for the LMT-poorer samples. This is most probably also an effect of the presence of comparably large primary lithium metatitanate dendrites, which contribute considerably more to the diffractogram as the LMT lamellae which are featured in the pearlite-like eutectic domains. For the amount of 'amorphous' lithium orthosilicate in the annealed samples, the difference between both grades of pebbles is relatively small. Yet, the amount is lower for the LMT-richer pebbles, which is somewhat unexpected if no other effects were present, but there is one observation that can explain the measured values of 'amorphous' content. The microstructure of the nominally 20 mol% LMT containing pebbles occasional shows the self-evident formation of lithium metatitanate precipitates (see Figure 4.25), which is not observable for the LMT-richer samples. Certainly, this may also be an issue of visibility, but since the precipitates were never observed for the nominally 30 mol% LMT containing pebbles, it is clear that this possibility is rather low. The formation of these precipitates requires the supersaturation of the lithium orthosilicate lattice by titanium. From the micrographs that are given in Figure 4.25, it is clear that the precipitates form along certain crystallographic planes and therefore, it is reasonable to assume that the metastable, dissolved titanium within the lithium orthosilicate lattice is already preordered along these planes. It is, from the present results, speculative whether lithium metatitanate precipitates already exist for shorter annealing durations and are simply not visible with the used preparation, which seems to be very likely. To have an effect on the diffractogram, however, there must be a change of the crystal orientation or the crystal structure.

The determined amount of 'amorphous' material suggests that the precipitates, if they are the cause for this observation, already exist after 4 days of annealing in such a form. In principle, the correlation of the lithium orthosilicate unit cell strain with the lithium surplus/deficit as shown in Figure 5.16 may also be (partially) an effect of the dissolved titanium in lithium orthosilicate. Yet, with the present results, this issue cannot be clarified. In any case, the evolution of the amorphous content as a function of annealing duration makes it clear, that apart from the rapid recovery of crystal defects that were generated by thermal stresses during fabrication, the coarsening of the microstructure is sluggish as a function of annealing duration as a considerable amount of amorphous material remains. This is in good accordance with the SEM micrographs of the cross sections (see Figure 4.24 and Figure 4.26) which show that in general a high degree of fine dispersion of both phases is maintained over 128 days of annealing, although some coarsening is observable.

It is, however, obvious that the coalescence of lithium metatitanate grains within a lithium orthosilicate matrix depends on the crystallographic orientation of both phases. While the initial lamellar structure of lithium metatitanate and lithium orthosilicate is preserved with little changes for some areas, spherical lithium metatitanate grains may have formed within the neighboring grain, i.e. former pearlite-like eutectic domain, from the lamellae. This results in a relatively inhomogeneous microstructure of the annealed pebbles. Most probably the absence of (semi-)coherent grain boundaries leads to the observed formation of spherical lithium metatitanate grains as (semi-)coherent grain boundaries generally imply notably lower interface energies compared to incoherent grain boundaries. As the reduction of the overall interface energy is achieved by minimizing the interfacial area of the lithium metatitanate grains, coarsening and the development of a near-spherical shape are inevitable consequences. For grains that share (semi-)coherent grain boundaries, the driving force, i.e. the reduction of the overall interface energy, is considerably lower. In the present case, it is evidently sufficiently low to prevent an excessive change of the morphology of a large number of grains as well as the grain size. The observation of lithium metatitanate precipitates after 128 days in the LMT-poorer samples is also a clear indication for the existence of coherent LOS-LMT grain boundaries. Yet, also for the cases when obviously incoherent grain boundaries are involved, the coarsening of the microstructure is very limited as described in section 4.3.3.

For significant coarsening to take place, a considerable diffusion flux between the respective grains of one phase is imperative. Because of the negligible solubility of lithium orthosilicate and lithium metatitanate for thermodynamic equilibrium conditions, the diffusion of lithium metatitanate through a lithium orthosilicate matrix should be relatively low (see also Knitter et al. [73]). As a result, the coarsening is in general not expected to be fast, even when the driving force for coarsening is high by the standards of this system. In principal, the coarsening of the microstructure as a whole is linked to the coarsening of the lithium metatitanate grains while the coarsening of the lithium orthosilicate matrix grains, which are significantly larger than the lithium metatitanate grains, is less readily observable. For the case of the amorphous content, as it is detected by the XRD measurements, the size of the lithium orthosilicate grains is not relevant but rather the spacing between the lithium metatitanate structures. This spacing is certainly a function of the coalescence of the lithium metatitanate grains. Yet, also the lithium orthosilicate grains, or for that matter the boundaries of the pearlite-like eutectic domains, grow as a function of annealing time as the microstructure images clearly show (see Figure 4.24 and Figure 4.26). Of course, for pebbles whose microstructure is dominated by lithium metatitanate dendrites, no grain growth of lithium orthosilicate grains is visible. For these pebbles, the coarsening of the grains seems to progress generally inconsiderably, most probably because the grain sizes of both phases are already large enough after fabrication so that the driving force for coalescence is too low for notable grain growth. In case of the lithium orthosilicate grains, regardless of the kind of microstructure and nominal composition, the grain growth seems clearly hampered by the presence of the finely dispersed lithium metatitanate. This is illustrated by the comparably rapid grain growth that pebbles, that lack a high number of lithium metatitanate grains, show. This is the case for the pebbles in which titanium is rather dissolved within the lithium orthosilicate lattice instead of forming separate lithium metatitanate grains (see Figure 4.25).

Evolution of the mechanical strength of the nominally 20 mol% LMT containing pebbles

With respect to the mechanical strength of the pebbles, especially for the nominally 20 mol% LMT containing ones, the 'amorphous' content of the pebbles apparently approaches an asymptotic value with increasing annealing duration as well and equally shows a relatively strong change after the first 4 days of annealing (see Figure 4.40 and Figure 4.41 for the mechanical strength). However, the 'amorphous' content of the pebbles does not seem to be dominantly affecting the mechanical strength as both quantities do not highly correlate. For the nominally 20 mol% LMT containing pebbles, Figure 4.44 and Figure 4.45 detail the development of the Weibull parameters of both pebble sizes that were investigated. Both figures show a similar behavior with a mildly lowered Weibull modulus as a result of the annealing and a significantly reduced characteristic strength. The relative constancy of the Weibull modulus suggests that the failure of the pebbles occurs by the same class of defects, and as already discussed in section 5.1.1, surface defects are causative for the failure. The micrographs of the surface of the pebbles (see Figure 4.12) as well as of the cross sections (see Figure 4.24) show that these cracks have grown considerably during the first 4 days of annealing, which in return reduces the critical stress for failure as explained in section 2.4. A similar correlation of the open porosity with the characteristic strength of the pebbles as for the biphasic emulsion method pebbles (see Figure 5.12) is not obtained for the KALOS pebbles.

This is already clear from the observed changes after 4 days of annealing. While the mechanical strength of the pebbles decreases as a result of significantly larger surface cracks, the open porosity of the samples decreases, although the value of open porosity should increase with increasing crack length (see Figure 4.37). Obviously, during the first 4 days of annealing a notable portion of the initially open porosity is transformed into closed pores as the simultaneous increase in closed porosity (see Figure 4.36) suggests. Yet, also the total porosity is considerably reduced during these 4 days of annealing (see Table C.15), which accords with the observed surface morphology of the pebbles that is visibly smoothed and lost the pearlite-like structures in exchange for spheroid lithium metatitanate grains that decorate the surface of lithium orthosilicate grains and their grain boundaries. This evolution of the surface morphology and the comparably high speed of transformation is, however, an effect of surface diffusion. As the diffusion coefficients of surface diffusion are by experience roughly two orders of magnitude higher than bulk diffusion, the smoothing of the pebble surface also quickly reaches a state at which no significant changes are visible. This is clearly visible for of both types of KALOS pebbles as the morphology after 4 days of annealing is virtually identical to the morphology that is obtained after 128 days of annealing. As a result, some form of bridge that at least punctually spans a pore is needed in order to close this pore via surface diffusion. Obviously, this is much more probable for small cracks and pores than for larger gaping cracks. In consequence, the smoothing and the transformation of open porosity into closed porosity by overgrowing small pores do not affect the mechanical strength if only surface defects of insignificant size are concerned. Yet, the smoothing by surface diffusion will also reduce the sharpness of the large crack tips and thus decrease the stress intensity in front of the cracks. Because of these reasons, the crack size cannot be estimated by the overall open porosity value as for the biphasic emulsion method pebbles.

Concerning the decrease of the characteristic strength of the pebbles, the growth of the critical cracks is consequentially not advanced by surface diffusion effects. The micrographs that are depicted in Figure 4.24 suggest that the largest cracks grow between domains of pearlite-like structure after 4 days of annealing, however not between all of them but between a few of them. It must therefore be assumed that significant stresses develop between these domains. One plausible explanation for the generation of these stresses is the phase transformation of γ- to β-lithium metatitanate. The transformation of the metastable γ-phase begins at comparably low temperatures of about 775 °C (see Figure 4.11) when the residual stresses within lithium orthosilicate are presumably not dissipated by creep or other processes. As the compressive stresses within lithium metatitanate are relieved during the phase transformation, the counteracting tensile stresses within lithium orthosilicate are consequentially released as well, which effectively results in a significant contraction of the domain in direction of the lamellae and a probably insignificant strain perpendicular to the direction of the lamellae. The temperature varied measurements of the as-received KA-30-LMT sample in section 4.2.3 supports this reasoning as the observed kink most probably represents the explained contraction of lithium orthosilicate. Of course, the reduced expansion of the unit cell volume is smaller than the strain that is eventually observed at room temperature, because the thermal expansion effectively reduces the amount of strain from the thermal expansion mismatch. Yet, the overall contraction of the LOS unit cell after cooling to room temperature is very close to the contraction that is observed for the annealed samples (see section 4.2.2).

The so generated contraction can lead to highly anisotropic strains which in return can certainly promote the growth of large cracks when two or more interfacing domains are unfavorably oriented, especially as a number of pores can be found at these boundaries. It is furthermore clear that this adjustment is a one-time occurrence and consequently, the crack extension is not increased by this effect with prolonged annealing. This perfectly agrees with the general observations of the LMT-poorer pebbles, independent of their size, in Figure 4.40 and Figure 4.41 that show stable mean contact pressure values for all annealing durations. Also the micrographs in Figure 4.12 and Figure 4.24 show no apparent increase in crack length as a function of annealing duration. Likewise the determined characteristic strength of the samples does not decrease significantly after the first 4 days of annealing. The differences in the characteristic strength between smaller and larger pebbles can more or less be explained by the size effect according to equation (5.2), although especially the Weibull moduli notably differ for most of the samples. Yet, a statistically attestable accordance can only be found for the as-received samples (see Figure D.2 in Appendix D). Presumably the effect of the contracting domains on the growth of cracks indirectly depends on the pebble size, as the in total more rapid cooling of the smaller pebbles leads to in average smaller domains, and consequentially for larger pebbles the cracks grow measurably larger in relative terms. Once the large cracks have formed, they should be somewhat mitigated by surface diffusion as already explained and as a result, the characteristic strength of the pebbles should increase, although probably not much.

This anticipated small increase in strength is counteracted by the thermal cycles that are involved in the annealing experiment during sampling. During each cooling from 900 °C to room temperature and consecutive heating-up of the samples, the pebbles are exposed to stresses that arise from the significant thermal expansion mismatch of lithium orthosilicate and β-lithium metatitanate, which is maximal around 700 °C (see Figure 4.10). But also the high anisotropy of the thermal expansion of lithium orthosilicate (see Figure 4.8) may generate stresses upon cooling that lead to the growth of existing cracks. Yet, these effects are clearly of subordinate importance as the biphasic LOS+LMT emulsion method samples do not suffer notably from this issue during cooling to room temperature after sintering (see section 5.1.1). Again the model of Grünberg [129] explains the involved tangential thermal stresses if one assumes that the cooling is fast enough compared to the thermal conduction of the pebble material. Because of the strong differences of lithium orthosilicate and lithium metatitanate in thermal expansion, also non-compressive radial stresses can be expected which should lead to a considerable weakening of the grain boundaries and thus facilitate the propagation of cracks along the grain boundaries.

Yet, obviously these processes do not dominate the major trend of the evolution of the mechanical strength of the pebbles. The generally visible discontinuity of the Weibull parameters of the annealed, nominally 20 mol% LMT containing samples of either size may, however, reflect another effect, although it is only clearly visible for the pebbles of 1000 μm diameter that were annealed for 32 days and 64 days. For both samples a number of data points deviate from the Weibull line and form an upward kink in the Weibull plot for low strengths (see Figure 4.46), similar to the as-received LMT-richer KALOS samples, although by far not as substantially. Also in the case of the nominally 20 mol% LMT containing pebbles, this observation is unambiguously linked to the presence of 'quasi-continuous failures', which are more common than it is apparent from the Weibull plots. The microstructure that is predestined to cause quasi-continuous failures, i.e. the pebble shows a high amount of large defects so that it disintegrates bit by bit, is that of the pebbles that show the incorporation of titanium into the lithium orthosilicate lattice (see Figure 4.25). The reason for the difficulties to identify these defects from the Weibull plot arises from the, in this case, relatively good blending of this failure mode into the Weibull plot of the classical failure from surface defects. As discussed in section 5.1.1 such failures can be regarded as unique failure mode of one subtype of pebbles with considerable justification. However, depending on the similarity of the Weibull parameters, both failure modes may be hardly distinguishable when both failure modes have an approximately equal chance of appearance. If the Weibull parameters of this failure mode change during annealing, these changes will impact the Weibull plot as a whole. Changes of the failure behavior of the basically single-phase pebbles are highly plausible when their cross section images are considered (see Figure 4.25). Already after 4 days of annealing a considerable amount of cracks forms between a number of grains. With increasing annealing duration the mean pore size of these pebbles increases as well. It is assumed that the moving grain boundaries purge the pebble from small pores as they are collected at it.

As a result between two interfacing grains a crack develops naturally. When the pores grow to a substantial size they may separate from the grain boundary and remain within the grain [200]. A similar effect should be observable for the lithium metatitanate grains which are either already present or which have precipitated and formed spherical grains during annealing. Both processes, the collection of pores and lithium metatitanate grains by the lithium orthosilicate grain boundary, have a clearly weakening effect. In conjunction with the increased grain size, which aggravates the influence of thermal expansion anisotropy, increasingly large crack networks develop within these pebbles as a function of annealing duration. As a result, the Weibull modulus and characteristic strength that are determined for the overall sample may be altered and therefore at a first glance unsystematic fitfulness of the determined Weibull parameters is observed. Principally the collection of pores at moving grain boundaries can also decrease the strength of the pebbles that show a microstructure with finely dispersed lithium metatitanate. Yet, this effect is far less obvious in these pebbles, probably because of the obstructed movement of the grain boundary. It therefore has to be of subordinate significance.

Evolution of the strength of the nominally 30 mol% LMT containing pebbles

For the nominally 30 mol% LMT containing pebbles, the mechanical strength develops considerably different than that of the nominally 20 mol% LMT containing ones. This is most visible for the pebbles that measure 1000 μm in diameter, as the failure behavior of the majority of these samples cannot be described satisfyingly by a single Weibull distribution. Yet, as described in section 5.1.1 and 5.2.1 the two sizes of the as-received samples of this type of pebbles behave in principle equally, although the fraction of the three identified failure types differ perceivably. For both sizes of the nominally 30 mol% LMT containing pebbles, the high number of apparent failures vanishes after annealing for 4 days, which is the exclusive reason for the observed increase in strength (see Figure 4.41). As already discussed, the presence of significant residual thermal stresses enables the quasi-plastic behavior of the pebbles, which most probably causes this failure mode. Once the residual thermal stresses are near zero, intragranular fracture may occur instead, according to Cai et al. [186]. Since these stresses are eliminated from the pebbles after 4 days of annealing, also the quasi-plastic 'apparent failure' mode should and does disappear in consequence. For the 1000 μm pebbles that are annealed for 4 days and 32 days, the principal failure behavior of the as-received pebbles is preserved without the appearance of a significant number of 'apparent failures'.

In effect, this means that respective probability p_1 or p_2 in equation (5.1) is zero and therefore the classical case of two non-competing failure modes is established as described by Abernethy [137]. In Figure 5.17 a comparison between the experimental values and the model is shown and it is clearly visible that the model represents the experimental data very well. In contrast to the Weibull parameters that were attributed to the volume defect failure of pebbles that basically exclusively show pearlite-like eutectic domains instead of primary lithium metatitanate dendrites (see Figure 5.4), the characteristic strength had to be lowered slightly to represent the experiment reasonably well. For the samples that were annealed for 4 days instead of 32 days, there was no need to change the characteristic strength (see Figure D.3). In consequence the size of the critical cracks does not change significantly after annealing for 32 days, which fits well to the images of the pebble cross sections (see Figure 4.26). Also with respect to the drawing probability there are moderate differences between these samples.

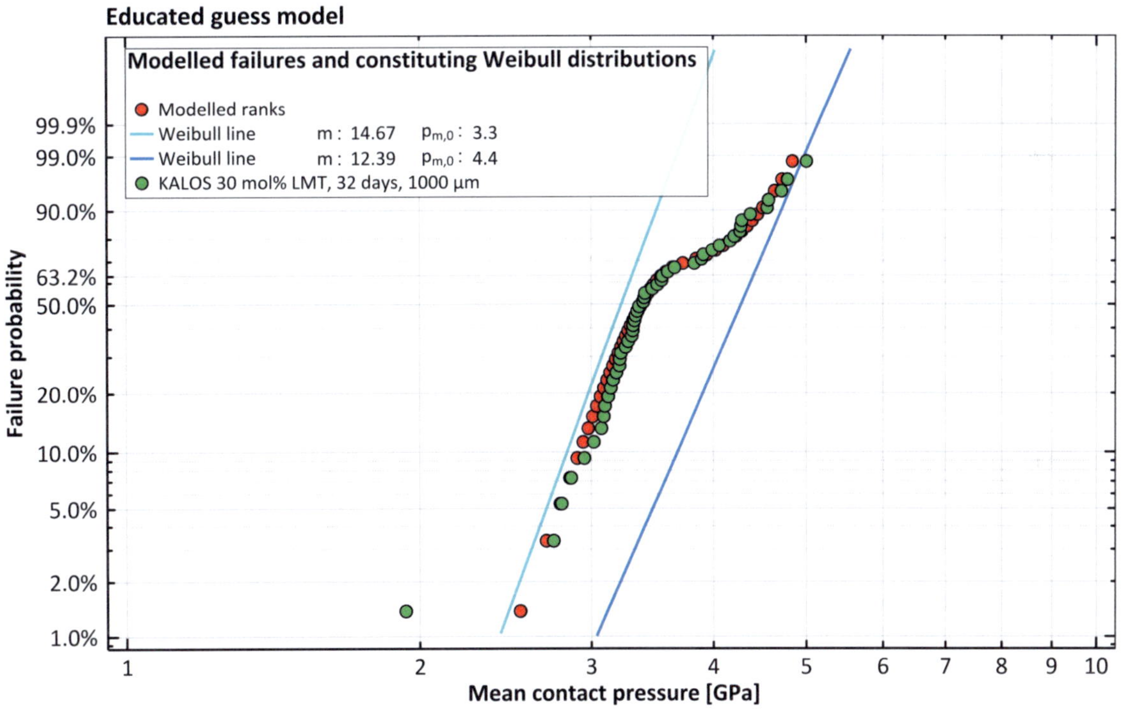

Figure 5.17: Modelled failure behavior for two non-competing failure modes with Weibull moduli of 14.67 and 12.39. The probability of testing a sample of latter failure mechanism is 35 %. The model is compared to the experimental data of nominally 30 mol% LMT containing KALOS pebbles of 1000 µm diameter that were annealed for 32 days.

About 75 % of the pebbles that were annealed for 4 days apparently fail by volume defects and feature a mainly pearlite-like microstructure or in other terms the microstructure of about 25 % of the pebbles is dominated by primary lithium metatitanate dendrites and fails from surface defects (again see Figure D.3). The latter value equals that of the as-received pebbles. For the samples that were annealed for 32 days, this value changes to about 35 %. Yet, the microstructure of the pebbles cannot change from being dominated by pearlite-like eutectic to being dominated by primary lithium metatitanate dendrites. In consequence the drawing probability should be relatively equal. The high similarity of the failure behavior of both types of pebbles, i.e. annealed for 4 and 32 days, is not only visible from Figure 4.41 and Figure 4.49, but also the microstructure has changed insignificantly during the annealing. Yet, the contraction of the lithium orthosilicate lamellae upon the γ to β phase transformation in lithium metatitanate, as discussed for the LMT poorer pebbles will also generate stresses that, in principle, can also lead to the growth of the cracks within the volume of the pebbles.

The growth of these cracks may, however, require significantly higher stresses to achieve the same extension as the interfaces between two domains are comparably intact with respect to the LMT-poorer samples and such stresses may not act significantly on the existing volume defects to extend them as the involved domains have to be oriented favorably. As a result, the strength of LMT-richer pebbles that show a pearlite-like microstructure does not degrade notably after 4 days of annealing in contrast to the LMT-poorer pebbles. The high-strength failure mechanism that is attributed to pebbles with a microstructure that is dominated by primary lithium metatitanate dendrites remains remarkably stationary during annealing. As the microstructure of the respective pebbles is also not changed perceivably, this observation is in good accordance with the mechanical data as well.

The Weibull plot of the 1000 μm pebbles, which nominally contain 30 mol% LMT, changes after annealing for 64 days. After annealing for this long, the fracture behavior is no longer reasonably described by two Weibull distributions but rather by one single Weibull distribution. The cross section images of Figure 4.26 clearly show that the microstructure of both types of pebbles is still significantly different and thus two Weibull distributions are still expected to be determinant for the overall fracture behavior. Yet, they are not clearly separable anymore. From the evolution of the surface morphology as well as from the cross section images (see Figure 4.13 and Figure 4.26), it is obvious that a significant amount of surface defects form in both types of pebbles, but significantly more severe in the ones that feature a mainly pearlite-like microstructure. As a consequence, the formerly volume defect dominated failure of the pebbles changes, as the failure from surface defects becomes significant, besides the still possible failure from volume defects for the pebbles with pearlite-like microstructure. An example for such a failure behavior is observable for the 10 mol% LAO containing emulsion method pebbles (see Figure 5.14). Yet, an unambiguous Weibull plot is logically not obtained as the failure of the pebbles whose microstructure is dominated by lithium metatitanate dendrites are also part of the compound failure probability distribution. As relatively few surface defects were perceivable in the pebbles that are dominated by lithium metatitanate dendrites and were annealed for up to 32 days, the appearance of some surface defects of considerable length will have a significant detrimental effect on the characteristic strength of the samples. As a result, the mean strength of the 1000 μm pebbles is significantly reduced after 64 days and even more after 128 days of annealing.

In principle, the slight deviations of the experimental data of the 1000 μm pebbles, that were annealed for 64 days and 128 days, from ideally linear behavior can still be modelled on the basis of two non-competing failure mechanisms, i.e. it is assumed that the contribution to failure by volume defects is insignificant (see Figure 5.18 and Figure D.4 in Appendix D). These models generally involve a substantial reduction of the characteristic strength of the samples that feature a lithium metatitanate dendrite dominated microstructure. In addition to that, also a significantly reduced Weibull modulus for the failure mechanism is assigned to the pebbles that are dominated by pearlite-like eutectic domains. These results are in good agreement with the observed microstructure of these pebbles. The resultant Weibull plot shows a slight S-shape, similar to the experimental data. If the low strength run-out values in the Weibull plots of the pebbles that were annealed for shorter times are also considered as meaningful values (see Figure 5.17 as one example), the S-shape of the Weibull plots is already foreshadowed by these samples.

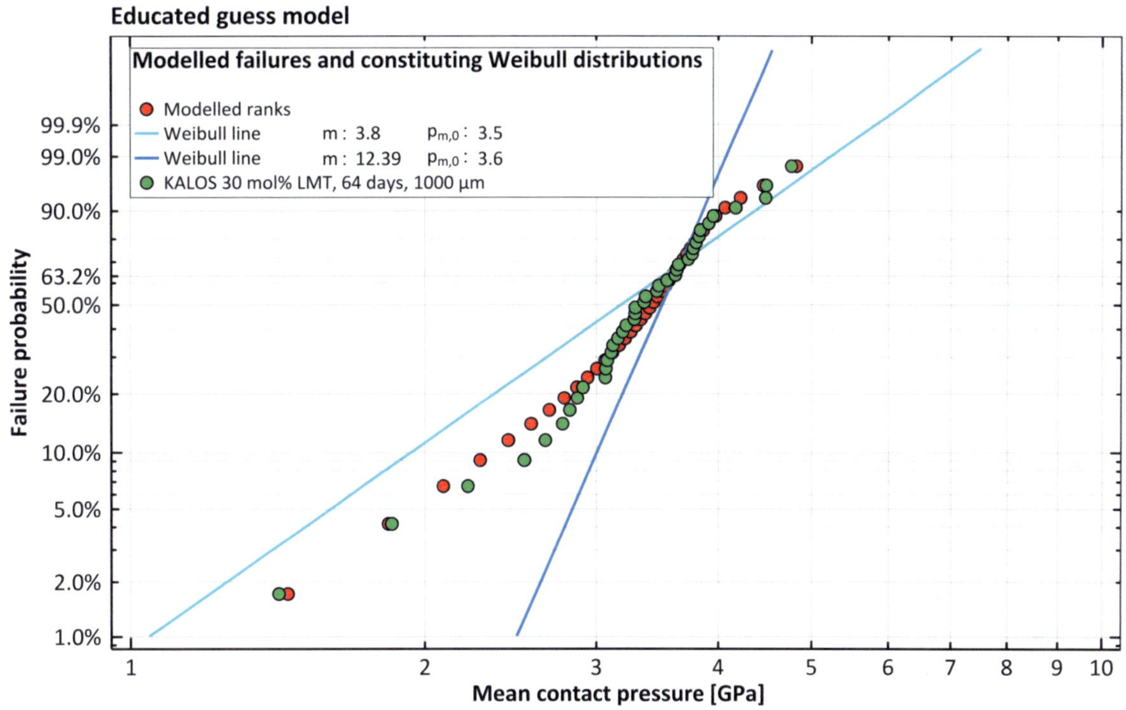

Figure 5.18: Modelled failure behavior for two non-competing failure modes with Weibull moduli of 3.8 and 12.39. The probability of testing a sample of latter failure mechanism is 50 %. The model is compared to the experimental data of nominally 30 mol% LMT containing KALOS pebbles of 1000 µm diameter that were annealed for 64 days.

This means in return that the failure by surface defects of pebbles that feature a microstructure, which is dominated by eutectic domains, is also taking place for pebbles that were annealed for short durations, yet to an exiguous extent. But this also shows that the more appropriate model would be, as Abernethy [137] calls it, 'Competing risk with a batch problem', since it comprises two simultaneously active failure modes in one subsample, i.e. the pebbles that are dominated by eutectic domains in this case, plus one non-competing failure mode in another subsample, i.e. the pebbles that are dominated by the lithium metatitanate dendrites. Because the deviations of the experimental data of the 1000 µm pebbles that were annealed for 64 days and 128 days from an ideal Weibull line are also statistically insignificant, the application of this model involves an unsatisfyingly high degree of arbitrariness and is thus not performed. Still, in consequence the description of the experimental data by a single Weibull line is most probably not exhaustive and a significant degree of discontinuity of the Weibull parameters is thus expected as a function of annealing duration when the parameters of the constituting Weibull distributions change. The microstructural changes, i.e. the formation and growth of surface cracks, which are reflected by the changing Weibull parameters, probably originate from the implicit thermal cycling during the sampling of the experiment. As already explained for the LMT-poorer pebbles, the thermal expansion mismatch between lithium orthosilicate and lithium metatitanate is substantial and thus the resultant stresses certainly as well. Obviously, tangential thermal stresses cause the generation of the cracks during the cooling of the pebbles since otherwise no cracks that initiate from the surface can be expected. It is not expected that the generated stresses are, in qualitative terms, significantly different from the stresses that are generated within the LMT-poorer pebbles during the thermal cycling that is effectively performed for sampling.

Yet, the already large surface cracks that are present within the LMT-poorer pebbles forestall the generation of large stresses, which results in considerably less severe degradation of the mechanical strength of the pebbles for longer annealing durations. Since both sets of samples are annealed without external stresses and since intrinsic stresses should be relieved at high temperatures, a subcritical crack growth is highly unlikely because the presence of constant or cyclic mechanical stress is required to extend the existing cracks [132]. As furthermore detailed for the LMT-poorer pebbles which fail quasi-continuously, significant grain growth may aggravate the stresses from thermal expansion anisotropy and thus may finally lead to the formation of the observed cracks. Yet, it is very difficult to determine the grain size of the lithium orthosilicate matrix of the KA-30-LMT pebbles and as a result, it is not possible to verify a possible significant grain growth after 64 days of annealing compared to less annealed pebbles.

For the smaller 500 µm pebbles of the LMT-richer sample, the evolution of the mechanical strength mimics that of the 1000 µm pebbles of the same grade (see Figure 4.40 and Figure 4.41). Yet, apart from the as-received state, which can easily compared to the larger 1000 µm pebbles as shown in section 5.2.1, the similarities are mostly not obvious. Generally one Weibull line sufficiently describes the failure behavior of the annealed pebbles. The so-identified failure mechanisms cannot be separated from each other with statistical significance as Figure 4.47 shows. Yet, beside the difference in size, the pebbles of the LMT-richer grade do not differ perceivably. Furthermore, the pebbles show a very similar failure behavior in the as-received state. It is therefore implausible that the annealed 500 µm pebbles should behave generally different than the 1000 µm pebbles and consequently the same failure mechanisms should dictate the mechanical behavior of the pebbles. Yet, it is certainly possible that the formation of surface defects is affecting the smaller pebbles more and after shorter annealing durations than the larger pebbles.

Therefore the failure from surface defects of pebbles with a microstructure that is dominated by pearlite-like eutectic domains can be notably more probable than for the larger pebbles for low annealing durations. Since the larger pebbles also show some indications of this failure mechanism for low annealing durations, this is the only reasonable explanation of the failure behavior of these 500 µm pebbles. In consequence the failure behavior may, similarly to the 1000 µm pebbles that were annealed for

64 and 128 days, appear as being governed by a single failure mechanism, which is in the end probably just a valid overall description of the failure probability as a function of the applied contact pressure. It is therefore, and as explained in the preceding paragraph when the discontinuity of the Weibull parameters is addressed, difficult to explain the changes of the determined Weibull parameters in the annealed state. That the small pebbles probably fail from more than one failure mechanism is supported by the SEM images of the fragments of annealed pebbles (see Figure 4.61).

In contrast to some ambiguity in the description of the mechanical behavior of the 500 µm pebbles in the annealed state, the transition from the as-received to the annealed state shows a clear picture. The annealing leads to an overall increase of the mechanical strength of the 500 µm pebbles, because the apparent failures disappear. The disappearing is certainly caused by the same reasons as for the 1000 µm pebbles. As Figure 4.47 shows, the Weibull moduli of the annealed 500 µm samples are very similar. Apparently, these pebbles reach the same state that the 1000 µm pebbles reach after 64 days of annealing already after 4 days of annealing and, similar to the large pebbles, remain in this state for the longer annealing durations.

5.3.2 Lithium excessive lithium metatitanate pebbles

In section 5.1.2 the properties of the as-received lithium excessive lithium metatitanate pebbles are discussed. From the results that were presented in section 4, it is evident that the annealing of these pebbles highly affects the pebble properties. Like the biphasic KALOS pebbles, the lithium excessive lithium metatitanate pebbles lose a portion of their lithium content during the annealing, which is clearly demonstrated by the ICP-OES results in Table 4.3 and Table 4.4. These losses of lithium by evaporation over time at elevated temperatures are not surprising as several authors have described this observation in the past [192–198]. The effects of these losses on the unit cell are generally mild as the XRD results show (see Figure 4.5). While the volume of the unit cell varies without visibly correlating with the observed lithium losses, the individual cell parameters show counteracting correlations as the cell expands in [100] direction with increasing surplus of lithium and simultaneously contracts in [010] direction as well as to a much lower extend also in [001] direction (see also Figure D.5 in Appendix D). As lithium ions fill the center space of each hexagonal Ti-O ring in the stoichiometric structure (see Figure 2.7) this hints at a modification of the ring structure in case of lithium non-stoichiometry.

Figure 5.19 exemplarily details the effect of the lithium content on the strain of the lithium metatitanate unit cell in [100] direction. In this case, the regression straight of the SG-LMT samples almost crosses the origin and is thus close to ideal behavior. Yet, this is not indicative for the other lattice parameters. The significant amount of impurity elements that both samples feature, especially the SG-LMT samples, certainly complicate the results and therefore the comparison of the absolute values is difficult. Yet, the overall trend that the lithium content primarily governs the observable strain of the unit cell is certainly represented reasonably well. As a function of annealing duration it is also clearly observable from the Rietveld analysis of the diffractograms that the preferred orientation of (001)-planes parallel to the sample surface increases substantially.

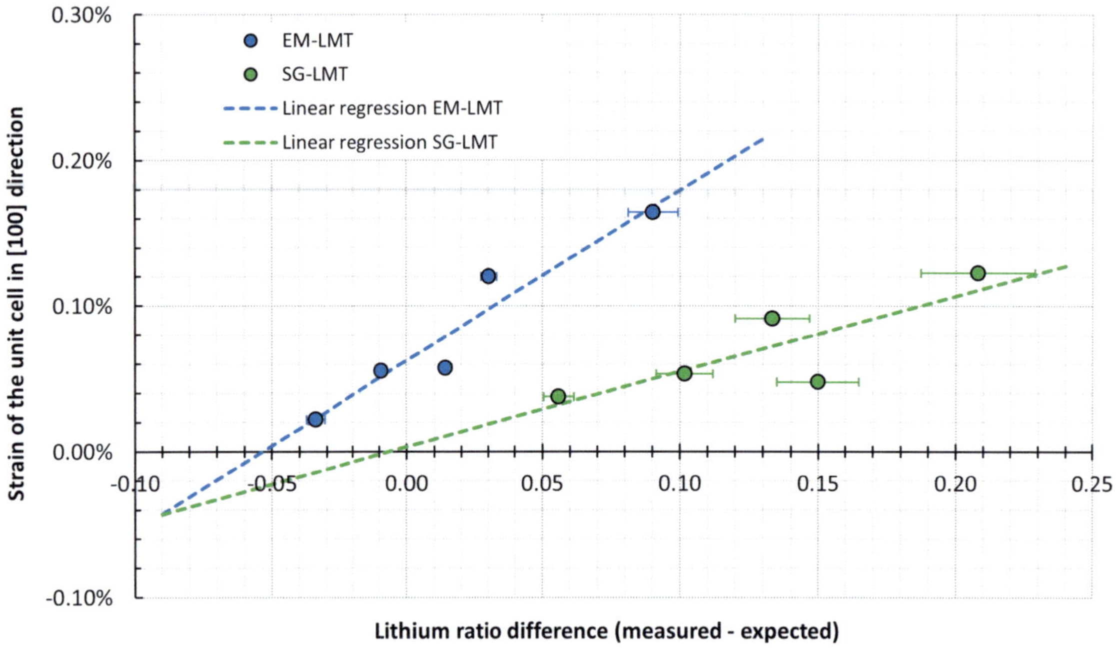

Figure 5.19: Strain of the β-lithium metatitanate unit cell in [100] direction as a function of the difference between the measured and expected (stoichiometric) lithium content. The error bars in x-direction show ±10 % on the obtained value, whereas the error that the refinement suggests is too low to be displayed in this graph. The linear regression shows a coefficient of determination of 0.924 and 0.676 for the EM-LMT and SG-LMT samples, respectively.

The degree of preferred orientation as given by equation (3.8) can reach relatively high values of about 35 % after 128 days of annealing, independent of the fabrication method. The value of preferred orientation correlates very well with the determined grain size for either grade of pebbles by a linear equation (see Figure D.6 in Appendix D). From the microstructure of the samples (see Figure 4.28 and Figure 4.29) it is observable that the grains are predominantly equiaxed while just a fraction of the grains appears to be elongated. The existence of slightly elongated grains is, furthermore, only apparent for long annealing durations of 64 days and 128 days. Consequently there would be no reason for the grains to align along a certain crystallographic direction. Yet, the samples are milled considerably during the preparation (see section 3.5.1).

It is therefore concluded that for one the (001)-planes are the glide planes for β-lithium metatitanate for which cleavage (during the preparatory milling) and plastic deformation are most preferential. The preferential milling of the grains along these planes generates platelets which are, because of their geometry, susceptible to preferred orientation. Moreover, it is logical that a larger initial grain size will lead to platelets with higher aspect ratios and thus to a higher tendency for preferred orientation. Secondly, the results also suggest that the initial grains, in accordance with the cross section images, have to have a relatively equiaxed shape since otherwise a high correlation with the determined grain size would not be expectable.

The considerable grain growth that both grades of pebbles show, is apparently facilitated by residual lithium carbonate as Hoshino details [97]. Although a clear verification of lithium carbonate for these pebbles was not possible for the examined pebble samples as neither the XRD analysis nor the SEM analysis of the pebbles' surface (see Figure 4.15 and Figure 4.16) showed signs of lithium carbonate impurities, it cannot be excluded that traces of lithium carbonate are present. However, some lithium orthotitanate was observed in the XRD analysis (see section 4.2.1) which might have a similar effect on the grain growth as lithium carbonate, but whether or not the growth of the lithium metatitanate grains is increased by an impurity phase remains unclear. The increase of the grain size as a function of annealing duration for the case of unobstructed, uniform, normal grain growth has been addressed by Mullins [201]. He detailed that the square of the grain size d at a given time t is proportional to the time by

$$d(t)^2 - d(0)^2 = Bt \tag{5.3}$$

with B as constant for the respective system. If the grain size in the as-received state is taken as grain size (0), the increase as a function of annealing duration follows the prediction of Mullins [201] very well as Figure 5.20 shows. This means that an overwhelming amount of the grains in both samples grows normally and that in consequence a homogenous microstructure should be expectable for all annealed samples, which excellently complies with the cross section images. If the observation of elongated grains is an indication of abnormal grain growth, this effect has not had a notable impact on the overall grain growth during the annealing for 128 days. Furthermore, the lithium surplus or deficit seems to have no significant effect on the grain growth. From Figure 5.20 it is also obvious that the evolution of the grain size for both grades of samples is highly similar. If an impurity phase contributed to the growth of the grains it has to be present in both grades of samples by a similar amount, which appears to be not too likely. It is therefore assumed that the growth of the lithium metatitanate grains in these two grades of pebbles is not or insignificantly enhanced by an impurity phase.

The annealing of the lithium excessive pebbles not only leads to a considerable grain growth but also to significant changes of the porosity of the samples as Figure 4.36 and Figure 4.37 show. Surprisingly, the principal development of the porosity of these pebbles as a function of annealing duration does not differ very much from the behavior of the KALOS pebbles, despite their considerable differences in microstructure and composition. The differences in the development of the porosity between these two groups of samples are mostly found in the magnitude of the porosity. The lithium metatitanate samples generally show a higher amount of closed pores, which is reasonable as sintering processes are generally prone to incomplete densification especially if the green density of the pebbles is relatively low as it must be assumed from the results of the biphasic emulsion method pebbles (see Figure 4.22). Since the sol-gel fabrication obviously leads to a higher closed porosity than that of the emulsion method pebbles, the same argument applies.

After 4 days of annealing, the closed porosity of both grades of lithium excessive pebbles significantly increases. For the EM-LMT samples, this can unambiguously be attributed to the closing of a fraction of initially open pores as the SEM images of the pebble surface clearly show (see Figure 4.15). This process is principally identical to the one which was described for the KALOS pebbles. Although some pores of the pebbles lose their access to the surrounding atmosphere, the total porosity of the sample is reduced (see Table C.15). This trend also holds for longer annealing durations and is mainly due to a reduction of the closed porosity with increasing annealing duration. The determined values of the sum of the open and imaginary porosity increase again after 32 days of annealing, which was also observed for the KALOS pebbles and is certainly caused by the same processes. For longer annealing durations, the open porosity value does not change significantly. As the reduction of the porosity equals an increase in pebble density, the pebbles must consequently decrease in size, which is clearly observable from the mean pebble diameter as detailed in Table 4.6.

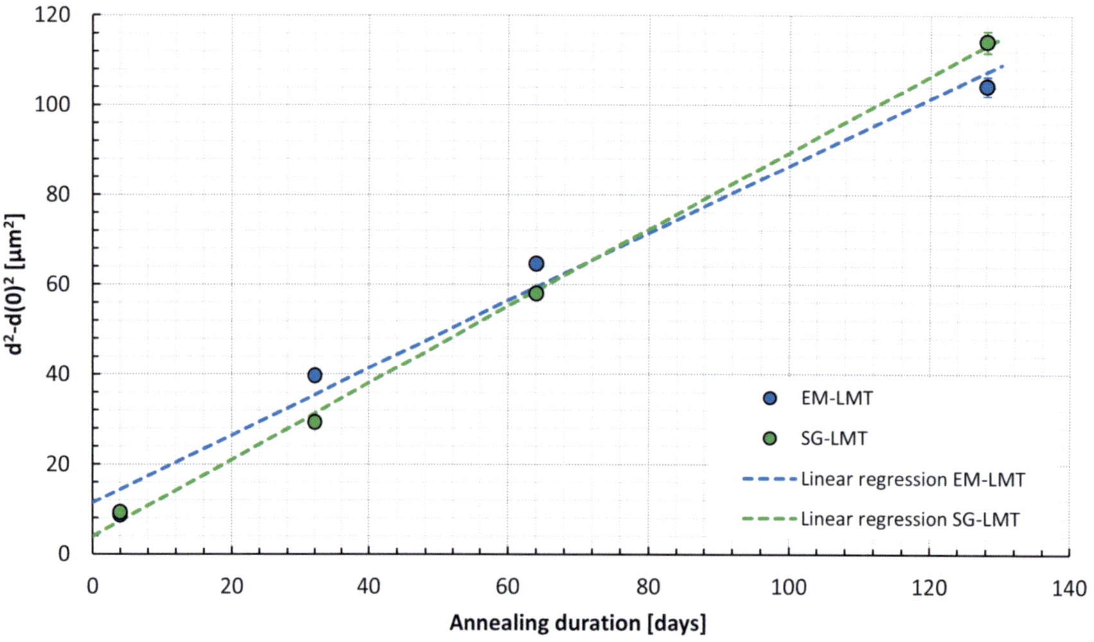

Figure 5.20: The squared grain size of the annealed lithium excessive pebbles is given as a function of the annealing duration according to equation (5.3). The linear regression shows a coefficient of determination of 0.982 and 0.999 for the EM-LMT and the SG-LMT samples, respectively. The estimated uncertainty of the grain size measurements is too low to be displayed in this plot.

For the SG-LMT samples, the closed porosity similarly increases after 4 days of annealing, although not as much as for the EM-LMT pebbles, but this increase is not accompanied by a decrease of the optically determined open porosity, which results in an increase of the total porosity of the sample. This increase in porosity accords with a slight increase in pebble size (see Table 4.6). Also the surface morphology (see Figure 4.16) does not suggest a reduction of the open porosity. The SEM images of the surfaces of the as-received pebbles show an already rather smooth surface which provides substantially less potential for the overgrowth of open pores as the surface of the EM-LMT samples. In effect, rather an increase in the open porosity is suggested by the surface morphology of the pebbles that were annealed for 4 days, as a large number of small dents in the surface has formed. Yet, the clearly visible cracks in the surface of the as-received SG-LMT pebbles, which constitute the majority of open pores for this sample, disappear after 4 days of annealing. Both processes are clearly counteracting each other in their effect on the determined open porosity value. Principally it may be possible that the dents have formed from the closed cracks, yet this hypothesis cannot be proven on the basis of the performed experiments. Nevertheless the results show that it is reasonable that the open porosity does not change much after 4 days of annealing. Yet, as little open porosity is transformed into closed porosity, the observed increase of the closed porosity has to originate from a different effect.

The cross section images of the SG-LMT pebbles most probably reveal the reason for this increase, namely the formation of a thick crust on some pebbles which occasionally detaches from the core of the pebble and thus forms large cavities between both parts (see Figure 4.29). The mechanism by which this crust forms is unclear. The crust itself appears in darker color in the SEM images, which suggests that it is composed of lighter atoms. As the XRD analysis (see section 4.2.1) suggests that small amounts of lithium orthotitanate are present in these pebbles, the crust may actually be composed by this phase. If that is the case, the majority of the lithium excess of these samples is not stored within the lattice of lithium metatitanate but in the crust of the pebbles. As lithium orthotitanate shows a significantly lower melting temperature compared to lithium metatitanate, the densification of such a crust has surely progressed much more than that of the core lithium metatitanate by the end of the sintering. As the EM-LMT samples show, the annealing leads to a notable shrinkage (see Table 4.6). If the core of the SG-LMT samples shrinks in the same way while the crust does not, because it is already close to 100 % density, the formation of cavities between the core and the crust would be the consequence. The significantly higher lithium surplus that the SG-LMT samples show would in principal favor the formation of a second phase.

Yet, the determined lithium surplus is still dissolvable within β-lithium metatitanate for which a maximum lithium surplus of 0.255 is implicitly reported, according to the phase diagram of Li_2O-TiO_2 (see Figure 2.10). With longer annealing durations the sum of the total and imaginary porosity of the samples stays relatively constant at a value of about 37 %, which is achieved after 4 days of annealing. Even if the crust of the pebbles does not compact visibly with increasing annealing duration, as the determined pebble diameters suggest (see Table 4.6), the core of the pebbles visibly densifies as the cross section images clearly display. These observations also support the proposed mechanism that generates the cavities between crust and core. If the crust of the pebbles were dense, the determined density of the pebbles should not change. Otherwise the densification of the core would also lead to a reduction of the observed closed porosity. Yet, only a fraction of pebbles shows the formation of crust and thus show a different behavior from the EM-LMT pebbles. As a result the observed decrease of the closed porosity with increasing annealing duration also has to be attributed to the densification of these pebbles. After 128 days of annealing the values of closed porosity of the KALOS pebbles and the lithium excessive lithium metatitanate pebbles reach similar values, although a moderately higher value is measured for the SG-LMT samples. This is most likely caused by the pebbles that show a gastight crust.

Effects on the strength of the lithium excessive samples

As opposed to the biphasic emulsion method pebbles, the strength of both grades of lithium excessive pebbles cannot be explained by the determined amount of porosity as both quantities develop conflictively as a function of annealing duration. This is found although the samples show a considerable amount of closed porosity which can reach similar levels as those of the LAO-containing samples. Yet, the size of the pores is notably larger for the LAO-containing samples, which is the more relevant property for explaining the strength of the samples. For the lithium excessive pebbles that were fabricated by the emulsion method, the Weibull modulus stays relatively constant as a function of annealing duration (see Figure 4.51) while the characteristic strength continuously decreases and thus, also supported by the SEM images of the fragments (see section 4.5.3), the failure mechanism is identical for all samples independent of the annealing duration. The only visible exception of this trend is the sample that was annealed for 4 days, which shows a notably higher Weibull modulus. As an increased Weibull modulus is an expression of a sharper defect size distribution, the observed increase after 4 days of annealing may reflect this property. The cross section images show that the pore structure changes from elongated cavities along large blocks of grains to spherical pores after 4 days of annealing (see Figure 4.28). Whether the size of the spherical pores varies as much as for the pores in the as-received pebbles is not obviously perceivable. In consequence, the available results do not explain the comparably high Weibull modulus of this particular set of samples.

Generally, one consequence of the transformation of the block-structure of the microstructure to a properly sintered microstructure with spherical pores is that, in terms of mechanical strength, the comparison between the as-received pebbles and the annealed pebbles is difficult. For the annealed pebbles, the grain size increases while the strength of the samples decreases. In the most general way, such a behavior is described by the so-called Hall-Petch equation

$$\sigma_c = \sigma_{res} + \frac{k_H}{\sqrt{d}} \tag{5.4}$$

where σ_{res} is a measure of the stress that is needed to create plastic deformation of a grain and k_H being a material specific constant [202,203]. The Hall-Petch model is regularly explained by the slip of a grain which leads to a dislocation pile-up in front of the grain boundary which in return raises the stress intensity at this point [203]. For β-lithium metatitanate, the slip along (001)-planes is highly likely as the presented results have shown and thus the Hall-Petch equation can reasonably be applied. Yet, the direct experimental proof for plasticity of β-lithium metatitanate by dislocation movement has, up to now, not been obtained. As Figure 5.21 shows, the determined characteristic mean contact pressure of both grades of annealed lithium excessive can be described very well by the increasing grain size that is established during the annealing (see Figure 5.20). Yet, as Figure 4.52 illustrates, the fraction of failures which are unambiguously explained by the fitted Weibull line decreases significantly with increasing annealing duration for the EM-LMT samples. Conversely, the number of apparent failures increases. As stated in section 2.3.1, the coalescence of cracks that originate from microplastic behavior during Hertzian loading is significantly dependent on the grain size, with a coarse grained material being significantly more susceptible to this behavior than fine grained material.

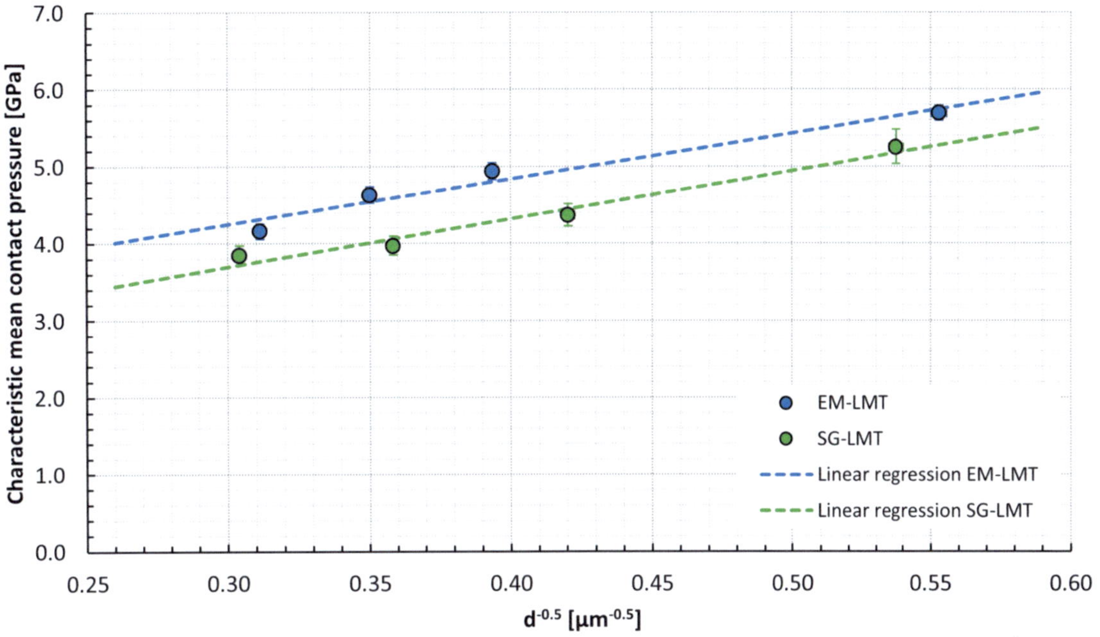

Figure 5.21: The characteristic mean contact pressure of the annealed lithium excessive pebbles is shown as a function of the grain size to the power of -0.5 according to the Hall-Petch relationship in equation (5.4). The linear regression shows a coefficient of determination of 0.957 and 0.972 for the EM-LMT and SG-LMT samples, respectively. The error bars in x-direction, which account for a measurement uncertainty of ±1 %, mostly vanish with the chosen scale. The error bars in y-direction reflect the one dimensional projections of the 90 % likelihood ratio confidence boundaries.

Therefore it is not surprising that an increasing amount of apparent failures is visible with increasing annealing duration as these failures were already associated with the coalescence of cracks below the contact area (see section 5.1.1). Furthermore, the coalescence of a substantial amount of the relevant cracks is a stochastic process as Lawn et al. [122] mention, which explains that a considerable number of pebbles still fails in a fully brittle way.

Yet, the grain size is more than doubled between the as-received state and the that after 128 days of annealing, which exceeds the span that Guiberteau et al. [124] observed, but it is not at all clear to which extent those results can be generalized. Of course, the increasing amount of apparent failures leads to a disproportionately high decrease of the mean mechanical strength of the EM-LMT pebbles with increasing annealing duration compared to the predictions of the Hall-Petch equation. In principle also the SG-LMT pebbles should show this behavior as the observed grain sizes are very similar. However, significantly less apparent failures are observed for these samples and also the characteristic deviation from Weibull behavior is not observed by the same degree, but overall a similar trend is observable with an increase of these failures for long annealing durations and in particular after 128 days of annealing (see Figure 4.54).

In general, the coalescence of cracks takes place in a limited volume below the mechanical contact, which is at or near the surface of the pebble. Any significant surface layer or a crust certainly changes the behavior of the pebbles. Especially since the formation of the cracks, that are associated with microplastic behavior, requires some form of plastic deformation of the grains or slip along the grain boundaries. The cross section images of the SG-LMT samples clearly show for a large number of pebbles that the microstructure near the surface features a considerably different structure compared to the bulk and the EM-LMT samples (see Figure 4.28, Figure 4.29 and Figure 4.30). With the assumption that this shell is composed of lithium orthotitanate instead of lithium metatitanate, it is expected that the sliding along certain crystallographic planes requires much higher stresses and consequently the reduced amount of 'apparent' failures should be reasonable. Figure 4.67 furthermore clearly illustrates that the shell of the pebbles is compressed during the uniaxial loading and does not break away before the measurements. Concerning the majority of the pebbles that can be represented by a Weibull line, the decrease of the characteristic strength with increasing annealing duration (see Figure 4.53) can be explained by the increasing grain size as described by the Hall-Petch equation (see Figure 5.21). In contrast to the EM-LMT pebbles, the annealing of the SG-LMT pebbles for 4 days increases the mean contact pressure at which the pebbles break compared to the as-received state (see Figure 4.42). In terms of the representation of the failure behavior, a single Weibull line can describe the measurements after 4 days of annealing instead of two. The two Weibull lines are probably the result of the transition from inter- to intragranular fracture as described in section 5.1.2. Since the cross section images of the SG-LMT pebbles suggest that intragranular fracture does not occur after 4 days of annealing (see Figure 4.67 as an example), it is fitting that a single Weibull line satisfyingly represents the obtained failure data.

The observation of transgranular fracture for the in comparison smaller grain size in the as-received state is unusual as a small grain size reduces the generated fracture surface area. Yet, the small grain size certainly reduces the intrinsic stresses that are generated by thermal expansion anisotropy, which is substantial for β-lithium metatitanate (see Figure 4.9). These stresses that form upon cooling after sintering certainly weaken the grain boundaries and therefore promote intergranular fracture. After annealing for 4 days or longer, the involved cooling will generate these thermal stresses again, although probably not as strongly because of the lower temperature difference. Consequently the probability of transgranular fracture is reduced further and evidently not observable anymore after 4 days of annealing, which closely resembles the fracture behavior of the EM-LMT. This is also true for the determined characteristic strength of the SG-LMT pebbles, which is much lower for either of the two subsegments in the as-received state. During this first annealing period of the samples, the relatively loose microstructure of the pebble core considerably increases in density as Figure 4.29 shows.

As a result also the mechanical strength of the pebbles has to be increased. With ongoing annealing, the strength of the SG-LMT pebbles develops in the same way as that of the EM-LMT samples (see Figure 5.21) and it is thus determined by the grain size, which naturally increases with the annealing duration. Yet, for the SG-LMT samples a perceivably lower Weibull modulus is determined. This is very probably a result of the overall non-uniform microstructure of the pebbles, as in principle two Weibull distributions are necessary to model the system correctly, however from the available data an unambiguous modeling with two Weibull distributions is not possible.

6 Conclusions

From the preceding sections it is unambiguously clear that the strength of advanced ceramic breeder pebbles is governed by the fabrication process, the material composition and the time for that these pebbles are annealed, i.e. the duration of usage. Consequently the strength and the probability of failure of an individual pebble is a combined result of all of these points. In this work the development of the mechanical strength of four grades of advanced ceramic breeder pebbles was investigated as a function of the annealing duration close to the maximum expected temperature during actual operation in a fusion blanket as a function of time. To complement the results of these investigations, two sets of samples were fabricated by the emulsion method by which a broad composition range was covered.

For the advanced ceramic breeder pebbles that were fabricated by the melt-based process KALOS, an important part of their mechanical strength is determined by several effects which are directly related to the fabrication as discussed in the preceding section. First and foremost, the solidification and subsequent cooling of the pebbles to room temperature involves high cooling rates and consequently high temperature gradients within the droplets and solid pebbles. In order to describe the mechanical stresses that are generated during the solidification, the model of Forgac et al. [128] was applied. It was shown that generally during the melt-based fabrication the liquid core of a solidifying droplet quickly experiences tensile forces while the solid surface is compressed. This state is maintained until the droplet has solidified completely. However, in most cases these forces grow too large so that the solid crust cannot withstand them, which results in the generation of dents and the droplets lose their initially virtually perfect sphericity. Yet, during the subsequent rapid cooling of the pebbles, eventually tensile stresses are generated in the surface of the samples, as Grünberg [129] showed, and the imperfections of the pebble surfaces facilitate the formation of cracks. In order to prevent the formation of such cracks, the KALOS process has to be adapted in such a way, that imperfections of the pebble surface are obviated. One option can be to increase the cooling of the solidifying droplets which leads to an increased shrinking of the solid crust onto the liquid core and therefore reduces the tensile forces within it. Of course, by contrast the subsequent cooling of the solid pebbles should be as mild as possible to decrease the temperature gradient between the cooling surface and the warmer pebble volume to decrease the tensile stresses within the crust.

Furthermore, the KALOS process evidently generates pebbles within a considerable composition range around the nominal value. The nominal composition is, nevertheless, realized with small deviations when the mean value of all pebbles is considered. It is additionally obvious that these variations in composition lead to the formation of distinctly different microstructures of the pebbles. Concerning the formation of surface cracks during cooling, the pebbles whose actual composition is close to the eutectic composition of the quasi-binary system Li_4SiO_4-Li_2TiO_3 are demonstrably more resilient and are virtually free from surface defects. This observation is explained by the formation of residual compressive surface stresses during crystallization, which counteracts the tensile stresses that are generated during cooling. These compressive stresses are expected to be most pronounced and best preserved when a relatively sharp solidification front is established as the secondary dendrite arms effectively lead to a premature solidification and a thus larger crust of the pebble. Consequently the fabrication of pebbles with a composition close to the eutectic composition can also enhance the overall mechanical strength considerably. Furthermore, such pebbles show a notably diminished probability of failing at low loads since volume defects lead to failure in their case and thus a significantly higher Weibull modulus is achieved.

With this exception, the as-received pebbles that were fabricated by the KALOS process fail from surface defects because comparably large surface cracks are formed by the stresses that are generated by the rapid cooling. Yet, also the lithium metatitanate content of the pebbles is without a doubt a significant factor with regard to the mechanical strength. Generally, a higher lithium metatitanate content leads to a higher mechanical strength when the pebbles fail ordinarily. This trend is observable from the LMT-poorer samples with a microstructure that is unambiguously assigned to a clearly hypoeutectic composition, which feature the lowest characteristic strength, to the pebbles that show an almost eutectic composition to the ones that are dominated by lithium metatitanate dendrites, and therefore a clearly hypereutectic composition, which show the highest characteristic strengths. This increase in strength as a function of the lithium metatitanate concentration is also determined for the biphasic emulsion method pebbles that are composed of lithium orthosilicate and lithium metatitanate. Although these pebbles show an entirely different microstructure, the strengthening effect of lithium metatitanate is clearly visible. Whether the microstructure has an influence on the strength of the pebbles and how large it is, is difficult to quantify on the basis of the results of this work. Yet, an R-curve behavior, i.e. an increasing resistance to crack propagation with increasing crack length, is not observed, also not for the pebbles that are dominated by lithium metatitanate dendrites for which such a behavior is thinkable.

Apart from the ordinary sudden failure of the pebbles, two different manners of failures were observed for the KALOS pebbles which do not lead to the immediate failure of the complete pebble. One of these two failure modes has significant influence on the mean pebble strength of the as-received pebbles and also changes the Weibull plot of these samples considerably for low loads. The most plausible explanation of this failure mode is the momentary coalescence of microcracks below the contact areas during uniaxial compression of the pebbles and the therefore abrupt deformation of the pebble as microplastically strained areas around the contact points indicate. As the residual thermal stresses, which originate from the significant thermal expansion mismatch between lithium orthosilicate and lithium metatitanate, can facilitate the formation of the relevant microcracks, an annealing step after fabrication easily removes this behavior. Yet, the expected temperatures during operation within a fusion blanket can certainly act as such an annealing as well. Furthermore, as the pebbles will not fail completely by these so-called 'apparent failures' their presence may even be tolerable in the end, if the pebble bed properties are not overly deteriorated until the residual stresses are relieved. From the presented results it is also visible that a sufficiently low size of the pearlite-like eutectic domains near the surface of the pebbles leads to the disappearing of this failure mode too. For the KALOS process this would require an increased cooling during the solidification of the droplets, which is also beneficial for the pebble morphology as already mentioned.

During the annealing in reference purge gas atmosphere, the KALOS pebbles significantly change, especially with regard to their mechanical strength. The pebbles with nominally hypoeutectic composition show a considerable drop in mechanical strength after 4 days of annealing, i.e. after the first sampling, which is associated with the contraction of the pearlite-like domains. This is a direct consequence of the phase transformation of metastable crystallized γ- to β-lithium metatitanate, which is the equilibrium phase. As also shown in this work, the phase transformation does not at all require 4 days to complete but much less time, so the associated changes will also be apparent much earlier. This phase transformation indubitably removes the compressive residual stresses within the lithium metatitanate lamellae and as a result the previously counteracting residual tensile stresses within the lithium orthosilicate lamellae are relieved by contraction as well. The already present surface cracks within the pebbles are thus extended and lead to the observed decline in strength as a one-time adjustment. In terms of the mean pebble strength, no notable changes take place with ongoing annealing. Yet, from the Weibull analysis it is obvious that a minority of these pebbles still change as pebbles that fail quasi-continuously appear after annealing for 32 days and their number increases as a function of annealing duration. The mode of quasi-continuous failure is assigned to pebbles which do not show the typical microstructure of mostly pearlite-like domains, but rather single grains of lithium orthosilicate in which a significant amount of titanium is dissolved. The grain growth within such pebbles is comparably unobstructed because of the absence of finely dispersed lithium metatitanate grains, which obviously limit grain growth effectively in all other KALOS pebbles. As a result, cracks of considerable length form along the grain boundaries between these large grains which are advanced by stresses from the significant thermal expansion anisotropy of the lithium orthosilicate lattice during sampling. Which conditions are necessary to generate this type of pebbles is not clear. Yet, it is self-evident that a higher lithium metatitanate content prevents the incorporation of the majority of the available titanium into the lithium orthosilicate lattice and in consequence the grain growth will not be as a serious issue. If, however, pebbles of this composition are fabricated, either a less rapid solidification of the pebbles has to be ensured to avoid the crystallization in a single phase, or the controlled precipitation of lithium metatitanate has to be accomplished after the fabrication. If this type of pebbles can be predominantly fabricated, the latter point clearly has the potential to tailor the pebble microstructure within certain limits.

For the nominally hypereutectic pebbles, the annealing first and foremost leads to an increase of the mean mechanical strength, which results from the annihilation of the residual thermal stresses and the consequential disappearing of low strength apparent failures. Of course, also this failure mode will disappear much sooner than after 4 days of annealing since the residual thermal stresses annihilate during the phase transformation within the lithium metatitanate grains. Until 32 days of annealing, the mechanical strength of the samples does not change notably but afterwards surface cracks of considerable length develop in both types of pebbles and alter the failure behavior. As a result, the failures of pebbles, that show a mostly pearlite-like microstructure, from volume defects is significantly reduced, which in consequence results in a reduced mean pebble strength after 64 days of annealing. As the surface cracks in pebbles whose microstructure is dominated by lithium metatitanate dendrites also grow considerably, the overall mean pebble strength is reduced significantly for longer annealing durations. The growing of these surface cracks is attributed to the anisotropic thermal expansion of lithium orthosilicate and the associated stresses which are generated during the sampling. However, also thermal cycles are expected during use in a fusion blanket. Whether the lithium orthosilicate matrix grains within the pebbles grow significantly and thus increase the crack lengths is, however, speculative although plausible. In principle, the generation of the highest stresses by the thermal expansion anisotropy is expected at temperatures during which lithium orthosilicate undergoes a phase transformation, as clearly demonstrated. If this phase transformation can be suppressed, e.g. by a suitable dopant, the degradation of the mechanical strength of these pebbles for long annealing durations can also be reduced significantly.

In the end, the mechanical strength of the KALOS pebbles is hard to estimate by a single comprehensive pebble property like e.g. the open porosity or the amorphous content. This is to a significant part due to the illustrated diversity of the pebble microstructures and the different manners of solidification, which are eventually the result of different pebble compositions. Also the different development of the mechanical strength as a function of annealing duration for the two sets of KALOS samples is directly linked to this cause. It goes without saying that the KALOS process should aim at producing pebbles that only marginally vary in their composition. Yet, this work cannot give insight into how to achieve this goal. For the KALOS pebbles, two different pebble sizes were characterized with regard to their mechanical strength. In the as-received state, the Weibull size-effect on the failure probability can reasonably explain the differences in the characteristic pebble strength. Yet, after annealing the larger pebbles are not as strong as theory predicts, which implies that the observed changes are generally pebble size dependent and more severe for the larger pebbles. It would therefore be sensible to reduce the mean size of the produced pebbles. The quantitative XRD analysis also reveals that the lithium orthosilicate lattice can take up a notable amount of excess lithium, which is unexpected from the literature data. Also the uptake of titanium is not reported in literature. Both observations are certainly attributed to the rapid cooling and the so increased probability that metastable states develop. Yet, the incorporation of impurities into the lithium orthosilicate lattice obviously influences the unit cell strain and thus complicates the reasonable estimation of the residual stresses within the samples. For the case of lithium, a reasonable correlation was found between the lithium content and the lithium orthosilicate unit cell strain. This is especially useful, as it was shown that a perceivable amount of lithium evaporates from the samples during annealing.

For the biphasic emulsion method pebbles that contain lithium metatitanate, the mechanical strength can, in contrast to the KALOS pebbles as well as the LAO containing pebbles, be estimated by one comprehensive property, namely the open porosity, if the pebbles fail from surface defects. Yet, in this case the open porosity is only a good estimator of the length of these cracks. Generally, pebbles of one composition only show one failure mode, as against the KALOS samples show multiple failure modes. From the results, it is also clear that the fracture toughness of the biphasic emulsion method pebbles is a function of the lithium metatitanate concentration, which should be identical for the KALOS pebbles. Yet, the exact scale of the increase of the fracture toughness is potentially obscured by differences of the surface morphology. For the lithium aluminate containing biphasic pebbles, the increase in fracture toughness can be determined much better, as the defect size or in this case pore size can properly be measured. The results prove that sizeable increases in fracture toughness can be realized by moderate second phase additions to the pebbles. For these samples a considerably larger increase in fracture toughness is determined in comparison to the lithium metatitanate containing emulsion method pebbles. Yet, further investigations are necessary to determine whether the strengthening effect of lithium aluminate is indeed significantly higher than that of lithium metatitanate.

To increase the strength of the LAO-containing emulsion method pebbles, the sintering program has to be adapted to release the forming gaseous species before they are encased within the pebble volume. In principle this also applies to the nominally 10 mol% lithium metatitanate containing pebble sample. However, as surface defects are defining the strength of these pebbles, this is not imperative. For the 10 mol% lithium aluminate containing sample, the vanishing of the closed pores will also just partially help to increase the mean pebble strength as eventually the surface defects determine the failure behavior when the failure from volume defects becomes increasingly unlikely with shrinking closed pores. Therefore, ideally also the surface defects should be reduced when the emulsion method is improved. As the observed seams represent critical surface defects for the biphasic emulsion method pebbles, an improvement has to focus on the drying and washing procedure of the emulsion method, since the seams' existence is attributed to the withdrawal of water from the swelled gel. For achieving this, either a slower removal of water from the gelled spheres has to be realized, or principally less water is added to the slurry from the beginning, if it is still possible to handle the slurry with a reduced amount of water. Generally, the densities that are achieved by the applied, non-adapted sintering program are reasonably high when the samples for which lithium orthosilicate dominates the densification behavior are concerned, i.e. for all but the 90 mol% lithium metatitanate containing pebbles. As a result only little improvements are possible for these pebbles in this regard. However, if the emulsion method is improved so that surface defects do not play a significant role, which is already the case for the 70 and 80 mol% lithium metatitanate containing samples, a tailored sintering program may further improve the mechanical strength of these pebbles.

For the lithium excessive lithium metatitanate pebbles that were fabricated by the emulsion method no principal differences to the biphasic pebbles that were fabricated by the same method are observable. Especially for the 90 mol% lithium metatitanate containing sample, the differences in the Weibull parameters are statistically not significant (compare Figure 4.51 and Figure 4.55). Yet, the considerable shrinkage of these pebbles after 4 days of annealing and the simultaneous increase in strength shows that the applied sintering program may not be optimal. This certainly reflects the difficulties to compare the Weibull moduli directly to the observed failure mode from volume defects for the lithium excessive pebbles in contrast to the failure from surface defects for the biphasic emulsion method pebbles. After achieving this state of high strength, further annealing significantly decreases the strength of the EM-LMT pebbles. The annealing obviously leads to homogeneous, unobstructed grain growth and the consequential reduction of the mechanical strength according to the Hall-Petch equation [202,203]. Because of the clearly observable plasticity of monoclinic lithium metatitanate along the (001)-planes, the Hall-Petch equation perfectly applies to this system. Yet, the growth of the lithium metatitanate grains also leads to the increasing observation of apparent failures, which can potentially be tolerated during use in a fusion blanket module as already mentioned, although this has not been verified until now. Generally the grain growth with ongoing annealing has to be reduced for these pebbles to avoid a considerable degradation in strength.

This can be achieved by an appropriate second phase like lithium orthosilicate, as the example of the KALOS pebbles clearly illustrates. Consequently, the fabricated biphasic emulsion method pebbles which feature high amounts of lithium metatitanate are highly likely to provide better long-term stability. In principle the same applies for the lithium excessive pebbles that were fabricated by the sol-gel method for which also unobstructed grain growth and the consequential decrease of the pebble strength according to the Hall-Petch equation are observed. However, in the as-received state the pebbles show a clearly different failure behavior compared to the emulsion method samples, which is associated with comparably low adherence between the individual lithium metatitanate grains. Yet, also for these pebbles, volume defects are obviously causative for failure. The reason for the scanty adherence is not clear, but as a result, the pebbles show inter- as well as intragranular fracture. The transition between both fracture modes leads to the observation of an R-curve behavior for these pebbles. Furthermore these pebbles develop a crust which presumably consists of lithium orthotitanate and detaches from the pebble core during annealing. For the mechanical strength this crust does not seem to have a perceivably negative effect. On the contrary, the presence of the lithium orthotitanate crust seemingly suppresses the appearance of apparent failures. The existence of this crust is attributed to an inhomogeneous distribution of the raw materials within the green pebbles, which is indisputably unwanted and the sol-gel method should be modified in this regard. Almost certainly this also involves a revision of the removal of water after the shaping of the pebbles similar to the emulsion method. For all initially lithium excessive pebbles, the lithium excess is steadily reduced with continued annealing and eventually a lithium deficit is established.

In this work, the presently available advanced ceramic breeder materials of the European Union and Japan were thoroughly characterized in terms of their mechanical strength and their failure mechanisms in the as-received state and also after a simulated use for up to 128 days. The presented results and models identify and explain the causes for the fracture of the pebbles, detail numerous insights into the mechanisms of their fracture and therefore provide several starting points for the improvement of the fabrication processes as well as the material composition. Moreover, the characterization of the failure behavior and mechanical strength of the current grades of material is essential for understanding the mechanical behavior of the pebble beds and for qualifying these materials for the use in a fusion power plant or its predecessor devices. The research on these topics will be carried out in the coming years in the process of increasing the maturity of the technical realization of fusion power to eventually realize commercial fusion power plants.

7 References

[1] P. Muehlich, M. Biberacher, H. Cabal Cuesta, U. Ciorba, C. Eherer, F. Gracceva, P. E. Gronheit, T. Hamacher, W. Han, Y. Lechon, A. Pina, D. Ward, The Potential Role for Fusion Power in Future Energy Markets, Proceedings of the 23rd IAEA Fusion Energy Conference (2010).

[2] V. Dornburg, D. van Vuuren, G. van de Ven, H. Langeveld, M. Meeusen, M. Banse, M. van Oorschot, J. Ros, G. van den Jan Born, H. Aiking, M. Londo, H. Mozaffarian, P. Verweij, E. Lysen, A. Faaij, Bioenergy revisited: Key factors in global potentials of bioenergy, Energy Environ. Sci. 3 (2010) 258.

[3] T.I.M. Beringer, W. Lucht, S. Sschaphoff, Bioenergy production potential of global biomass plantations under environmental and agricultural constraints, GCB Bioenergy 3 (2011) 299–312.

[4] D.P. van Vuuren, J. van Vliet, E. Stehfest, Future bio-energy potential under various natural constraints, Energy Policy 37 (2009) 4220–4230.

[5] M. Rogovin, G.T. Frampton, Three Mile Island: a report to the commissioners and to the public, United States, 1980.

[6] The Chernobyl Forum 2003-2005, Chernobyl's Legacy: Health, Environmental and Socio-Economic Impacts and Recommendations to the Governments of Belarus, the Russian Federation and Ukraine., Vienna, Austria, 2006.

[7] Z.A. Medvedev, The legacy of Chernobyl, 1st ed., W.W. Norton & Co, New York, 1992, ©1990.

[8] The Great East Japan Earthquake Expert Mission, IAEA international fact finding expert mission of the Fukushima Dai-ichi NPP accident following the Great East Japan Earthquake and Tsunami, 2011.

[9] R. Köcher, Atemberaubende Wende, in: FAZ-Monatsberichte des Instituts für Demoskopie.

[10] P.L. Kunsch, J. Friesewinkel, Nuclear energy policy in Belgium after Fukushima, Energy Policy 66 (2014) 462–474.

[11] M. Siegrist, V.H.M. Visschers, Acceptance of nuclear power: The Fukushima effect, Energy Policy 59 (2013) 112–119.

[12] Y. Kim, M. Kim, W. Kim, Effect of the Fukushima nuclear disaster on global public acceptance of nuclear energy, Energy Policy 61 (2013) 822–828.

[13] Z. Ming, L. Yingxin, O. Shaojie, S. Hui, L. Chunxue, Nuclear energy in the Post-Fukushima Era: Research on the developments of the Chinese and worldwide nuclear power industries, Renewable and Sustainable Energy Reviews 58 (2016) 147–156.

[14] N. Lior, Sustainable energy development: The present (2009) situation and possible paths to the future, Energy 35 (2010) 3976–3994.

[15] M. Ichimiya, The Status of Generation IV Sodium-Cooled Fast Reactor Technology Development and its Future Project, Energy Procedia 7 (2011) 79–87.

[16] D. Turnbull, A. Glaser, R.J. Goldston, Investigating the value of fusion energy using the Global Change Assessment Model, Energy Economics 51 (2015) 346–353.

[17] J.K. Shultis, R.E. Faw, Fundamentals of nuclear science and engineering, 2nd ed., CRC Press, Boca Raton, 2008.

[18] A.A. Harms, Principles of fusion energy: An introduction to fusion energy for students of science and engineering, World Scientific, Singapore, London, 2005.

[19] G.M. McCracken, P.E. Stott, Fusion: The energy of the universe, 2nd ed., Academic Press, Waltham, MA, Oxford, UK, 2013.

[20] S. Mertens, T. Lasserre, S. Groh, G. Drexlin, F. Glück, A. Huber, A.W.P. Poon, M. Steidl, N. Steinbrink, C. Weinheimer, Sensitivity of next-generation tritium beta-decay experiments for keV-scale sterile neutrinos, J. Cosmol. Astropart. Phys. 2015 (2015) 20.

[21] L.V. Boccaccini, J.-F. Salavy, R. Lässer, A. Li Puma, R. Meyder, H. Neuberger, Y. Poitevin, G. Rampal, The European test blanket module systems: Design and integration in ITER, Fusion Engineering and Design 81 (2006) 407–414.

[22] J. Reimann, S. Hermsmeyer, Thermal conductivity of compressed ceramic breeder pebble beds, Fusion Engineering and Design 61-62 (2002) 345–351.

[23] M.H.H. Kolb, R. Knitter, U. Kaufmann, D. Mundt, Enhanced fabrication process for lithium orthosilicate pebbles as breeding material, Fusion Engineering and Design 86 (2011) 2148–2151.

[24] R. Knitter, P. Chaudhuri, Y.J. Feng, T. Hoshino, I.-K. Yu, Recent developments of solid breeder fabrication, Journal of Nuclear Materials 442 (2013) S420-S424.

[25] ITER Organization, Illustration of the ITER tokamak, 2014, https://www.iter.org/doc/all/content/com/gallery/media/7%20-%20technical/2009_04_29%20machine.jpg, accessed 8 May 2018.

[26] S. Hermsmeyer, L.V. Boccaccini, U. Fischer, C. Köhly, J. Rey, D. Ward, Reactor integration of the helium cooled pebble bed blanket for DEMO, Fusion Engineering and Design 75-79 (2005) 779–783.

[27] L.V. Boccaccini, L. Giancarli, G. Janeschitz, S. Hermsmeyer, Y. Poitevin, A. Cardella, E. Diegele, Materials and design of the European DEMO blankets, Journal of Nuclear Materials 329-333 (2004) 148–155.

[28] L.V. Boccaccini, G. Aiello, J. Aubert, C. Bachmann, T. Barrett, A. Del Nevo, D. Demange, L. Forest, F. Hernandez, P. Norajitra, G. Porempovic, D. Rapisarda, P. Sardain, M. Utili, L. Vala, Objectives and status of EUROfusion DEMO blanket studies, Fusion Engineering and Design 109-111 (2016) 1199–1206.

[29] M. Dalle Donne, U. Fischer, M. Küchle, A Helium-Cooled, Poloidal Blanket with Ceramic Breeder and Beryllium Multiplier for the Next European Torus, Nuclear Technology 71 (1985) 15–28.

[30] J.E. Vetter, Breeding blanket development for NET, Fusion Engineering and Design 11 (1989) 101–114.

[31] Mohamed A. Abdou, A. René Raffray, Zinovy R. Gorbis, Mark S. Tillack, Yoichi Watanabe, Alice Y. Ying, Mahmoud Z. Youssef, Kaoru Fujimura, A Helium-Cooled Solid Breeder Concept for the Tritium-Producing Blanket of the International Thermonuclear Experimental Reactor, Fusion Technology (1989) 166–182.

[32] E. Proust, L. Anzidei, M. Dalle Donne, U. Fischer, T. Kuroda, Solid breeder blanket design and tritium breeding, Fusion Engineering and Design 16 (1991) 73–84.

[33] P.H. Sager, E.R. Hager, Design considerations for access to Tokamak fusion reactors, Nuclear Engineering and Design 39 (1976) 153–164.

[34] F. Carre, Chevereau G., Gervaise F., Proust E., Status of fusion reactor blanket evaluation studies in France, Fusion Technology 1985 491–500.

[35] Dale L. Smith, Charles C. Baker, Dai Kai Sze, Grover D. Morgan, M. A. Abdou, Steven J. Piet, K. R. Schultz, Ralph W. Moir, James D. Gordon, Overview of the Blanket Comparison and Selection Study, Fusion Sci. Technol (1985).

[36] L. Giancarli, M. Dalle Donne, W. Dietz, Status of the European breeding blanket technology, Fusion Engineering and Design 36 (1997) 57–74.

[37] G. Aiello, J. Aktaa, F. Cismondi, G. Rampal, J.-F. Salavy, F. Tavassoli, Assessment of design limits and criteria requirements for Eurofer structures in TBM components, Journal of Nuclear Materials 414 (2011) 53–68.

[38] M.E. Sawan, M.A. Abdou, Physics and technology conditions for attaining tritium self-sufficiency for the DT fuel cycle, Fusion Engineering and Design 81 (2006) 1131–1144.

[39] S. Hermsmeyer, S. Malang, U. Fischer, S. Gordeev, Lay-out of the He-cooled solid breeder model B in the European power plant conceptual study, Fusion Engineering and Design 69 (2003) 281–287.

[40] A. Ciampichetti, F.S. Nitti, A. Aiello, I. Ricapito, K. Liger, D. Demange, L. Sedano, C. Moreno, M. Succi, Conceptual design of Tritium Extraction System for the European HCPB Test Blanket Module, Fusion Engineering and Design 87 (2012) 620–624.

[41] F. Hernández, F. Cismondi, B. Kiss, Thermo-mechanical analyses and assessment with respect to the design codes and standards of the HCPB-TBM Breeder Unit, Fusion Engineering and Design 87 (2012) 1111–1117.

[42] R. Knitter, M.H.H. Kolb, F. Cismondi, A. Abou-Sena, L.V. Boccaccini, R. Annabattula, M. Kamlah, Magielsen L., S. van Til, A. Fedorov, J.-F. Salavy, C. Fausser, J. Quinones, F4E-2009-GRT-030 (Action-1), Task-1 Final Report (0-D3): Development / Qualification / Procurement Plan for Ceramic Breeder Pebbles (2011).

[43] F. Cismondi, B. Kiss, F. Hernandez, E. NDiaye, G. Legradi, J. Reimann, M. Ilic, The fundamental role of fluid dynamic analyses in the design of the solid EU Test Blanket Module, Fusion Engineering and Design 87 (2012) 1123–1129.

[44] Y. Gan, F. Hernandez, D. Hanaor, R. Annabattula, M. Kamlah, P. Pereslavtsev, Thermal Discrete Element Analysis of EU Solid Breeder Blanket Subjected to Neutron Irradiation, Fusion Sci. Technol 66 (2014).

[45] A. Donato, A critical review of Li_2O ceramic breeder material properties correlations and data, Fusion Engineering and Design 38 (1998) 369–392.

[46] E.T. Cheng, Concentration limits of natural elements in low activation fusion materials, Journal of Nuclear Materials 258-263 (1998) 1767–1772.

[47] E.T. Cheng, Waste management aspect of low activation materials, Fusion Engineering and Design 48 (2000) 455–465.

[48] M. Dalle Donne, E. Bojarsky, U. Fischer, M. Küchle, P. Norajitra, G. Reimann, H. Reiser, H.D. Baschek, E. Bogusch, The Karlsruhe solid breeder blanket and the test module to be irradiated in ITER/NET, Fusion Engineering and Design 17 (1991) 87–94.

[49] M. Zmitko, Y. Poitevin, L. Boccaccini, J.-F. Salavy, R. Knitter, A. Möslang, A.J. Magielsen, J.B.J. Hegeman, R. Lässer, Development and qualification of functional materials for the EU Test Blanket Modules: Strategy and R&D activities, Journal of Nuclear Materials 417 (2011) 678–683.

[50] D. Carloni, L.V. Boccaccini, F. Franza, S. Kecskes, Requirements for helium cooled pebble bed blanket and R&D activities, Fusion Engineering and Design 89 (2014) 1341–1345.

[51] L.M. Giancarli, M. Abdou, D.J. Campbell, V.A. Chuyanov, M.Y. Ahn, M. Enoeda, C. Pan, Y. Poitevin, E. Rajendra Kumar, I. Ricapito, Y. Strebkov, S. Suzuki, P.C. Wong, M. Zmitko, Overview of the ITER TBM Program, Fusion Engineering and Design 87 (2012) 395–402.

[52] W. Pannhorst, V. Geiler, G. Rake, B. Speit, D. Sprenger, Production process of lithium orthosilicate pebbles, Association Euratom-CEA Cadarache, France, 1998.

[53] R. Hoppe, K. Bernet, A. Möller, Einkristall-Synthese hochschmelzender Oxyde bei niederer Temperatur: Γ-$Li_4[SiO_4]$ — ohne Fehlordnung, isotyp mit $Na_4[SnO_4]$. (Was heißt 'Isotypie'?), Z. anorg. allg. Chem. 629 (2003) 1285–1293.

[54] H. Kleykamp, Enthalpy, heat capacity, second-order transitions and enthalpyof fusion of Li_4SiO_4 by high-temperature calorimetry, Thermochimica Acta 287 (1996) 191–201.

[55] B. Konar, M.-A. van Ende, I.-H. Jung, Critical evaluation and thermodynamic optimization of the Li-O, and Li_2O-SiO_2 systems, Journal of the European Ceramic Society (2017).

[56] Y. Deng, C. Eames, J.-N. Chotard, F. Lalere, V. Seznec, S. Emge, O. Pecher, C.P. Grey, C. Masquelier, M.S. Islam, Structural and Mechanistic Insights into Fast Lithium-Ion Conduction in Li_4SiO_4-Li_3PO_4 Solid Electrolytes, Journal of the American Chemical Society 137 (2015) 9136–9145.

[57] B.H.W.S. de Jong, D. Ellerbroek, A.L. Spek, Low-temperature structure of lithium nesosilicate, Li_4SiO_4 and its Li 1s and O 1s X-ray photoelectron spectrum, Acta Crystallogr B Struct Sci 50 (1994) 511–518.

[58] O. Götzmann, Thermodynamics of ceramic breeder materials for fusion reactors, Journal of Nuclear Materials 167 (1989) 213–224.

[59] H. Zimmermann, Mechanische Eigenschaften von Lithiumsilikaten für Fusionsreaktor-Brutblankets, Kernforschungszentrum, Karlsruhe, 1989.

[60] M.C. Billone, W.T. Grayhack, Summary of mechanical properties data and correlations for Li_2O, Li_4SiO_4, $LiAlO_2$, and Be, United States, 1988.

[61] K. Bär, C.Y. Chu, J.P. Singh, K.C. Goretta, J.L. Routbort, M.C. Billone, R.B. Poeppel, Mechanical properties and deformation of polycrystalline lithium orthosilicate, Fusion Engineering and Design 8 (1989) 371–374.

[62] J. Ortiz-Landeros, L. Martínez-dlCruz, C. Gómez-Yáñez, H. Pfeiffer, Towards understanding the thermoanalysis of water sorption on lithium orthosilicate (Li_4SiO_4), Thermochimica Acta 515 (2011) 73–78.

[63] V. Schauer, G. Schumacher, Study of adsorption and desorption of water on Li_4SiO_4, Journal of Nuclear Materials 167 (1989) 225–230.

[64] B. Alcántar-Vázquez, P.R. Díaz Herrera, A. Barrera González, Y. Duan, H. Pfeiffer, Analysis of the CO_2–H_2O Chemisorption in Lithium Silicates at Low Temperatures (30–80 °C), Ind. Eng. Chem. Res. 54 (2015) 6884–6892.

[65] Q. Wang, J. Luo, Z. Zhong, A. Borgna, CO_2 capture by solid adsorbents and their applications: Current status and new trends, Energy Environ. Sci. 4 (2011) 42–55.

[66] T. Hoshino, K. Tsuchiya, K. Hayashi, M. Nakamura, H. Terunuma, K. Tatenuma, Preliminary test for reprocessing technology development of tritium breeders, Journal of Nuclear Materials 386-388 (2009) 1107–1110.

[67] R. Knitter, B. Löbbecke, Reprocessing of lithium orthosilicate breeder material by remelting, Journal of Nuclear Materials 361 (2007) 104–111.

[68] G. Schumacher, M. Dalle Donne, S. Dorner, Properties of lithium orthosilicate spheres, Journal of Nuclear Materials 155-157 (1988) 451–454.

[69] R. Knitter, B. Alm, G. Roth, Crystallisation and microstructure of lithium orthosilicate pebbles, Journal of Nuclear Materials 367-370 (2007) 1387–1392.

[70] R. Knitter, P. Risthaus, Fabrication of Lithium Orthosilicate Pebbles by Melt-Spraying - Reproducibility and Yield, Germany, 2004.

[71] L. Wondraczek, J. Deubener, S.T. Misture, R. Knitter, Crystallization Kinetics of Lithium Orthosilicate Glasses, J American Ceramic Society 89 (2006) 1342–1346.

[72] P. Waibel, J. Matthes, O. Leys, M. Kolb, H.B. Keller, R. Knitter, High-Speed Camera-Based Analysis of the Lithium Ceramic Pebble Fabrication Process, Chem. Eng. Technol. 37 (2014) 1654–1662.

[73] R. Knitter, M.H.H. Kolb, U. Kaufmann, A.A. Goraieb, Fabrication of modified lithium orthosilicate pebbles by addition of titania, Journal of Nuclear Materials 442 (2013) S433-S436.

[74] O. Leys, C. Odemer, U. Maciejewski, M.H.H. Kolb, R. Knitter, Microstructure Analysis of Melt-based Lithium Orthosilicate/Metatitanate Pebbles, PM 50 (2013) 196–204.

[75] R. Knitter, M.H.H. Kolb, O. Leys, S. Pupeschi, Specific Grant 01 of the FPA-380-A3, Final Technical Report D3.2: Development and Characterisation of Li-ceramic pebbles/pebble beds (2016).

[76] T. Hoshino, K. Mukai, M.H.H. Kolb, R. Knitter, Cutting edge of R and D in solid tritium breeder materials toward the 'Solid' blankets, Journal of Plasma Physics and Fusion Technology (2017) 83–90.

[77] L.E. Tanner, H. Okamoto, The Pt-Si (Platinum-Silicon) system, JPE 12 (1991) 571–574.

[78] K. Kataoka, Y. Takahashi, N. Kijima, H. Nagai, J. Akimoto, Y. Idemoto, K.-i. Ohshima, Crystal growth and structure refinement of monoclinic Li2TiO3, Materials Research Bulletin 44 (2009) 168–172.

[79] A. Laumann, H. Boysen, M. Bremholm, K.T. Fehr, M. Hoelzel, M. Holzapfel, Lithium Migration at High Temperatures in $Li_4Ti_5O_{12}$ Studied by Neutron Diffraction, Chem. Mater. 23 (2011) 2753–2759.

[80] J. Morales, J. Santos-Peña, R. Trócoli, S. Franger, Electrochemical activity of rock-salt-structured $LiFeO_2/Li_{4/3}Ti_{2/3}O_2$ nanocomposites in lithium cells, J Nanopart Res 10 (2008) 217–226.

[81] F.F. Barblan, Untersuchungen zur Kristallchemie von Fe_2O_3 und Ti O_2, sowie ihrer Alkaliverbindungen, Schweizerische Mineralogische und Petrographische Mitteilungen (1943) 295–352.

[82] H. Kleykamp, Phase equilibria in the Li–Ti–O system and physical properties of Li_2TiO_3, Fusion Engineering and Design 61-62 (2002) 361–366.

[83] G. Izquierdo, A.R. West, Phase equilibria in the system Li_2O-TiO_2, Materials Research Bulletin 15 (1980) 1655–1660.

[84] M. Hara, Y. Togashi, M. Matsuyama, Y. Oya, K. Okuno, Crystal structure change of $Li_{2+x}TiO_{3+y}$ tritium breeder under moist air, Journal of Nuclear Materials 404 (2010) 217–221.

[85] N. Roux, Compilation of blanket-relevant properties data for Li_2TiO_3 pebbles and pebble beds, in: M. Yamawaki (Ed.), CBBI-9: Proceedings of the ninth international workshop on ceramic breeder blanket interactions, Japan, 2001, p. 253.

[86] P. Gierszewski, H. Hamilton, J. Miller, J. Sullivan, R. Verrall, J. Earnshaw, D. Ruth, R. Macauley-Newcombe, G. Williams, Canadian ceramic breeder technology: Recent results, Fusion Engineering and Design 27 (1995) 297–306.

[87] P. Gierszewski, Review of properties of lithium metatitanate, Fusion Engineering and Design 39-40 (1998) 739–743.

[88] T. Hoshino, Development of methods for reprocessing and reuse of tritium breeder materials in broader approach activities, Journal of Nuclear Materials 442 (2013) S425-S428.

[89] K. Tsuchiya, H. Kawamura, K. Fuchinoue, H. Sawada, K. Watarumi, Fabrication development and preliminary characterization of Li_2TiO_3 pebbles by wet process, Journal of Nuclear Materials 258-263 (1998) 1985–1990.

[90] T. Hoshino, F. Oikawa, Trial fabrication tests of advanced tritium breeder pebbles using sol–gel method, Fusion Engineering and Design 86 (2011) 2172–2175.

[91] T. Hoshino, M. Nakamichi, Development of fabrication technologies for advanced breeding functional materials For DEMO reactors, Fusion Engineering and Design 87 (2012) 486–492.

[92] T. Hoshino, Development of fabrication technologies for advanced tritium breeder pebbles by the sol–gel method, Fusion Engineering and Design 88 (2013) 2264–2267.

[93] A.N. Enyashin, T.A. Denisova, A.L. Ivanovskii, Structural, electronic properties and stability of metatitanic acid (H_2TiO_3) nanotubes, Chemical Physics Letters 484 (2009) 44–47.

[94] K. Tsuchiya, H. Kawamura, Development of wet process with substitution reaction for the mass production of Li_2TiO_3 pebbles, Journal of Nuclear Materials 283-287 (2000) 1380–1384.

[95] T. Hoshino, Trial examination of direct pebble fabrication for advanced tritium breeders by the emulsion method, Fusion Engineering and Design 89 (2014) 1431–1435.

[96] M.H.H. Kolb, K. Mukai, R. Knitter, T. Hoshino, Li_4SiO_4 based breeder ceramics with Li_2TiO_3, $LiAlO_2$ and $Li_XLa_YTiO_3$ additions, part I: Fabrication, Fusion Engineering and Design 115 (2017) 39–48.

[97] T. Hoshino, Optimization of sintering conditions of advanced tritium breeder pebbles fabricated by the emulsion method, Fusion Engineering and Design 98-99 (2015) 1788–1791.

[98] T. Nishitani, T. Yamanishi, H. Tanigawa, M. Nakamichi, T. Nozawa, T. Hoshino, K. Ochiai, Blanket material and technology developments toward DEMO under the Broader Approach framework, Fusion Engineering and Design 89 (2014) 1699–1703.

[99] W. Dienst, H. Zimmermann, Investigation of the mechanical properties of ceramic breeder materials, Journal of Nuclear Materials 155-157 (1988) 476–479.

[100] N.S. Kulkarni, T.M. Besmann, K.E. Spear, Thermodynamic Optimization of Lithia-Alumina, Journal of the American Ceramic Society 91 (2008) 4074–4083.

[101] C. Denuziere, N. Roux, Data and properties of lithium aluminate γ $LiAlO_2$, France, 1988.

[102] M. Marezio, J.P. Remeika, High-Pressure Synthesis and Crystal Structure of α-$LiAlO_2$, The Journal of Chemical Physics 44 (1966) 3143–3144.

[103] M. Marezio, The crystal structure and anomalous dispersion of γ-$LiAlO_2$, Acta Cryst 19 (1965) 396–400.

[104] J. Charpin, F. Botter, M. Briec, B. Rasneur, E. Roth, N. Roux, J. Sannier, Investigation of γ lithium aluminate as tritium breeding material for a fusion reactor blanket, Fusion Engineering and Design 8 (1989) 407–413.

[105] S. Claus, H. Kleykamp, W. Smykatz-Kloss, Phase equilibria in the Li_4SiO_4-Li_2SiO_3 region of the pseudobinary Li_2O-SiO_2 system, Journal of Nuclear Materials 230 (1996) 8–11.

[106] J.A. Mergos, C.T. Dervos, Structural and dielectric properties of Li_2O-doped TiO_2, Materials Characterization 60 (2009) 848–857.

[107] R.J. Cava, D.W. Murphy, S. Zahurak, A. Santoro, R.S. Roth, The crystal structures of the lithium-inserted metal oxides $Li_{0.5}TiO_2$ anatase, $LiTi_2O_4$ spinel, and $Li_2Ti_2O_4$, Journal of Solid State Chemistry 53 (1984) 64–75.

[108] D.A.H. Hanaor, M.H.H. Kolb, Y. Gan, M. Kamlah, R. Knitter, Solution based synthesis of mixed-phase materials in the Li_2TiO_3–Li_4SiO_4 system, Journal of Nuclear Materials 456 (2015) 151–161.

[109] Heinrich Hertz, Ueber die Berührung fester elastischer Körper, Journal für die reine und angewandte Mathematik (Crelle's Journal) 1882 (1882).

[110] M.T. Huber, Zur Theorie der Berührung fester elastischer Körper, Ann. Phys. 319 (1904) 153–163.

[111] K.L. Johnson, Contact mechanics, 1st ed., Cambridge University Press, Cambridge [Cambridgeshire], New York, 1987.

[112] S. Timoshenko, J.N. Goodier, Theory of elasticity, 3rd ed., McGraw-Hill, New York, 1970.

[113] B.R. Lawn, Indentation of Ceramics with Spheres: A Century after Hertz, Journal of the American Ceramic Society 81 (1998) 1977–1994.

[114] B.R. Lawn, S.K. Lee, I.M. Peterson, S. Wuttiphan, Model of Strength Degradation from Hertzian Contact Damage in Tough Ceramics, Journal of the American Ceramic Society 81 (1998) 1509–1520.

[115] R.J. Verrall, A sphere compression test for measuring the mechanical properties of dental composite materials, Journal of Dentistry 4 (1976) 11–14.

[116] P.H. Shipway, I.M. Hutchings, Fracture of brittle spheres under compression and impact loading. I. Elastic stress distributions, Philosophical Magazine A 67 (1993) 1389–1404.

[117] Y. Hiramatsu, Y. Oka, Determination of the tensile strength of rock by a compression test of an irregular test piece, International Journal of Rock Mechanics and Mining Sciences & Geomechanics Abstracts 3 (1966) 89–90.

[118] G. Wijk, Some new theoretical aspects of indirect measurements of the tensile strength of rocks, International Journal of Rock Mechanics and Mining Sciences & Geomechanics Abstracts 15 (1978) 149–160.

[119] K.T. Chau, X.X. Wei, R.H.C. Wong, T.X. Yu, Fragmentation of brittle spheres under static and dynamic compressions: experiments and analyses, Mechanics of Materials 32 (2000) 543–554.

[120] S. Zhao, Y. Gan, M. Kamlah, Spherical ceramic pebbles subjected to multiple non-concentrated surface loads, International Journal of Solids and Structures 49 (2012) 658–671.

[121] W.G. Luscher, J.R. Hellmann, A.E. Segall, D.L. Shelleman, B.E. Scheetz, A Critical Review of the Diametral Compression Method for Determining the Tensile Strength of Spherical Aggregates, J. Test. Eval. 35 (2007) 1–6.

[122] B.R. Lawn, N.P. Padture, F. Guiberteau, H. Cai, A model for microcrack initiation and propagation beneath hertzian contacts in polycrystalline ceramics, Acta Metallurgica et Materialia 42 (1994) 1683–1693.

[123] H. Cai, S.M.A. Kalceff, B.R. Lawn, Deformation and fracture of mica-containing glass-ceramics in Hertzian contacts, J. Mater. Res. 9 (1994) 762–770.

[124] F. Guiberteau, N.P. Padture, B.R. Lawn, Effect of Grain Size on Hertzian Contact Damage in Alumina, Journal of the American Ceramic Society 77 (1994) 1825–1831.

[125] P. Kudinov, V. Kudinova, T.-N. Dinh, Molten Oxidic Particle Fracture during Quenching in Water, 7th International Conference on Multiphase Flow - ICMF 2010 Proceedings (2010) P2.32.

[126] F.E. Genceli, M. Rodriguez Pascual, S. Kjelstrup, G.-J. Witkamp, Coupled Heat and Mass Transfer during Crystallization of $MgSO_4 \cdot 7H_2O$ on a Cooled Surface, Crystal Growth & Design 9 (2009) 1318–1326.

[127] J.M. Forgac, J.C. Angus, Solidification of metal spheres, MTB 12 (1981) 413–416.

[128] J.M. Forgac, T.P. Schur, J.C. Angus, Solidification of a Sphere: The Effects of Thermal Contraction and Density Change Upon Freezing, J. Appl. Mech. 46 (1979) 83.

[129] G. Grünberg, Über den in einer isotropen Kugel durch ungleichförmige Erwärmung erregten Spannungszustand, Z. Physik (Zeitschrift für Physik) 35 (1926) 548–555.

[130] W. Weibull, A statistical distribution function of wide applicability, Journal of Applied Mechanics (1951) 293–297.

[131] A.A. Griffith, The theory of rupture, in: C.B. Biezeno, J.M. Burgers (Eds.), Proceedings of the first International Congress for Applied Mechanics, Delft, 1924, pp. 56–63.

[132] D. Munz, T. Fett, Ceramics: Mechanical Properties, Failure Behaviour, Materials Selection, Springer Berlin Heidelberg, Berlin, Heidelberg, 1999.

[133] S. Zhao, Y. Gan, M. Kamlah, T. Kennerknecht, R. Rolli, Influence of plate material on the contact strength of Li_4SiO_4 pebbles in crush tests and evaluation of the contact strength in pebble–pebble contact, Engineering Fracture Mechanics 100 (2013) 28–37.

[134] R.K. Annabattula, Y. Gan, S. Zhao, M. Kamlah, Mechanics of a crushable pebble assembly using discrete element method, Journal of Nuclear Materials 430 (2012) 90–95.

[135] S. Zhao, Y. Gan, M. Kamlah, Failure initiation and propagation of Li_4SiO_4 pebbles in fusion blankets, Fusion Engineering and Design 88 (2013) 8–16.

[136] R. Danzer, P. Supancic, J. Pascual, T. Lube, Fracture statistics of ceramics – Weibull statistics and deviations from Weibull statistics, Engineering Fracture Mechanics 74 (2007) 2919–2932.

[137] R.B. Abernethy, The new Weibull handbook: Reliability & statistical analysis for predicting life, safety, risk, support costs, failures, and forecasting warranty claims, substantiation and accelerated testing, using Weibull, Log normal, Crow-AMSAA, Probit, and Kaplan-Meier models, 5th ed., R.B. Abernethy, North Palm Beach, Fla., 2008.

[138] M. Tiryakioğlu, Weibull Analysis of Mechanical Data for Castings II: Weibull Mixtures and Their Interpretation, Metall and Mat Trans A 46 (2015) 270–280.

[139] ReliaSoft Corporation, Life Data Analysis Reference Book, Available under a a Creative Commons Attribution-NonCommercial-ShareAlike 4.0 International License, 2015.

[140] H. Hirose, Bias correction for the maximum likelihood estimates in the two-parameter Weibull distribution, IEEE Trans. Dielect. Electr. Insul. 6 (1999) 66–68.

[141] R. Ross, Bias and standard deviation due to Weibull parameter estimation for small data sets, IEEE Trans. Dielect. Electr. Insul. 3 (1996) 28–42.

[142] National Physical Laboratory (Great Britain), A guide to the measurement of humidity, Institute of Measurement and Control, [London], 1996.

[143] K.N. Marsh, Recommended reference materials for the realization of physicochemical properties, Blackwell Scientific Publications, Oxford [Oxfordshire], Boston, 1987.

[144] I. Ricapito, A. Ciampichetti, P. Agostini, G. Benamati, Tritium processing systems for the helium cooled pebble bed test blanket module, Fusion Engineering and Design 83 (2008) 1461–1465.

[145] R Core Team, R: A Language and Environment for Statistical Computing, R Foundation for Statistical Computing, Vienna, Austria (2014).

[146] Jürgen Symynck, abrem: Abernethy Reliability Methods, R package version 0.1.17/r102 (2014).

[147] Z. Hashin, S. Shtrikman, A variational approach to the theory of the elastic behaviour of multiphase materials, Journal of the Mechanics and Physics of Solids 11 (1963) 127–140.

[148] M. Wang, N. Pan, Elastic property of multiphase composites with random microstructures, Journal of Computational Physics 228 (2009) 5978–5988.

[149] R.W. Zimmerman, Hashin-Shtrikman bounds on the poisson ratio of a composite material, Mechanics Research Communications 19 (1992) 563–569.

[150] F. Yan, D.-h. Han, Theoretical validation of fluid substitution by Hashin-Shtrikman bounds, in: SEG Technical Program Expanded Abstracts 2011, pp. 2251–2255.

[151] A.C. Larson, R.B. von Dreele, General Structure Analysis System (GSAS), Los Alamos National Laboratory Report LAUR (2000) 86–748.

[152] R.W. Cheary, A.A. Coelho, J.P. Cline, Fundamental parameters line profile fitting in laboratory diffractometers, Journal of Research of the National Institute of Standards and Technology 109 (2004) 1–25.

[153] H. Berger, Study of the Kα emission spectrum of copper, X-Ray Spectrom. 15 (1986) 241–243.

[154] W.A. Dollase, Correction of intensities for preferred orientation in powder diffractometry: Application of the March model, J Appl Crystallogr 19 (1986) 267–272.

[155] A. March, Mathematische Theorie der Regelung nach der Korngestah bei affiner Deformation, Zeitschrift für Kristallographie - Crystalline Materials 81 (1932).

[156] E. Zolotoyabko, Determination of the degree of preferred orientation within the March–Dollase approach, J Appl Crystallogr 42 (2009) 513–518.

[157] A. Boulineau, L. Croguennec, C. Delmas, F. Weill, Structure of Li_2MnO_3 with different degrees of defects, Solid State Ionics 180 (2010) 1652–1659.

[158] M.I. Aroyo, J.M. Perez-Mato, C. Capillas, E. Kroumova, S. Ivantchev, G. Madariaga, A. Kirov, H. Wondratschek, Bilbao Crystallographic Server: I. Databases and crystallographic computing programs, Zeitschrift für Kristallographie 221 (2006).

[159] M.I. Aroyo, A. Kirov, C. Capillas, J.M. Perez-Mato, H. Wondratschek, Bilbao Crystallographic Server. II. Representations of crystallographic point groups and space groups, Acta crystallographica. Section A, Foundations of crystallography 62 (2006) 115–128.

[160] M.I. Aroyo, J.M. Perez-Mato, D. Orobengoa, E. Tasci, G. de La Flor, A. Kirov, Crystallography online: Bilbao crystallographic server, Bulgarian Chemical Communications 43 (2011) 183–197.

[161] A.A. Coelho, Whole-profile structure solution from powder diffraction data using simulated annealing, J Appl Crystallogr 33 (2000) 899–908.

[162] P.W. Stephens, Phenomenological model of anisotropic peak broadening in powder diffraction, J Appl Crystallogr 32 (1999) 281–289.

[163] D.L. Bish, S.A. Howard, Quantitative phase analysis using the Rietveld method, J Appl Crystallogr 21 (1988) 86–91.

[164] R.J. Hill, C.J. Howard, Quantitative phase analysis from neutron powder diffraction data using the Rietveld method, J Appl Crystallogr 20 (1987) 467–474.

[165] P. Riello, Quantitative Analysis of Amorphous Fraction in the Study of the Microstructure of Semi-crystalline Materials, in: R. Hull, R.M. Osgood, J. Parisi, H. Warlimont, E.J. Mittemeijer, P. Scardi (Eds.), Diffraction Analysis of the Microstructure of Materials, Springer Berlin Heidelberg, Berlin, Heidelberg, 2004, pp. 167–184.

[166] I.C. Madsen, N.V.Y. Scarlett, A. Kern, Description and survey of methodologies for the determination of amorphous content via X-ray powder diffraction, Zeitschrift für Kristallographie 226 (2011) 944–955.

[167] L. Spiess, L. Spiess, H. Behnken, C. Genzel, R. Schwarzer, G. Teichert, Moderne Röntgenbeugung: Röntgendiffraktometrie für Materialwissenschaftler, Physiker und Chemiker, 2nd ed., Vieweg+Teubner Verlag / GWV Fachverlage, Wiesbaden, Wiesbaden, 2009.

[168] D.S. Young, B.S. Sachais, L.C. Jefferies, The Rietveld Method, 1993.

[169] R.G. Munro, Evaluated Material Properties for a Sintered alpha-Alumina, Journal of the American Ceramic Society 80 (1997) 1919–1928.

[170] Y. Okada, Y. Tokumaru, Precise determination of lattice parameter and thermal expansion coefficient of silicon between 300 and 1500 K, J. Appl. Phys. 56 (1984) 314–320.

[171] E. Macherauch, H.-W. Zoch (Eds.), Praktikum in Werkstoffkunde, Vieweg+Teubner, Wiesbaden, 2011.

[172] Y. Idemoto, J.W. Richardson, N. Koura, S. Kohara, C.-K. Loong, Crystal structure of $(Li_xK_{1-x})_2CO_3$ (x = 0, 0.43, 0.5, 0.62, 1) by neutron powder diffraction analysis, Journal of Physics and Chemistry of Solids 59 (1998) 363–376.

[173] H. Völlenkle, Verfeinerung der Kristallstrukturen von Li_2SiO_3 und Li_2GeO_3, Zeitschrift für Kristallographie - Crystalline Materials 154 (1981).

[174] Z.K. Heiba, K. El-Sayed, Structural and anisotropic thermal expansion correlation of Li_2ZrO_3 at different temperatures, J Appl Crystallogr 35 (2002) 634–636.

[175] L. Zhao, X. Long, X. Chen, C. Xiao, Y. Gong, Q. Guan, J. Li, L. Xie, X. Chen, S. Peng, Design, synthesis and characterization of the advanced tritium breeder: $Li_{4+x}Si_{1-x}Al_xO_4$ ceramics, Journal of Nuclear Materials 467 (2015) 911–916.

[176] J.H. Hubbell, S.M. Seltzer, Tables of X-Ray Mass Attenuation Coefficients and Mass Energy-Absorption Coefficients from 1 keV to 20 MeV for Elements Z = 1 to 92 and 48 Additional Substances of Dosimetric Interest, 1996, http://www.nist.gov/pml/data/xraycoef/, accessed 15 June 2015.

[177] S.L. Mair, The electron distribution of the hydroxide ion in lithium hydroxide, Acta Cryst A 34 (1978) 542–547.

[178] R.P. Gunawardane, J.G. Fletcher, M.A.K.L. Dissanayake, R.A. Howie, A.R. West, Crystal Structure Refinement of Li_4TiO_4 Containing Tetrahedrally Coordinated Ti^{4+} and Tetragonally Packed Oxide Ions, Journal of Solid State Chemistry 112 (1994) 70–72.

[179] K.V.K. Rao, S.V.N. Naidu, L. Iyengar, Thermal Expansion of Rutile and Anatase, J American Ceramic Society 53 (1970) 124–126.

[180] Y.S. Touloukian, R.K. Kirby, E.R. Taylor, T.Y.R. Lee, Thermal Expansion - Nonmetallic Solids, in: Y.S. Touloukian (Ed.), Thermophysical Properties of Matter - the TPRC Data Series, 1977.

[181] M.E. Payton, M.H. Greenstone, N. Schenker, Overlapping confidence intervals or standard error intervals: What do they mean in terms of statistical significance?, Journal of Insect Science 3 (2003) 1–6.

[182] J.R. Lanzante, A Cautionary Note on the Use of Error Bars, J. Climate 18 (2005) 3699–3703.

[183] M.H.H. Kolb, R. Knitter, T. Hoshino, Li_4SiO_4 based breeder ceramics with Li_2TiO_3, $LiAlO_2$ and $Li_XLa_YTiO_3$ additions, part II: Pebble properties, Fusion Engineering and Design 115 (2017) 6–16.

[184] P.H. Shipway, I.M. Hutchings, Fracture of brittle spheres under compression and impact loading. II. Results for lead-glass and sapphire spheres, Philosophical Magazine A 67 (1993) 1405–1421.

[185] A.G. Duba, H.C. Heard, The brittle-ductile transition in rocks: The Heard volume, American Geophysical Union, Washington, D.C, op. 1990.

[186] H. Cai, N.P. Padture, B.M. Hooks, B.R. Lawn, Flaw tolerance and toughness curves in two-phase particulate composites: SiC/glass system, Journal of the European Ceramic Society 13 (1994) 149–157.

[187] L. Kogut, I. Etsion, Elastic-Plastic Contact Analysis of a Sphere and a Rigid Flat, J. Appl. Mech. 69 (2002) 657.

[188] R.O. Ritchie, Mechanisms of Fatigue-Crack Propagation in Ductile and Brittle Solids, International Journal of Fracture 100 (1999) 55–83.

[189] J.J. Kruzic, R.L. Satet, M.J. Hoffmann, R.M. Cannon, R.O. Ritchie, The Utility of R-Curves for Understanding Fracture Toughness-Strength Relations in Bridging Ceramics, J American Ceramic Society 91 (2008) 1986–1994.

[190] D.A. Gorham, A.D. Salman, The failure of spherical particles under impact, Wear 258 (2005) 580–587.

[191] O. Leys, T. Bergfeldt, M.H.H. Kolb, R. Knitter, A.A. Goraieb, The reprocessing of advanced mixed lithium orthosilicate/metatitanate tritium breeder pebbles, Fusion Engineering and Design 107 (2016) 70–74.

[192] S. Ogawa, Y. Masuko, H. Kato, H. Yuyama, Y. Sakai, E. Niwa, T. Hashimoto, K. Mukai, T. Hosino, K. Sasaki, Li vaporization property of two-phase material of Li_2TiO_3 and Li_2SiO_3 for tritium breeder, Fusion Engineering and Design 98-99 (2015) 1859–1863.

[193] T. Hoshino, M. Dokiya, T. Terai, Y. Takahashi, M. Yamawaki, Non-stoichiometry and its effect on thermal properties of Li_2TiO_3, Fusion Engineering and Design 61-62 (2002) 353–360.

[194] H. Kashimura, M. Nishikawa, K. Katayama, S. Matsuda, M. Shimozori, S. Fukada, T. Hoshino, Mass loss of Li_2TiO_3 pebbles and Li_4SiO_4 pebbles, Fusion Engineering and Design 88 (2013) 2202–2205.

[195] K. Mukai, K. Sasaki, T. Terai, A. Suzuki, T. Hoshino, Vaporization property and crystal structure of lithium metatitanate with excess Li, Journal of Nuclear Materials 442 (2013) S447-S450.

[196] T. Hoshino, M. Yasumoto, K. Tsuchiya, K. Hayashi, H. Nishimura, A. Suzuki, T. Terai, Vapor species evolved from Li_2TiO_3 heated at high temperature under various conditions, Fusion Engineering and Design 81 (2006) 555–559.

[197] T. Hoshino, M. Yasumoto, K. Tsuchiya, K. Hayashi, H. Nishimura, A. Suzuki, T. Terai, Non-stoichiometory and vaporization characteristic of $Li_{2.1}TiO_{3.05}$ in hydrogen atmosphere, Fusion Engineering and Design 82 (2007) 2269–2273.

[198] R. Yamamoto, K. Katayama, T. Hoshino, T. Takeishi, S. Fukada, Li mass loss from Li_2TiO_3 with excess Li pebbles fabricated by optimized sintering condition, Fusion Engineering and Design (2017).

[199] G. Piazza, J. Reimann, E. Günther, R. Knitter, N. Roux, J.D. Lulewicz, Behaviour of ceramic breeder materials in long time annealing experiments, Fusion Engineering and Design 58-59 (2001) 653–659.

[200] R.J. Brook, Pore-Grain Boundary Interactions and Grain Growth, J American Ceramic Society 52 (1969) 56–57.

[201] W.W. Mullins, The statistical self-similarity hypothesis in grain growth and particle coarsening, J. Appl. Phys. 59 (1986) 1341–1349.

[202] C.B. Carter, M.G. Norton, Ceramic materials: Science and engineering, 2nd ed., Springer, New York, NY, 2013.

[203] M. Kato, Hall-Petch Relationship and Dislocation Model for Deformation of Ultrafine-Grained and Nanocrystalline Metals, Mater. Trans. 55 (2014) 19–24.

[204] N.L. Johnson, S. Kotz, N. Balakrishnan, Continuous univariate distributions, 2nd ed., Wiley, New York, 1994-1995.

[205] F.F. Ferreira, E. Granado, W. Carvalho, S.W. Kycia, D. Bruno, R. Droppa, X-ray powder diffraction beamline at D10B of LNLS: Application to the Ba_2FeReO_6 double perovskite, Journal of synchrotron radiation 13 (2006) 46–53.

[206] S. Pillet, M. Souhassou, C. Lecomte, K. Schwarz, P. Blaha, M. Rérat, A. Lichanot, P. Roversi, Recovering experimental and theoretical electron densities in corundum using the multipolar model: IUCr Multipole Refinement Project, Acta crystallographica. Section A, Foundations of crystallography 57 (2001) 290–303.

[207] D.M. Többens, N. Stüßer, K. Knorr, H.M. Mayer, G. Lampert, E9: The New High-Resolution Neutron Powder Diffractometer at the Berlin Neutron Scattering Center, MSF 378-381 (2001) 288–293.

List of abbreviations

CAD	Computer Aided Design
DEMO	Demonstration Power Plant
DIN	Deutsche Industrie Norm
D-T	Deuterium-Tritium
EDX	Energy Dispersive X-Ray spectroscopy
EM	Emulsion Method
EN	European Standard
EU	European Union
EUROFER	European Reduced Activation Ferritic/Martensitic Steel
FPA	Fundamental Parameters Approach
HCPB	Helium Cooled Pebble Bed Blanket
ICDD	International Centre for Diffraction Data
ICP-AES	Inductively Coupled Plasma Atomic Emission Spectrometry
ICP-OES	Inductively Coupled Plasma Optical Emission Spectrometry
ICSC	International Chemical Safety Card
ICSD	Inorganic Crystal Structure Database
ITER	International Thermonuclear Experimental Reactor
JAEA	Japan Atomic Energy Agency
KALOS	Karlsruhe LithiumOrthosilicate
KIT	Karlsruhe Institute of Technology
LAO	Lithium Aluminate
LCO	Lithium Carbonate
LMT	Lithium Metatitanate
LOS	Lithium Orthosilicate
LOT	Lithium Orthotitanate
LTA	Long-term annealing
MLE	Maximum Likelihood Estimation
PC	Personal Computer
PO	Preferred Orientation
PVA	Polyvinyl Alcohol
QST	National Institutes for Quantum and Radiological Science and Technology
RBA	Reduced Bias Adjustment
SEM	Scanning Electron Microscopy
SG	Sol-Gel method
TBM	Test Blanket Module
TBR	Tritium Breeding Ratio
TDMS	Technical Data Management Streaming
XRD	X-Ray Diffraction
XRF	X-Ray Fluorescence spectroscopy
DOC	Degree of Crystallinity

List of figures

List of tables

Appendix A

In this appendix, further information from the literature review is given. For the most part, the Weibull distribution in general as well as in view of fracture mechanics is addressed.

Tentative phase diagram Li-Ti-O

In Figure A.1, the tentative isothermal section of the system Li-Ti-O according to Kleykamp [82] is shown. It is clear, that several phases exist in which the valence of titanium is reduced with respect to the most commonly observed state of +4.

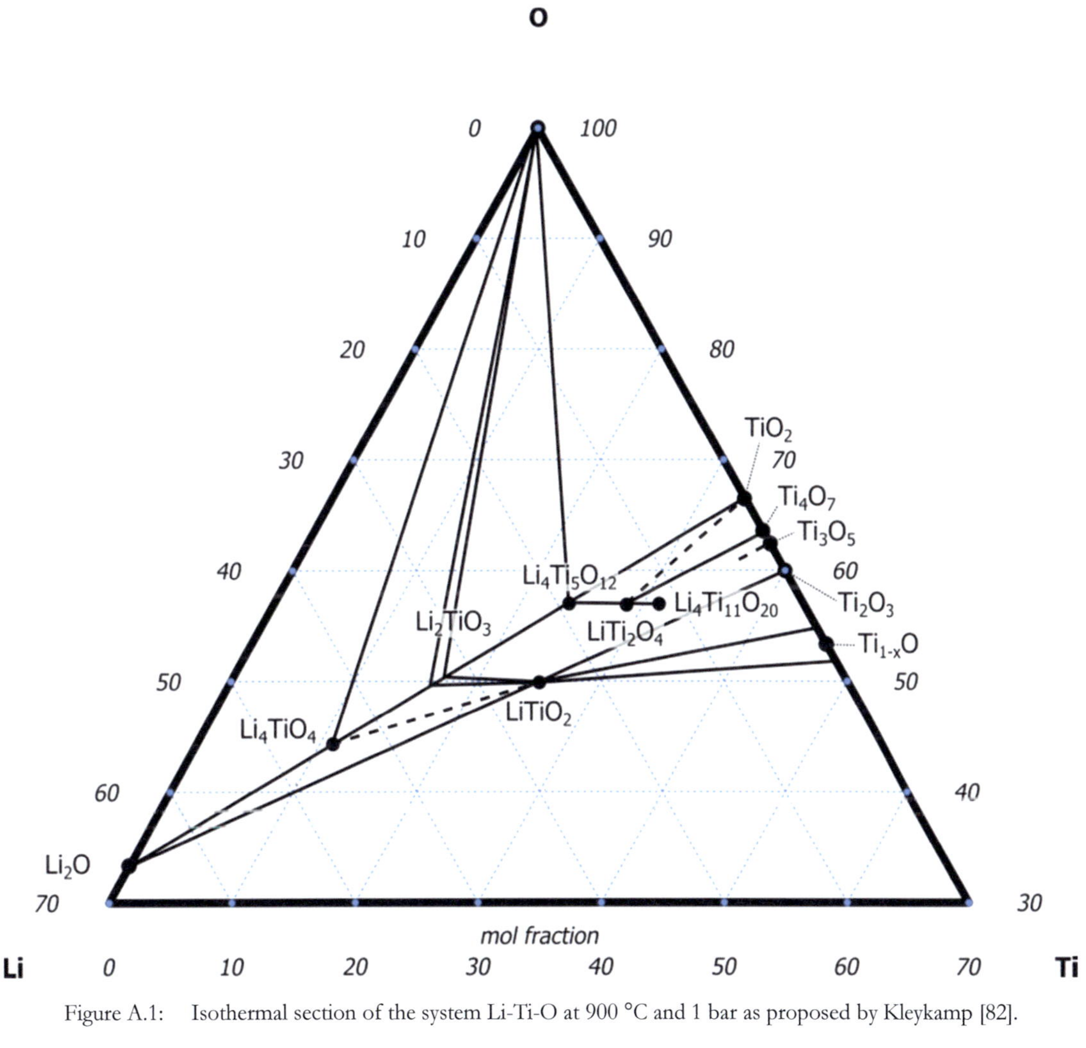

Figure A.1: Isothermal section of the system Li-Ti-O at 900 °C and 1 bar as proposed by Kleykamp [82].

Derivation of the Weibull distribution

Weibull defines the probability of an occurring failure as

$$F(x) = 1 - e^{-\varphi(x)} \tag{A.1}$$

which conveniently deals with 'weakest-link-in-a-chain' failure as briefly outlined by Weibull [130] and which is reproduced in the following. If P_n is defined as failure probability of a chain of n elements, the nonfailure of the chain relates to the nonfailure of its individual segments as

$$(1 - P_n) = (1 - P)^n \tag{A.2}$$

with P as the failure probability of an individual link. In case the failure probability P can be described by equation (A.1), it follows that

$$(1 - P)^n = e^{-n\varphi(x)}. \tag{A.3}$$

Therefore, the failure of the whole chain is simply described by

$$P_n = 1 - e^{-n\varphi(x)}. \tag{A.4}$$

Weibull specifies the function (x) as the simplest function to be positive and non-decreasing, while vanishing at a given value, which can be different from zero.

$$\varphi(x) = \frac{(x - x_u)^m}{x_0} \tag{A.5}$$

with the parameters m, x_u and x_0 satisfies the requests. The so-derived equation of the failure probability is usually written as

$$F(t) = 1 - e^{-\left(\frac{t-t_0}{\eta}\right)^{\beta}} \tag{A.6}$$

and is referred to as 3-parameter (cumulative) Weibull distribution function. Here t denotes a time or age parameter of any possible sort which is related to the failure probability of the investigated system. t_0 represents a shift in origin which is usually attributed to a preexisting aging of the specimens if t_0 is negative. In case t_0 is positive, the specimens show a period in which no failure occurs.

The other two parameters β and η are determinative for the characteristic of the Weibull distribution. η, which is called scale parameter or Characteristic Life. The shape of the distribution is determined by β, which is thus also called shape parameter. It defines whether the hazard function or instantaneous failure rate

$$h(t) = \left(\frac{\beta}{\eta}\right)\left(\frac{t - t_0}{\eta}\right)^{\beta-1} \tag{A.7}$$

increases, decreases or is constant with increasing age.

Properties of the Weibull distribution

The expected value of failure stress, and in this case also its mean value, is given by

$$E(\sigma_c) = \mu_{\sigma_c} = \sigma_0 \Gamma\left(1 + \frac{1}{m}\right) \tag{A.8}$$

for a continuous random variable σ_c distributed according to equation (2.17) [204], with Γ denoting the gamma function

$$\Gamma(u) = \int_0^\infty x^{u-1} e^{-u} dx. \tag{A.9}$$

For values of m larger than 1 the mean value of the Weibull distribution is predominantly determined by σ_0, whereas it is only marginally influenced by m as illustrated in Figure A.2. This behavior is reversed when m is lower than 1 and tends to zero. For m equal to 1, the mean value of the Weibull distribution is identical to σ_0.

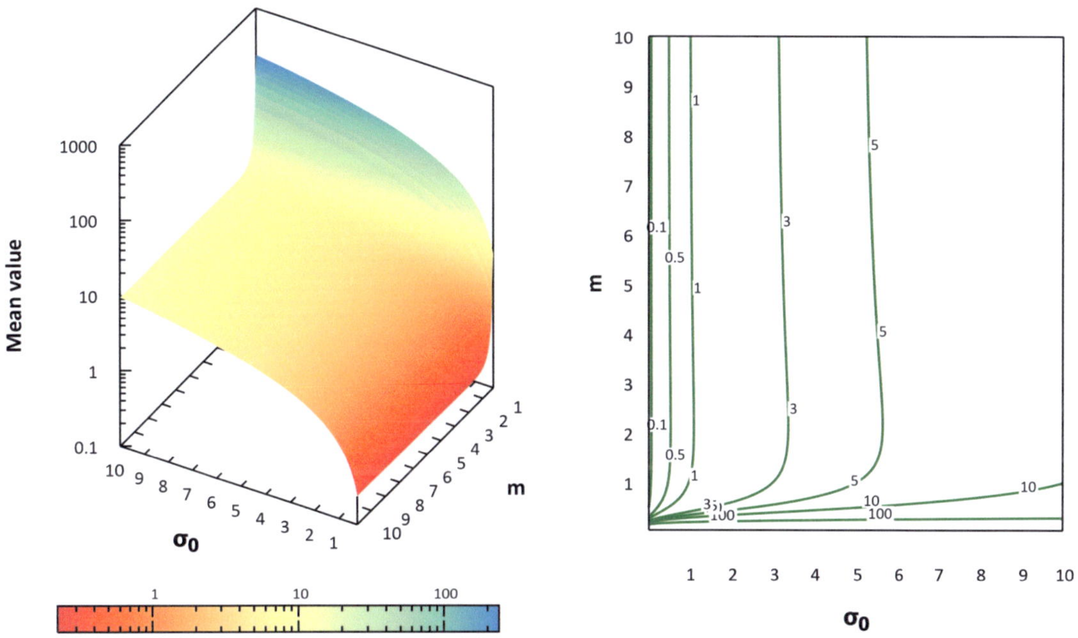

Figure A.2: The mean value of Weibull distributions as a function of the Weibull parameters m and σ_0. On the left hand side, the 3 dimensional surface of the equation (A.8) is illustrated, while on the right hand side iso-value contours are plotted.

The variance of σ_c,

$$Var(\sigma_c) = \sigma_{\sigma_c}^2 = \sigma_0{}^2\left(\Gamma\left(1+\frac{2}{m}\right)-\left(\Gamma\left(1+\frac{1}{m}\right)\right)^2\right) \tag{A.10}$$

equals the square of the standard deviation $\sigma_{\sigma c}$ for a continuous random variable [204]. As opposed to the expected value function (A.8) both Weibull parameters generally contribute significantly to the resultant variance. The function of equation (A.10) is plotted in Figure A.3.

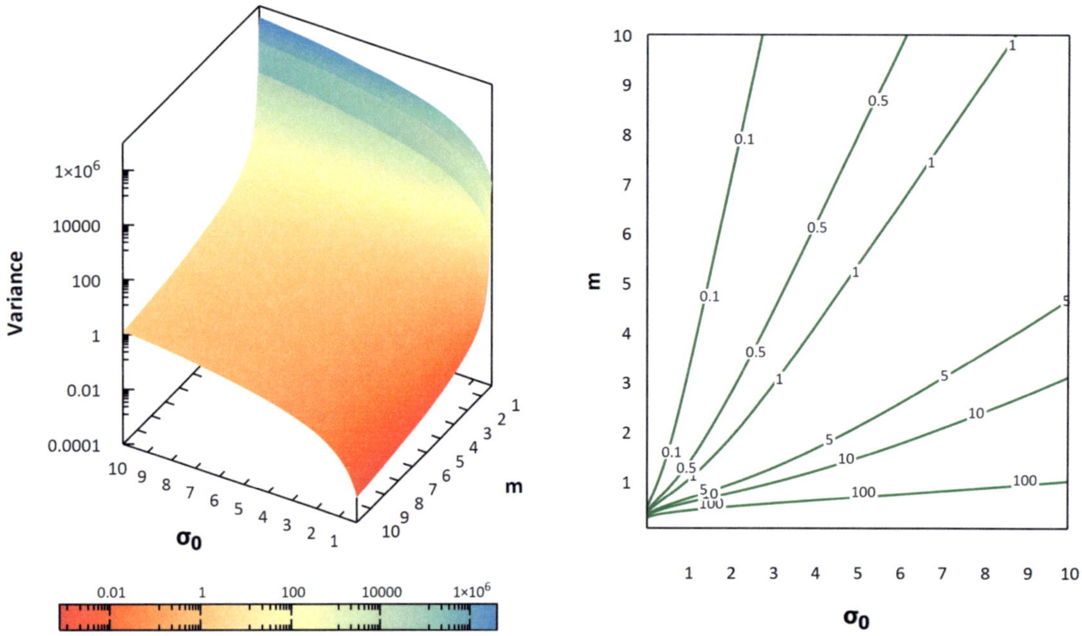

Figure A.3: The variance of Weibull distributions as a function of the Weibull parameters m and σ_0. On the left hand side, the 3 dimensional surface of the equation (A.10) is illustrated, while on the right hand side iso-value contours are plotted.

Derivation of the volume-dependent form of the Weibull distribution

Generally, equation (2.17) is only strictly valid for samples of the same volume. Usually a reference volume V_0 is defined to address different sample volumes. If a sample with ν-times the reference volume is measured, the probability of surviving a given stress is easily given by

$$1 - F(\sigma_c, V) = \left(1 - F(\sigma_c, V_0)\right)^{\nu} = \left(1 - F(\sigma_c, V_0)\right)^{\frac{V}{V_0}}. \tag{A.11}$$

After some rearranging and inserting of equation (2.17), the principal volume dependence of the Weibull statistic

$$1 - F(\sigma_c, V) = e^{\frac{V}{V_0} ln\left(1 - F(\sigma_c, V_0)\right)} = e^{\frac{V}{V_0} ln\left(1 - 1 + e^{-\left(\frac{\sigma_c}{\sigma_v}\right)^m}\right)} = e^{-\frac{V}{V_0}\left(\frac{\sigma_c}{\sigma_v}\right)^m} \tag{A.12}$$

is reached.

Appendix B

In this appendix additional information on the experimental approach and the used parameters is provided. This mostly concerns the models of the Rietveld analysis and the implementation of the maximum likelihood method for the estimation of the Weibull parameters of the samples.

Parameters of the mechanical testing

The exact amount of mechanically tested pebbles, as described in section 3.4, is given in Table B.1 with respect to the sample type and the pebble size.

Table B.1: The number of pebbles, which were mechanically tested in uniaxial compression experiments, is given for each specific sample. Please note that in case of the pebbles which were fabricated by the emulsion or the sol-gel method, the pebble size may deviate significantly from 1000 µm as explained in section 3.4.

Sample	Annealing dur./nominal second phase conc. [d]/[mol%]	Number of tested 500 µm pebbles	Number of tested (≈)1000 µm pebbles
KA-20-LMT: He/H_2	0	50	50
	4	40	40
	32	50	50
	64	40	40
	128	50	50
KA-30-LMT: He/H_2	0	50	50
	4	40	40
	32	50	50
	64	40	40
	128	50	50
EM-LMT: He/H_2	0	–	50
	4	–	40
	32	–	40
	64	–	50
	128	–	40
SG-LMT: He/H_2	0	–	50
	4	–	40
	32	–	40
	64	–	40
	128	–	50

LOS/LMT EM	10	–	50
	20	–	50
	30	–	50
	40	–	50
	50	–	50
	60	–	50
	70	–	50
	80	–	40
	90	–	40
LOS/LAO EM	10	–	40
	20	–	40
	30	–	40

In Table B.2 the estimated Young's moduli and Poisson's ratios are listed. These parameters were determined according to the procedure that is detailed in section 3.4.

Table B.2: Estimated mechanical parameters of all samples.

Sample	Annealing dur./nominal second phase conc.	Young's modulus	Poisson's ratio
	[d]/[mol%]	[GPa]	[-]
KA-20-LMT: He/H_2	0	129.97	0.23
	4	99.25	0.25
	32	108.93	0.25
	64	105.28	0.25
	128	116.43	0.24
KA-30-LMT: He/H_2	0	129.94	0.24
	4	115.86	0.25
	32	133.18	0.24
	64	132.08	0.24
	128	119.04	0.25
EM-LMT: He/H_2	0	230.98	0.29
	4	189.49	0.27
	32	247.11	0.29
	64	265.08	0.30
	128	259.66	0.30
SG-LMT: He/H_2	0	217.82	0.28
	4	206.14	0.28
	32	226.67	0.29
	64	245.07	0.29
	128	239.47	0.29
LOS/LMT EM	10	88.41	0.26
	20	108.56	0.25
	30	108.98	0.26
	40	117.97	0.26
	50	130.36	0.26
	60	126.98	0.28
	70	139.83	0.29
	80	139.54	0.30
	90	147.87	0.31
LOS/LAO EM	10	102.84	0.22
	20	105.81	0.23
	30	108.86	0.24

MLE fitting

The fitting of the failure data is performed by a number of functions written in R, which are detailed in the following.

```
##Hirose-Ross debias function
hrbu<-function(Qx,Qs=NULL)  {
        r<-Qx
        n<-r+Qs
        if (length(Qs)==0||Qs==0) {
        ## This is the Hirose Beta Unbias factor for complete failure samples
        BU<-1/( 1.0115+1.278/r+2.001/r^2+20.35/r^3-46.98/r^4 )
        }else{
                ## This is the Ross Beta Unbias factor for samples with suspensions
                BU<- 1/(1+1.37/(r-1.92)*sqrt(n/r))
                }
        BU
}

##Abernethy debias function
rba<-function(Qx, dist="weibull", basis="median")  {
        if(Qx<3)  {
                stop("insufficient data points")
        }
        if(Qx>343) return(1.0)

        ## factorial(x) is simply gamma(1+x)
        num<-gamma(1+(Qx-2)/2)
        den<-gamma(1+(Qx-3)/2)
        C4<-sqrt(2/(Qx-1))*num/den

                if(dist=="weibull")  {
                        if(basis=="median")  {
                                return(C4^3.5)
                        }
                        if(basis=="mean")  {
                                return(C4^6)
                        }
                }
                if(dist=="lognormal")  {
                return(sqrt(Qx/(Qx-1))/C4)
                }
}

## Function for the calculation of a 2-parameter Weibull fit
mlefitM<-function(Init.Beta  =  2,  Init.Eta  =  1,  minmeth="L-BFGS-B",  x,  s=NULL,  inter-
val=NULL, debmeth="rba-median", debias.Eta=TRUE)  {

        ##Function to minimize
        negLL<-function(x,s,interval,w.vals)  {
                Beta <- w.vals[1]
                Eta  <- w.vals[2]
                suscomp<-0
                intcomp<-0
                failcomp<-sum(dweibull(x,Beta,Eta,log=TRUE))
                if(length(s)>0)  {
                        suscomp<-sum(pweibull(s,Beta,Eta,lower.tail=FALSE,log.p=TRUE))
                }
                if(length(interval)>0)  {
                        right <- interval$right
                        left  <- interval$left
                        intcomp<-
log(prod(pweibull(right,Beta,Eta,lower.tail=TRUE,log.p=FALSE)-
```

```
pweibull(left,Beta,Eta,lower.tail=TRUE,log.p=FALSE)))
                }
                value<- -(failcomp+suscomp+intcomp)
        }
        w.vals <- c(Init.Beta, Init.Eta)

        ##Actual minimization
        y.mle = optim(c(w.vals[1], w.vals[2]), negLL, gr=NULL, method = minmeth, lower =
c(0.1, 0.5), x=x,s=s, interval=interval, hessian = TRUE)

        ##Output
        outvec <- c(y.mle$par[2],y.mle$par[1],-y.mle$value[1],y.mle$convergence[1])

        ##Debias
        negLL.1p<-function(x,s,interval,w.val,beta.val)  {
                Beta <- beta.val
                Eta  <- w.val
                suscomp<-0
                intcomp<-0
                failcomp<-sum(dweibull(x,Beta,Eta,log=TRUE))
                if(length(s)>0)  {
                        suscomp<-sum(pweibull(s,Beta,Eta,lower.tail=FALSE,log.p=TRUE))
                }
                if(length(interval)>0)  {
                        right <- interval$right
                        left  <- interval$left
                        intcomp<-
log(prod(pweibull(right,Beta,Eta,lower.tail=TRUE,log.p=FALSE)-
pweibull(left,Beta,Eta,lower.tail=TRUE,log.p=FALSE)))
                }
                value<- -(failcomp+suscomp+intcomp)
        }
        w.val <- y.mle$par[2]

        Qx<-length(x)+length(interval)
        Qs<-length(s)
                if (debmeth == "hrbu") {
                        outvec[2] <- outvec[2]*hrbu(Qx,Qs)
                        if (debias.Eta == TRUE) {
                                y.mle.encore  =  optim(w.val,  negLL.1p,  gr=NULL,  method  =
minmeth,  lower  =  c(0.1,  0.5),  x=x,s=s,  interval=interval,  beta.val=outvec[2],  hessian  =
TRUE)
                                outvec[1] <- y.mle.encore$par[1]
                        }
                }
                if (debmeth == "rba-mean") {
                        outvec[2] <- outvec[2]*rba((Qx+Qs), dist="weibull", basis="mean")
                        if (debias.Eta == TRUE) {
                                y.mle.encore  =  optim(w.val,  negLL.1p,  gr=NULL,  method  =
minmeth,  lower  =  c(0.1,  0.5),  x=x,s=s,  interval=interval,  beta.val=outvec[2],  hessian  =
TRUE)
                                outvec[1] <- y.mle.encore$par[1]
                        }
                }
                if (debmeth == "rba-median") {
                        outvec[2] <- outvec[2]*rba((Qx+Qs), dist="weibull", basis="median")
                        if (debias.Eta == TRUE) {
                                y.mle.encore  =  optim(w.val,  negLL.1p,  gr=NULL,  method  =
minmeth,  lower  =  c(0.1,  0.5),  x=x,s=s,  interval=interval,  beta.val=outvec[2],  hessian  =
TRUE)
                                outvec[1] <- y.mle.encore$par[1]
                        }
                }

        names(outvec) <- c("Eta", "Beta", "LogL", "Convergence code")
        return(outvec)
```

```
}

## Function for the calculation of the likelihood contours for a 2-parameter Weibull fit
MLEw2pContourM<-function(x,s=NULL,                                                              inter-
val=NULL,CL=0.9,DF=1,MLEfit=NULL,ptDensity=100,RadLimit=1e-5,   debias=TRUE,   debmeth="rba-
median", show=FALSE)  {
        ## limits for accuracy of parameter determination
        ## RadLimit=1.0e-5


        ## Internal functions
        LL<-function(x,s,interval,Beta,Eta)  {
                suscomp<-0
                intcomp<-0
                failcomp<-sum(dweibull(x,Beta,Eta,log=TRUE))
                if(length(s)>0)  {
                        suscomp<-sum(pweibull(s,Beta,Eta,lower.tail=FALSE,log.p=TRUE))
                }
                if(length(interval)>0)  {
                        right <- interval$right
                        left  <- interval$left
                        intcomp<-
log(prod(pweibull(right,Beta,Eta,lower.tail=TRUE,log.p=FALSE)-
pweibull(left,Beta,Eta,lower.tail=TRUE,log.p=FALSE)))
                }
                value<-failcomp+suscomp+intcomp
        }


##start of main procedure
        if(missing(MLEfit))  {
                MLEfit<-MLEw2p_cpp(x,s)
        }
        Beta_hat <- MLEfit[2]
        Eta_hat  <- MLEfit[1]

        MLLx<-LL(x,s,interval,Beta_hat,Eta_hat)
        FF<-1
        if(debias==TRUE)  {
        Nf<-length(x)
        FF<-(Nf-1)/(Nf+0.618)
        }

        ##Initial iteration
                if (length(x)>=40) {
                                shape.llim  <- 0.7*Beta_hat
                                shape.ulim  <- 1.3*Beta_hat
                                scale.llim  <- 0.7*Eta_hat
                                scale.ulim  <- 1.3*Eta_hat
                                } else {
                                shape.llim  <- (0.7/(40/length(x))**2.15)*Beta_hat
                                shape.ulim  <- (1.3*(40/length(x))**2.15)*Beta_hat
                                scale.llim  <- (0.7/(40/length(x))**2.15)*Eta_hat
                                scale.ulim  <- (1.3*(40/length(x))**2.15)*Eta_hat
                                if (scale.ulim > 150) {
                                        scale.ulim <- 150
                                        }
                                }
                                shape.limits <- c(shape.llim,shape.ulim)
                                scale.limits <- c(scale.llim,scale.ulim)

                ContourGrid = matrix(0.0, nrow = ptDensity, ncol = ptDensity)
                shape.vals = seq(shape.limits[1], shape.limits[2], len = ptDensity)
                scale.vals = seq(scale.limits[1], scale.limits[2], len = ptDensity)
                for(i in seq(along = shape.vals)){
                        for(j in seq(along = scale.vals)){
```

```
            ContourGrid[i,j] = LL(x,s,interval,shape.vals[i],scale.vals[j])
                    }
            }
            Contourtemp <- contourLines(shape.vals, scale.vals, ContourGrid, levels =
MLLx-qchisq(CL,DF)/2)
            Contour<-data.frame(Contourtemp[[1]]$y,Contourtemp[[1]]$x)

        ##Final iteration

            maxBeta<-max(Contour[,2])
            minBeta<-min(Contour[,2])
            minEta<-min(Contour[,1])
            maxEta<-max(Contour[,1])

            shape.llim  <- 0.8*minBeta
            shape.ulim  <- 1.2*maxBeta
            scale.llim  <- 0.8*minEta
            scale.ulim  <- 1.2*maxEta

            shape.limits <- c(shape.llim,shape.ulim)
            scale.limits <- c(scale.llim,scale.ulim)

            ContourGrid = matrix(0.0, nrow = ptDensity, ncol = ptDensity)
            shape.vals = seq(shape.limits[1], shape.limits[2], len = ptDensity)
            scale.vals = seq(scale.limits[1], scale.limits[2], len = ptDensity)
            for(i in seq(along = shape.vals)){
                    for(j in seq(along = scale.vals)){
            ContourGrid[i,j] = LL(x,s,interval,shape.vals[i],scale.vals[j])
                    }
            }
            Contourtemp <- contourLines(shape.vals, scale.vals, ContourGrid, levels =
MLLx-qchisq(CL,DF)/2)
            Contour<-data.frame(Contourtemp[[1]]$y,Contourtemp[[1]]$x)
            names(Contour) <- c("Eta", "Beta")



        U.rba<-1
        if(debias==TRUE)  {

        ##Debias
        Qx<-length(x)+length(interval)
        Qs<-length(s)
        if (length(debmeth)>0) {
                if (debmeth == "hrbu") {
                        U.rba <- hrbu(Qx,Qs)
                        }
                if (debmeth == "rba-mean") {
                        U.rba <- rba((Qx+Qs), dist="weibull", basis="mean")
                        }
                if (debmeth == "rba-median") {
                        U.rba <- rba((Qx+Qs), dist="weibull", basis="median")
                        }
                }
        }
        Contour[2]<-Contour[2]*U.rba

        if(show==TRUE)  {
                maxBeta<-max(Contour[,2])
                minBeta<-min(Contour[,2])
                minEta<-min(Contour[,1])
                maxEta<-max(Contour[,1])

                ylo<-floor(minBeta)
                yhi<-floor(maxBeta)+1
```

```
            EtaDec<-10^(floor(log(minEta)/log(10))-1)
            xlo<-EtaDec*(floor(minEta/EtaDec)-1)
            xhi<-EtaDec*(  floor(maxEta/EtaDec)+1    )

            plot(Eta_hat,Beta_hat*rba,xlim=c(xlo,xhi),ylim=c(ylo,yhi))
            lines(Contour)
    }
Contour
}

## Function for the calculation of the confidence boundaries for a 2-parameter Weibull fit
MLEw2pBoundsM<-function(x,              s=NULL,              interval=NULL,            CL=0.9,
Blives=(c(1,5,10,20,30,40,50,60,80,90,95,99)/100),     MLEcontour=NULL,   minmeth="L-BFGS-B",
debmeth="rba-median", debias=FALSE,show=FALSE)  {

        ## calculate contour if missing
        if(missing(MLEcontour))  {
                MLEcontour<-MLEw2pContour(x, s, CL,debias=debias)
        }

        ypts<-log(qweibull(Blives,1,1))
                j=1

        yval<-c(Blife=ypts)

        Eta<-MLEcontour[j,1]
        Beta<-MLEcontour[j,2]
                xvals=NULL
        for(k in 1:length(ypts) ) {

                xval<-ypts[k]/Beta+log(Eta)
                xvals<-c(xvals,xval)
                names(xvals)<-NULL
        }

        outmat<-rbind(yval,xlo=xvals, Eta=rep(Eta,length(ypts)),Beta=rep(Beta,length(ypts)),

                xhi=xvals, Eta=rep(Eta,length(ypts)),Beta=rep(Beta,length(ypts)))


                        clen=length(MLEcontour[,1])
                for(j in 1:clen)  {
                        Eta<-MLEcontour[j,1]
                        Beta<-MLEcontour[j,2]
                                xvals=NULL
                        for(k in 1:length(ypts) ) {

                                xval<-ypts[k]/Beta+log(Eta)

                                if(xval<outmat[2,k])  {
                                        outmat[2,k]=xval
                                        outmat[3,k]=Eta
                                        outmat[4,k]=Beta
                                }
                                if(xval>outmat[5,k])  {
                                        outmat[5,k]=xval
                                        outmat[6,k]=Eta
                                        outmat[7,k]=Beta
                                }
                        }
                }


        ## calculate the Datum vector
        MLEfit  <-  mlefitM(Init.Beta  =  2,  Init.Eta  =  1,  minmeth=minmeth,x=x,s=s,  inter-
val=interval, debmeth=debmeth)
```

```
        Eta<-MLEfit[1]
        Beta<-MLEfit[2]
                xvals=NULL
        for(k in 1:length(ypts) ) {

                xval<-ypts[k]/Beta+log(Eta)
                xvals<-c(xvals,xval)
                names(xvals)<-NULL
        }

        outDF<-data.frame(ypts=ypts,Lower=outmat[2,],Datum=xvals, Upper=outmat[5,])


        if(show==TRUE)  {
                plot(xvals,ypts,type="l")
                lines(outmat[2,],outmat[1,],col="red")
                lines(outmat[5,],outmat[1,],col="blue")
        }

outDF
}
```

Rietveld analysis

In the following supplemental information for the experimental approach concerning the performed Rietveld analysis is presented.

Sample preparation using a swing mill

To obtain high quality samples for powder X-ray diffraction, the particle size distribution of the powder has to be reasonably narrow as well as the particle size has to be reasonably small so that the powder samples qualifies as a random arrangement of particles. Yet, particle sizes below 1 µm are not anticipated as this leads to visible peak broadening. In Figure B.1 the resultant particles of a relevant test sample of KALOS pebbles after milling in the swing mill is detailed.

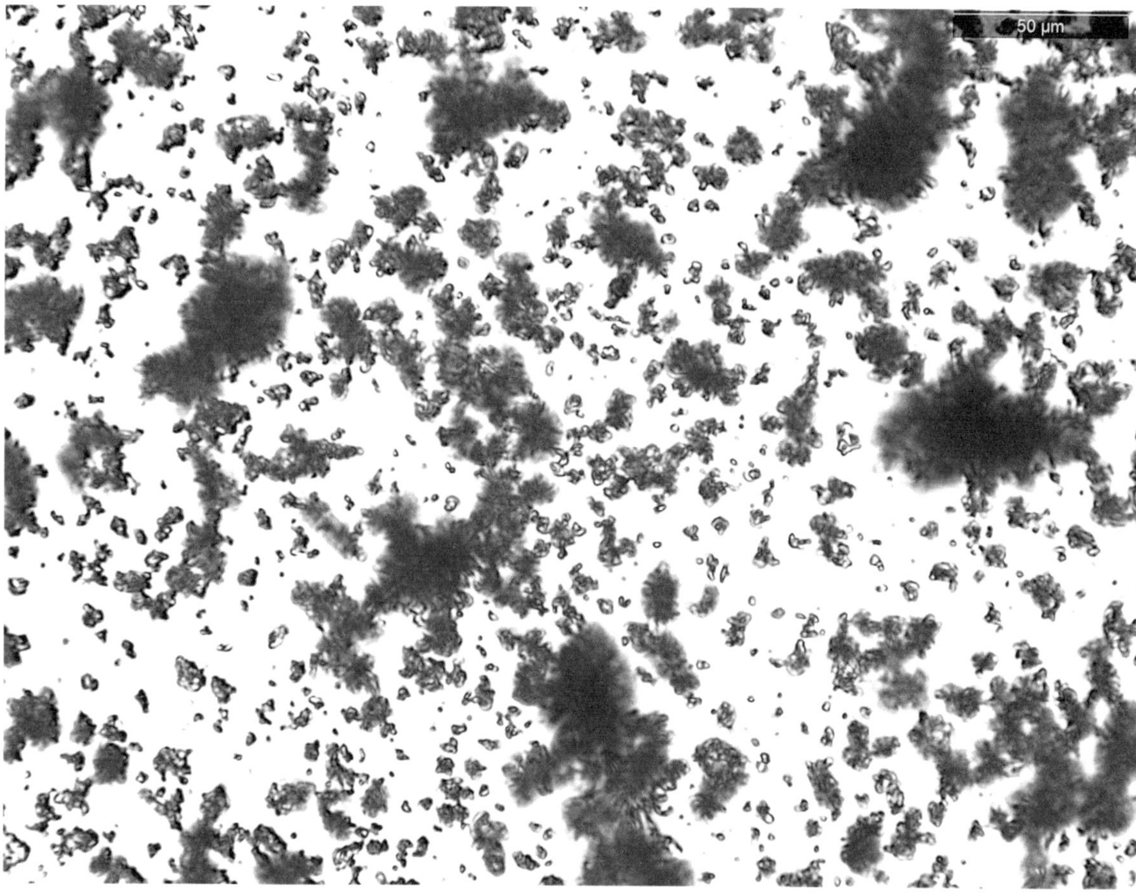

Figure B.1: Optical micrograph of a milled KALOS test pebble sample with a nominal LMT content of 20 mol%. The milling was performed within the swing mill according to the parameters given in section 3.5.1.

General refinement settings

When the fundamental parameters approach is applied for a Rietveld analysis, the experimental setup has to be sufficiently described with respect to its impact on the diffracted X-ray beam. The following TOPAS macros are shown as an example for the performed Rietveld analysis. In this particular case, the Bruker-AXS D8 configured for room temperature measurements is addressed by the code.

```
r_wp   19.0246547 r_exp   0.732864758 r_p   15.2636549 r_wp_dash   20.4895564 r_p_dash
21.3006393 r_exp_dash  0.789295471 weighted_Durbin_Watson  0.0209718363 gof  25.9592981

'------------------------------------------------------------
'General information about refinement here
'------------------------------------------------------------

iters 5000
chi2_convergence_criteria 0.0001
do_errors

'------------------------------------------------------------
'General settings and instrument convolutions
'------------------------------------------------------------

   x_calculation_step = Yobs_dx_at(Xo);
   convolution_step 3

   lam ymin_on_ymax 0.001

   Lam_recs
   {
      0.0159  1.534753  3.6854
      0.5691  1.540596  0.4370
      0.0762  1.541058  0.6000
      0.2517  1.544410  0.5200
      0.0871  1.544721  0.6200
   }
   LP_Factor(!unpolarized, 0)

   Radius(280, 280)

   axial_conv
      filament_length  12
      sample_length  25
      receiving_slit_length  16
      primary_soller_angle  2.5
      secondary_soller_angle  2.5
      axial_n_beta  30

   lpsd_th2_angular_range_degrees  2.9434395
       lpsd_equitorial_divergence_degrees  0.6
       lpsd_equitorial_sample_length_mm 25


   Specimen_Displacement(@,0)
```

Supercell model for LMT stacking faults

The anisotropic peak broadening which results from stacking faults is modeled by the following TOPAS script which utilizes the so-called supercell approach. For the supercell, the unit cell of β-lithium metatitanate was stacked 29 times on top of each other in (001)-direction. During the refinement, only the purely lithium containing layers were kept constant, while the other two layers may be moved in (100)- or (010)-direction or a combination of both.

```
#define Stack_Generate_Dummy_Atoms
macro AA   { @ }
   str
      MVW( 25461.350, 12397.655`, 100.000`)
      scale @  1.81859038e-005`

      space_group P1

      macro XX( x)  { x }
      macro YY( y)  { y }
      macro ZZ(& z) { = z / Nc; }

      load site x y z occ beq layer {

         Li1     XX( 0.23830 )    YY( 0.07950 )    ZZ( 0.00000 )    Li+1 1   1.16066548    A
         Li1002  XX( 0.76170 )    YY( 0.07950 )    ZZ( 0.50000 )    Li+1 1   1.16066548    A
         Li1003  XX( 0.76170 )    YY( 0.92050 )    ZZ( 0.00000 )    Li+1 1   1.16066548    A
         Li1004  XX( 0.23830 )    YY( 0.92050 )    ZZ( 0.50000 )    Li+1 1   1.16066548    A
         Li1005  XX( 0.73830 )    YY( 0.57950 )    ZZ( 0.00000 )    Li+1 1   1.16066548    A
         Li1006  XX( 0.26170 )    YY( 0.57950 )    ZZ( 0.50000 )    Li+1 1   1.16066548    A
         Li1007  XX( 0.26170 )    YY( 0.42050 )    ZZ( 0.00000 )    Li+1 1   1.16066548    A
         Li1008  XX( 0.73830 )    YY( 0.42050 )    ZZ( 0.50000 )    Li+1 1   1.16066548    A
         Li2     XX( 0.25000 )    YY( 0.25000 )    ZZ( 0.50000 )    Li+1 1   1.27120505    A
         Li2002  XX( 0.75000 )    YY( 0.25000 )    ZZ( 0.00000 )    Li+1 1   1.27120505    A
         Li2003  XX( 0.75000 )    YY( 0.75000 )    ZZ( 0.50000 )    Li+1 1   1.27120505    A
         Li2004  XX( 0.25000 )    YY( 0.75000 )    ZZ( 0.00000 )    Li+1 1   1.27120505    A
         Li3     XX( 0.00000 )    YY( 0.08520 )    ZZ( 0.25000 )    Li+1 1   0.315827341   B
         Li3002  XX( 0.00000 )    YY( 0.91480 )    ZZ( 0.75000 )    Li+1 1   0.315827341   C
         Li3003  XX( 0.50000 )    YY( 0.58520 )    ZZ( 0.25000 )    Li+1 1   0.315827341   B
         Li3004  XX( 0.50000 )    YY( 0.41480 )    ZZ( 0.75000 )    Li+1 1   0.315827341   C
         Ti1     XX( 0.00000 )    YY( 0.41749 )    ZZ( 0.25000 )    Ti+4 1   0.458739213   B
         Ti1002  XX( 0.00000 )    YY( 0.58251 )    ZZ( 0.75000 )    Ti+4 1   0.458739213   C
         Ti1003  XX( 0.50000 )    YY( 0.91749 )    ZZ( 0.25000 )    Ti+4 1   0.458739213   B
         Ti1004  XX( 0.50000 )    YY( 0.08251 )    ZZ( 0.75000 )    Ti+4 1   0.458739213   C
         Ti2     XX( 0.00000 )    YY( 0.74971 )    ZZ( 0.25000 )    Ti+4 1   0.454001802   B
         Ti2002  XX( 0.00000 )    YY( 0.25029 )    ZZ( 0.75000 )    Ti+4 1   0.454001802   C
         Ti2003  XX( 0.50000 )    YY( 0.24971 )    ZZ( 0.25000 )    Ti+4 1   0.454001802   B
         Ti2004  XX( 0.50000 )    YY( 0.75029 )    ZZ( 0.75000 )    Ti+4 1   0.454001802   C
         O1      XX( 0.14116 )    YY( 0.26356 )    ZZ( 0.13720 )    O-2  1   0.42636691    B
         O1002   XX( 0.85884 )    YY( 0.26356 )    ZZ( 0.36280 )    O-2  1   0.42636691    B
         O1003   XX( 0.85884 )    YY( 0.73644 )    ZZ( 0.86280 )    O-2  1   0.42636691    C
         O1004   XX( 0.14116 )    YY( 0.73644 )    ZZ( 0.63720 )    O-2  1   0.42636691    C
         O1005   XX( 0.64116 )    YY( 0.76356 )    ZZ( 0.13720 )    O-2  1   0.42636691    B
         O1006   XX( 0.35884 )    YY( 0.76356 )    ZZ( 0.36280 )    O-2  1   0.42636691    B
         O1007   XX( 0.35884 )    YY( 0.23644 )    ZZ( 0.86280 )    O-2  1   0.42636691    C
         O1008   XX( 0.64116 )    YY( 0.23644 )    ZZ( 0.63720 )    O-2  1   0.42636691    C
         O2      XX( 0.10555 )    YY( 0.58444 )    ZZ( 0.13719 )    O-2  1   0.433473025   B
         O2002   XX( 0.89445 )    YY( 0.58444 )    ZZ( 0.36281 )    O-2  1   0.433473025   B
         O2003   XX( 0.89445 )    YY( 0.41556 )    ZZ( 0.86281 )    O-2  1   0.433473025   C
         O2004   XX( 0.10555 )    YY( 0.41556 )    ZZ( 0.63719 )    O-2  1   0.433473025   C
         O2005   XX( 0.60555 )    YY( 0.08444 )    ZZ( 0.13719 )    O-2  1   0.433473025   B
         O2006   XX( 0.39445 )    YY( 0.08444 )    ZZ( 0.36281 )    O-2  1   0.433473025   B
         O2007   XX( 0.39445 )    YY( 0.91556 )    ZZ( 0.86281 )    O-2  1   0.433473025   C
         O2008   XX( 0.60555 )    YY( 0.91556 )    ZZ( 0.63719 )    O-2  1   0.433473025   C
```

```
      O3     XX( 0.13958 )    YY( 0.90588 )    ZZ( 0.13454 )    O-2  1  0.442158277   B
      O3002  XX( 0.86042 )    YY( 0.90588 )    ZZ( 0.36546 )    O-2  1  0.442158277   B
      O3003  XX( 0.86042 )    YY( 0.09412 )    ZZ( 0.86546 )    O-2  1  0.442158277   C
      O3004  XX( 0.13958 )    YY( 0.09412 )    ZZ( 0.63454 )    O-2  1  0.442158277   C
      O3005  XX( 0.63958 )    YY( 0.40588 )    ZZ( 0.13454 )    O-2  1  0.442158277   B
      O3006  XX( 0.36042 )    YY( 0.40588 )    ZZ( 0.36546 )    O-2  1  0.442158277   B
      O3007  XX( 0.36042 )    YY( 0.59412 )    ZZ( 0.86546 )    O-2  1  0.442158277   C
      O3008  XX( 0.63958 )    YY( 0.59412 )    ZZ( 0.63454 )    O-2  1  0.442158277   C
   }


   ' Values on left are from reference
   a   LP (5.069935   ,  5.068459` )
   b   LP (8.775487   ,  8.782927` )
   prm lpc  9.75370` min = 9.740 - .1; max = 9.740 + .1;
   c   = lpc Nc;
   al  90
   be  LPA(100.211998 ,  100.07060` )
   ga  90

   CS_L(, 10000.00000)
   CS_G(, 10000.00000)

   Strain_L(@, 0.17071`)
   Strain_G(@, 0.00013`)

   'Atom anti-bump script. Did not provide an improved result and is thus not used
   'atomic_interaction Libump = IF R < 1.9 THEN (R-1.9)^2 ELSE 0 ENDIF;
   'ai_radius 2.5 ai_sites_1 O* ai_sites_2 Li* penalty = Libump;
   '
   'atomic_interaction Tibump = IF R < 1.8 THEN (R-1.8)^2 ELSE 0 ENDIF;
   'ai_radius 2.5 ai_sites_1 O* ai_sites_2 Ti* penalty = Tibump;
   '
   'atomic_interaction LTbump = IF R < 2.8 THEN (R-2.8)^2 ELSE 0 ENDIF;
   'ai_radius 3.5 ai_sites_1 Li* ai_sites_2 Ti* penalty = LTbump;

   macro StackA(ze, sxc, sxv, syc, syv, szc, szv)
      {
         #m_argu sxc
         #m_argu syc
         #m_argu szc
         If_Prm_Eqn_Rpt(szc, szv, min = Val - .3; max = Val + .3; val_on_continue = Val
+ Rand(-.05, .05) T;)

         stack A
            sx = (((ze + CeV(szc, szv)) / Nc)*Cos(Get(be)*Pi/180))/Get(a);
            sy syc syv VV(b)
            sz = (ze + CeV(szc, szv)) / Nc;

            #ifdef Stack_Generate_Dummy_Atoms
               #m_ifarg ze "0" #m_else
                  generate_these "*"
                     generate_name_append _##ze
               #m_endif
            #endif
      }
   macro StackB(ze, sxc, sxv, syc, syv, szc, szv)
      {
         #m_argu sxc
         #m_argu syc
         #m_argu szc
         If_Prm_Eqn_Rpt(szc, szv, min = Val - .1; max = Val + .1; val_on_continue = Val
+ Rand(-.01, .01) T;)

         stack B
            sx sxc sxv VV(a)
```

```
            sy syc syv VV(b)
            sz = (ze + CeV(szc, szv)) / Nc;

            #ifdef Stack_Generate_Dummy_Atoms
               #m_ifarg ze "0" #m_else
                  generate_these "*"
                     generate_name_append _##ze
               #m_endif
            #endif
      }
   macro StackC(ze, sxc, sxv, syc, syv, szc, szv)
      {
         #m_argu sxc
         #m_argu syc
         #m_argu szc
         If_Prm_Eqn_Rpt(szc, szv, min = Val - .1; max = Val + .1; val_on_continue = Val
+ Rand(-.01, .01) T;)

         stack C
            sx sxc sxv VV(a)
            sy syc syv VV(b)
            sz = (ze + CeV(szc, szv)) / Nc;

            #ifdef Stack_Generate_Dummy_Atoms
               #m_ifarg ze "0" #m_else
                  generate_these "*"
                     generate_name_append _##ze
               #m_endif
            #endif
      }

   prm !Nc 29

   StackA(0,     ,0,  ,0,  ,0)
   StackA(1,     ,0,  ,0,  ,0)
   StackA(2,     ,0,  ,0,  ,0)
   StackA(3,     ,0,  ,0,  ,0)
   StackA(4,     ,0,  ,0,  ,0)
   StackA(5,     ,0,  ,0,  ,0)
   StackA(6,     ,0,  ,0,  ,0)
   StackA(7,     ,0,  ,0,  ,0)
   StackA(8,     ,0,  ,0,  ,0)
   StackA(9,     ,0,  ,0,  ,0)
   StackA(10,    ,0,  ,0,  ,0)
   StackA(11,    ,0,  ,0,  ,0)
   StackA(12,    ,0,  ,0,  ,0)
   StackA(13,    ,0,  ,0,  ,0)
   StackA(14,    ,0,  ,0,  ,0)
   StackA(15,    ,0,  ,0,  ,0)
   StackA(16,    ,0,  ,0,  ,0)
   StackA(17,    ,0,  ,0,  ,0)
   StackA(18,    ,0,  ,0,  ,0)
   StackA(19,    ,0,  ,0,  ,0)
   StackA(20,    ,0,  ,0,  ,0)
   StackA(21,    ,0,  ,0,  ,0)
   StackA(22,    ,0,  ,0,  ,0)
   StackA(23,    ,0,  ,0,  ,0)
   StackA(24,    ,0,  ,0,  ,0)
   StackA(25,    ,0,  ,0,  ,0)
   StackA(26,    ,0,  ,0,  ,0)
   StackA(27,    ,0,  ,0,  ,0)
   StackA(28,    ,0,  ,0,  ,0)

   StackB(0,     ,0,  ,0,  ,0)
   StackB(1,     @,0, @,0,  ,0)
   StackB(2,     @,0, @,0,  ,0)
```

```
StackB(3,     @,0, @,0, ,0)
StackB(4,     @,0, @,0, ,0)
StackB(5,     @,0, @,0, ,0)
StackB(6,     @,0, @,0, ,0)
StackB(7,     @,0, @,0, ,0)
StackB(8,     @,0, @,0, ,0)
StackB(9,     @,0, @,0, ,0)
StackB(10,    @,0, @,0, ,0)
StackB(11,    @,0, @,0, ,0)
StackB(12,    @,0, @,0, ,0)
StackB(13,    @,0, @,0, ,0)
StackB(14,    @,0, @,0, ,0)
StackB(15,    @,0, @,0, ,0)
StackB(16,    @,0, @,0, ,0)
StackB(17,    @,0, @,0, ,0)
StackB(18,    @,0, @,0, ,0)
StackB(19,    @,0, @,0, ,0)
StackB(20,    @,0, @,0, ,0)
StackB(21,    @,0, @,0, ,0)
StackB(22,    @,0, @,0, ,0)
StackB(23,    @,0, @,0, ,0)
StackB(24,    @,0, @,0, ,0)
StackB(25,    @,0, @,0, ,0)
StackB(26,    @,0, @,0, ,0)
StackB(27,    @,0, @,0, ,0)
StackB(28,    @,0, @,0, ,0)

StackC(0,     ,0, ,0, ,0)
StackC(1,     @,0, @,0, ,0)
StackC(2,     @,0, @,0, ,0)
StackC(3,     @,0, @,0, ,0)
StackC(4,     @,0, @,0, ,0)
StackC(5,     @,0, @,0, ,0)
StackC(6,     @,0, @,0, ,0)
StackC(7,     @,0, @,0, ,0)
StackC(8,     @,0, @,0, ,0)
StackC(9,     @,0, @,0, ,0)
StackC(10,    @,0, @,0, ,0)
StackC(11,    @,0, @,0, ,0)
StackC(12,    @,0, @,0, ,0)
StackC(13,    @,0, @,0, ,0)
StackC(14,    @,0, @,0, ,0)
StackC(15,    @,0, @,0, ,0)
StackC(16,    @,0, @,0, ,0)
StackC(17,    @,0, @,0, ,0)
StackC(18,    @,0, @,0, ,0)
StackC(19,    @,0, @,0, ,0)
StackC(20,    @,0, @,0, ,0)
StackC(21,    @,0, @,0, ,0)
StackC(22,    @,0, @,0, ,0)
StackC(23,    @,0, @,0, ,0)
StackC(24,    @,0, @,0, ,0)
StackC(25,    @,0, @,0, ,0)
StackC(26,    @,0, @,0, ,0)
StackC(27,    @,0, @,0, ,0)
StackC(28,    @,0, @,0, ,0)


Preferred_Orientation(@, 1,, 0 0 1)

view_structure

' Helper macros
macro LP(& l0, l)
   {
      AA l
```

```
            min = 10 - .1;
            max = 10 + .1;
        }
    macro LPA(& 10, 1)
        {
            AA 1
            min = 10 - 2;
            max = 10 + 2;
        }
    macro VV(a)
        {
            min = Val - .65/Constant(Get(a));
            max = Val + .65/Constant(Get(a));
            val_on_continue = Val + Rand(-1, 1) T / Constant(Get(a));
        }
```

Refinement of the annealed yttria standard

For the determination of the background function that originates from the experimental setup, an annealed yttria powder with 99.99 % purity (REacton®, Alfa Aesar) was used. The measurement as well as the Rietveld analysis is given in Figure B.2.

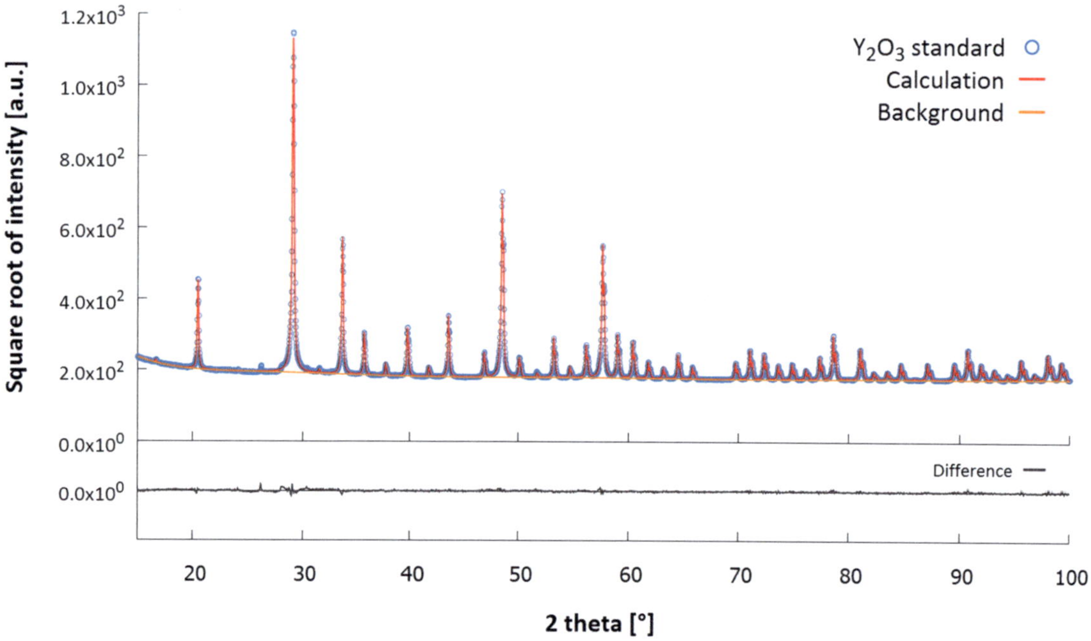

Figure B.2: Rietveld refinement of the yttria standard that was used to determine the background that is related to the experimental setup (Bruker AXS D8). The crystal structure of yttria was taken from Ferreira et al. [205].

Refinement of the measurement background

The refinement of the measurement background, when the amorphous content of a sample is estimated, has to be performed with care. The following TOPAS macro code details the approach. The scale parameter of the 1 over x function is identical to the first Chebyshev coefficient and is thus passed on. The other Chebyshev coefficients were determined from the analysis of the yttria standard (see also Figure B.2).

```
#ifdef bkgr
   macro Notbkgr {    }
#else
   macro Notbkgr { ! }
#endif


#ifdef constbkgr
   macro Notcbkgr {    }
#else
   macro Notcbkgr { ! }
#endif


One_on_X(Notbkgr##BKG, 447116.96570)
bkg !BKG  24190.4614  2043.22043  1105.95532 -1332.23015  1455.6495 -1286.73035  1064.20983
-911.707441  734.35593 -338.487499 -65.6616817
prm Notcbkgr##constbkg -23935.05324 min = -1e5; max = 1e6;
fit_obj = constbkg;
```

Comparison of hand-ground and milled pebbles

In Figure B.3 two XRD data of nominally 20 mol% LMT containing KALOS pebbles which were annealed for 32 days are compared. One important difference of these two samples is different preparation before the actual measurement. One sample was hand-ground in an agate mortar, while the other one was milled in a swing mill (see section 3.5.1). It is evident, that the sample which was milled in the swing mill shows lithium carbonate as impurity phase, while the hand-ground sample does not.

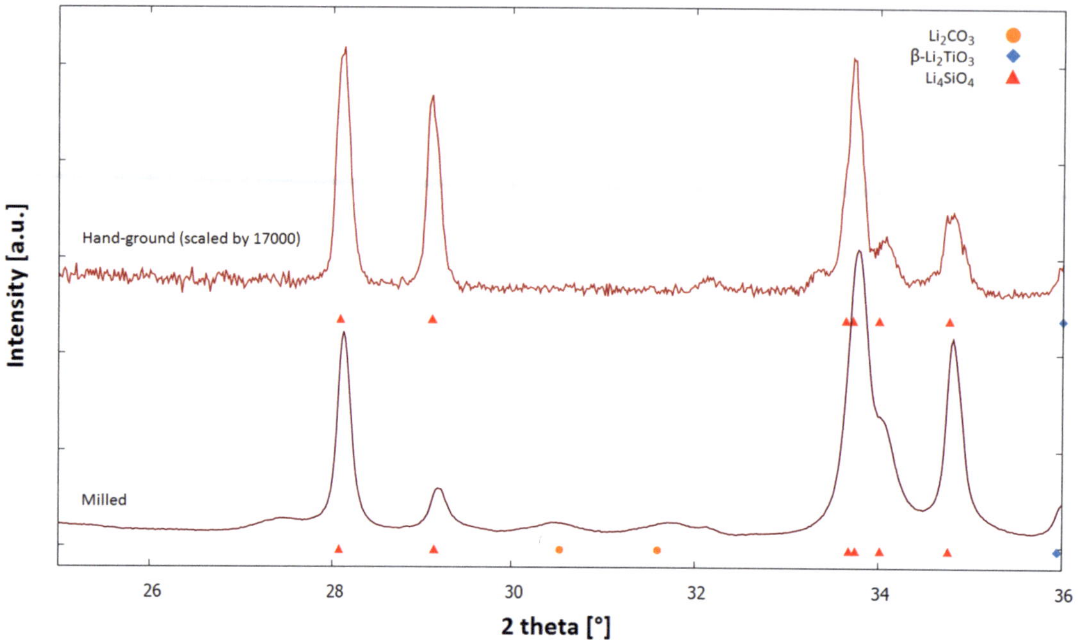

Figure B.3: Excerpt of two XRD data that were acquired from nominally 20 mol% LMT containing KALOS pebbles, which were annealed for 32 days. Because of different measurement conditions, the intensities of the hand-ground sample had to be scaled by a factor of 17000. Peaks are indicated for indentified phases if they are significantly strong. The lattice parameters of all phases were refined. The equilibrium crystal structures of Li_2CO_3, β-Li_2TiO_3 and Li_4SiO_4 are given by Idemoto et al. [172], Kataoka et al. [78] and Deng et al. [56], respectively. Both samples were prepared differently for the measurements. One was hand-ground in an agate mortar, while the other one was milled in a swing mill (see section 3.5.1).

Temperature deviation of the XRD heating chamber

The actual temperature of the powder samples within the heating chamber of the Bruker AXS D8 as a function of the anticipated temperature is given in Figure B.4.

For the determination of the temperature the well-known thermal expansion coefficients of silicon and alumina, as reported in literature by Munro as well as Okada and Tokumaro [169,170], were used. For the measurements a powder sample with a mixture of 80 wt.% alumina powder and 20 wt.% silicon powder was prepared. In total 39 diffractograms at different temperatures between 50 °C and 1200 °C were obtained and simultaneously refined. For the actual sample temperature as a function of the anticipated temperature, a polynomial of fourth degree is assumed and the respective coefficients are refined directly.

The structure of alumina is reported by Pillet et al. [206] and the structure of silicon by Többens et al. [207]. The temperature at the sample is always lower than the anticipated temperature as shown in Figure B.4. At about 600 °C, the difference between ideal and actual temperature reaches 20 K which remains constant for higher temperatures and decreases with lower anticipated temperatures.

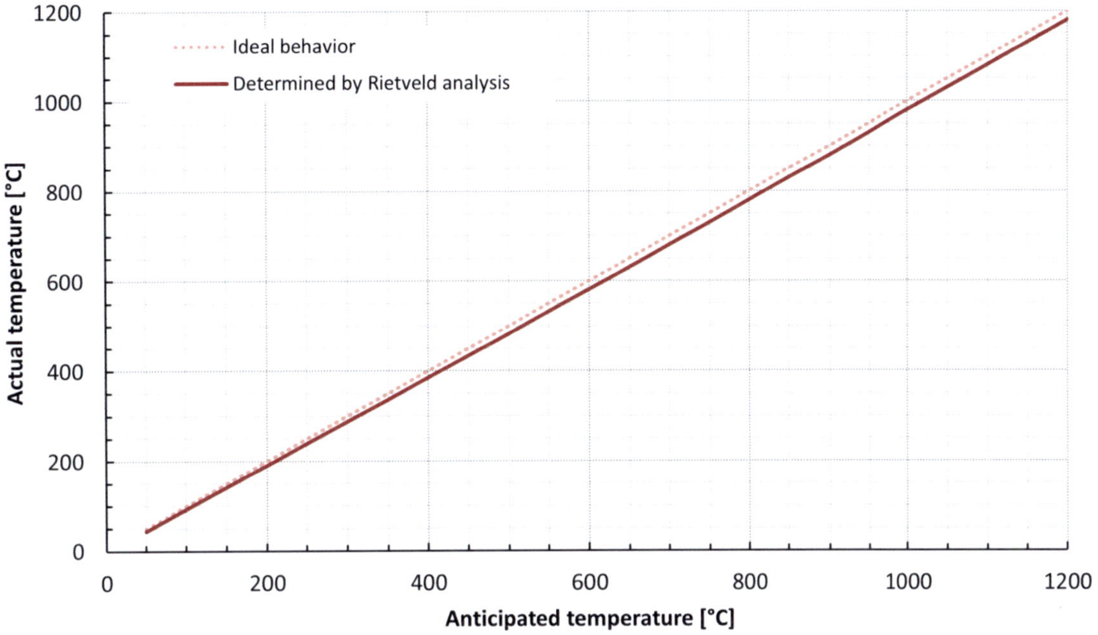

Figure B.4: Temperature of the powder sample holder within the heating chamber of the Bruker AXS D8 as a function of the anticipated temperature. For the determination of the actual temperature the known thermal expansion of alumina and silicon was used.

Appendix C

In the following additional and more detailed results concerning four of the five main topics of section 4, namely section 4.1, Phase analysis, Pebble morphology and microstructure as well as Porosity, are presented.

Detailed elemental composition

In the following, the determined molar elemental compositions of the various pebble samples are given. Also the detection limit of the method that was used to analyze the sample for each element is given. For a number of sample, the impurity concentration of certain elements lies below the detection limit, which is abbreviated by the term 'b.d.l.' (below detection limit). In such cases, the impurity concentration of the respective element may have any value below the detection limit.

Table C.1: Molar elemental composition of the nominally 20 mol% LMT containing KALOS samples that were annealed in He/H_2 atmosphere as determined by XRF (Si and Ti) and ICP-OES (rest).

Element	Detection limit	KA-20-LMT	KA-20-LMT	KA-20-LMT	KA-20-LMT	KA-20-LMT
		0 d	4 d	32 d	64 d	128 d
Li [%]	0.18	42.86	43.19	43.11	43.10	42.97
Si [%]	0.05	9.11	9.29	9.32	9.36	9.38
Ti [%]	0.03	2.13	2.09	2.17	2.14	2.18
Al [ppm]	5.2	281.3	289.6	284.2	284.1	280.5
Au [ppm]	1.4	b.d.l.	b.d.l.	b.d.l.	b.d.l.	b.d.l.
Ir [ppm]	0.7	b.d.l.	b.d.l.	b.d.l.	b.d.l.	b.d.l.
Mg [ppm]	–	–	–	–	–	–
Na [ppm]	30.3	b.d.l.	b.d.l.	b.d.l.	b.d.l.	b.d.l.
Pt [ppm]	1.4	4.1	3.8	4.3	4.3	4.2
Rh [ppm]	2.7	7.6	7.1	7.3	7.3	7.6

Table C.2: Molar elemental composition of the nominally 30 mol% LMT containing KALOS samples that were annealed in He/H_2 atmosphere as determined by XRF (Si and Ti) and ICP-OES (rest).

Element	Detection limit	KA-30-LMT	KA-30-LMT	KA-30-LMT	KA-30-LMT	KA-30-LMT
		0 d	4 d	32 d	64 d	128 d
Li [%]	0.42	41.09	41.28	41.50	41.32	40.38
Si [%]	0.05	8.45	8.39	8.50	8.50	8.36
Ti [%]	0.03	3.92	3.98	4.01	3.98	4.14
Al [ppm]	5.4	236.8	237.7	242.6	235.2	254.0
Au [ppm]	1.5	b.d.l.	b.d.l.	b.d.l.	b.d.l.	b.d.l.
Ir [ppm]	0.8	b.d.l.	b.d.l.	b.d.l.	b.d.l.	b.d.l.
Mg [ppm]	–	–	–	–	–	–
Na [ppm]	31.9	b.d.l.	b.d.l.	b.d.l.	b.d.l.	b.d.l.
Pt [ppm]	1.5	3.7	3.4	3.4	4.0	4.6
Rh [ppm]	2.9	6.5	6.6	6.8	6.5	6.7

Table C.3: Molar elemental composition of the lithium excessive pebbles that were fabricated by the emulsion method and were annealed in He/H_2 atmosphere as determined by XRF (Ti) and ICP-OES (rest).

Element	Detection limit	EM-LMT	EM-LMT	EM-LMT	EM-LMT	EM-LMT
		0 d	4 d	32 d	64 d	128 d
Li [%]	0.53	33.01	33.03	32.81	32.93	32.77
Si [%]	0.33	b.d.l.	b.d.l.	b.d.l.	b.d.l.	b.d.l.
Ti [%]	0.04	15.79	16.27	16.29	16.54	16.67
Al [ppm]	6.8	45.5	10.1	10.8	16.3	27.9
Au [ppm]	2.8	b.d.l.	b.d.l.	b.d.l.	b.d.l.	b.d.l.
Ir [ppm]	1.0	b.d.l.	b.d.l.	b.d.l.	b.d.l.	b.d.l.
Mg [ppm]	22.6	b.d.l.	b.d.l.	b.d.l.	b.d.l.	b.d.l.
Na [ppm]	39.9	147.5	220.8	b.d.l.	b.d.l.	b.d.l.
Pt [ppm]	1.9	b.d.l.	b.d.l.	b.d.l.	b.d.l.	b.d.l.
Rh [ppm]	3.6	b.d.l.	b.d.l.	b.d.l.	b.d.l.	b.d.l.

Table C.4: Molar elemental composition of the lithium excessive pebbles that were fabricated by the sol-gel method and were annealed in He/H_2 atmosphere as determined by XRF (Ti) and ICP-OES (rest).

Element	Detection limit	SG-LMT	SG-LMT	SG-LMT	SG-LMT	SG-LMT
		0 d	4 d	32 d	64 d	128 d
Li [%]	0.52	33.34	33.40	33.61	33.26	33.20
Si [%]	0.32	b.d.l.	b.d.l.	b.d.l.	b.d.l.	b.d.l.
Ti [%]	0.04	15.10	15.66	15.63	15.83	16.15
Al [ppm]	6.7	337.1	340.4	345.2	339.6	365.2
Au [ppm]	2.8	b.d.l.	b.d.l.	b.d.l.	b.d.l.	b.d.l.
Ir [ppm]	0.9	b.d.l.	b.d.l.	b.d.l.	b.d.l.	b.d.l.
Mg [ppm]	22.4	564.0	480.7	494.7	437.8	560.0
Na [ppm]	39.5	390.3	286.9	b.d.l.	b.d.l.	b.d.l.
Pt [ppm]	1.9	b.d.l.	b.d.l.	b.d.l.	b.d.l.	b.d.l.
Rh [ppm]	3.5	b.d.l.	b.d.l.	b.d.l.	b.d.l.	b.d.l.

Additional details of the results that were obtained by XRD

In the following additional details and aspects of the quantitative analysis results by X-ray diffraction are given.

Refined parameters of the quantitative XRD analysis

The relevant crystallographic parameters that were obtained from the Rietveld analysis of the XRD data for all examined samples are shown in the following tables. Also, as an indicator of the goodness of the refinement, the parameter R_{wp} is given (see Young et al. [168]). For addressing the preferred orientation of the lithium metatitanate crystallites in (001)-direction in the refinements the March parameter (see equation (3.8)) is given. The amount of determined lithium carbonate, which originates from the milling of the samples to a fine powder by the swing mill, is also given for the concerned samples.

Table C.5: Determined parameters of the nominally 20 mol% LMT containing KALOS samples that were annealed in He/H_2 atmosphere.

KA-20-LMT	As-received	4 days	32 days	64 days	128 days
Lithium orthosilicate					
a [Å]	11.590	11.586	11.580	11.581	11.578
b [Å]	6.132	6.129	6.125	6.124	6.124
c [Å]	16.818	16.809	16.797	16.797	16.792
ß [°]	99.258	99.202	99.160	99.155	99.136
Density [g/cm³]	2.362	2.365	2.369	2.369	2.370
Weight fraction	88.92 %	85.97 %	84.92 %	83.53 %	83.19 %
Molar fraction	91.64 %	88.25 %	87.15 %	85.58 %	85.19 %
Lithium metatitanate					
a [Å]	4.140	5.061	5.062	5.062	5.072
b [Å]	4.140	8.785	8.782	8.786	8.765
c [Å]	4.140	9.752	9.751	9.751	9.745
ß [°]	90	100.104	100.147	100.105	99.903
Density [g/cm³]	3.424	3.415	3.417	3.415	3.416
Weight fraction	7.43 %	10.48 %	11.47 %	12.89 %	13.24 %
March param.	1.000	0.992	0.968	0.990	0.984
Molar fraction	8.36 %	11.75 %	12.85 %	14.42 %	14.81 %
Lithium carbonate					
Weight fraction	3.65 %	3.55 %	3.61 %	3.58 %	3.57 %
Amorphous content					
DOC	71.53 %	75.31 %	79.48 %	80.09 %	80.11 %
Weight fract. LOS	16.16 %	15.52 %	12.04 %	12.91 %	12.98 %
Weight fract. LMT	12.32 %	9.17 %	8.48 %	7.00 %	6.91 %
Refinement quality					
R_{wp}	7.362	6.932	6.718	6.411	6.281

Table C.6: Determined parameters of the nominally 30 mol% LMT containing KALOS samples that were annealed in He/H_2 atmosphere.

KA-30-LMT	As-received	4 days	32 days	64 days	128 days
Lithium orthosilicate					
a [Å]	11.568	11.553	11.556	11.555	11.554
b [Å]	6.118	6.110	6.111	6.111	6.110
c [Å]	16.768	16.751	16.754	16.753	16.748
ß [°]	99.079	99.040	99.036	99.042	99.036
Density [g/cm³]	2.377	2.386	2.384	2.385	2.386
Weight fraction	68.97 %	64.26 %	65.01 %	63.09 %	65.44 %
Molar fraction	70.74 %	65.37 %	65.89 %	64.07 %	65.20 %
Lithium metatitanate					
a [Å]	4.142	5.062	5.063	5.063	5.072
b [Å]	4.142	8.783	8.785	8.786	8.767
c [Å]	4.142	9.749	9.752	9.751	9.746
ß [°]	90	100.092	100.118	100.095	99.913
Density [g/cm³]	3.419	3.417	3.414	3.414	3.415
Weight fraction	26.13 %	31.18 %	30.82 %	32.39 %	31.98 %
March param.	1.000	0.984	0.980	0.997	0.989
Molar fraction	29.26 %	34.63 %	34.11 %	35.93 %	34.80 %
Lithium carbonate					
Weight fraction	4.90 %	4.56 %	4.17 %	4.51 %	2.58 %
Amorphous content					
DOC	65.48 %	83.94 %	85.01 %	84.89 %	86.60 %
Weight fract. LOS	21.79 %	11.95 %	11.00 %	12.58 %	9.90 %
Weight fract. LMT	12.73 %	4.12 %	3.99 %	2.53 %	3.50 %
Refinement quality					
R_{wp}	5.759	4.143	4.489	4.418	4.221

Table C.7: Determined parameters of the EM-LMT pebbles that were annealed in He/H_2 atmosphere.

EM-LMT	As-received	4 days	32 days	64 days	128 days
Lithium metatitanate					
a [Å]	5.071	5.068	5.065	5.065	5.063
b [Å]	8.778	8.782	8.789	8.794	8.793
c [Å]	9.749	9.754	9.756	9.757	9.757
ß [°]	99.913	100.057	100.103	100.117	100.117
Density [g/cm³]	3.410	3.410	3.410	3.408	3.409
Weight fraction	100 %	100 %	100 %	100 %	100 %
March param.	0.919	0.836	0.774	0.713	0.666
Amorphous content					
DOC	91.64 %	92.31 %	92.70 %	93.39 %	92.34 %
Weight fract. LMT	8.36 %	7.69 %	7.30 %	6.61 %	7.66 %
Refinement quality					
R_{wp}	3.699	4.034	4.402	4.274	5.175

Table C.8: Determined parameters of the SG-LMT pebbles that were annealed in He/H_2 atmosphere.

SG-LMT	As-received	4 days	32 days	64 days	128 days
Lithium metatitanate					
a [Å]	5.069	5.067	5.065	5.065	5.064
b [Å]	8.783	8.783	8.788	8.791	8.792
c [Å]	9.754	9.753	9.756	9.757	9.756
ß [°]	100.071	100.088	100.143	100.146	100.127
Density [g/cm³]	3.410	3.412	3.411	3.409	3.409
Weight fraction	100 %	100 %	100 %	100 %	100 %
March param.	0.911	0.809	0.775	0.769	0.710
Amorphous content					
DOC	90.14 %	92.09 %	93.03 %	93.75 %	92.79 %
Weight fract. LMT	9.86 %	7.91 %	6.97 %	6.25 %	7.21 %
Refinement quality					
R_{wp}	3.953	4.343	4.303	4.009	4.150

Table C.9: Determined parameters of the biphasic lithium orthosilicate samples that contain lithium metatitanate as second phase. The samples were fabricated by the emulsion method. In the table, the nominal LMT content is given.
*: taken from the nominally 80 mol% LMT containing sample as the refinement error was too large for reliable values.

EM-X-LMT	10 mol%	20 mol%	30 mol%	40 mol%	50 mol%	60 mol%	70 mol%	80 mol%	90 mol%
Lithium orthosilicate									
a [Å]	11.547	11.545	11.546	11.543	11.542	11.542	11.545	11.544	11.544*
b [Å]	6.105	6.106	6.107	6.112	6.112	6.111	6.108	6.107	6.107*
c [Å]	16.737	16.738	16.741	16.742	16.744	16.746	16.746	16.746	16.746*
ß [°]	99.034	99.031	99.027	99.018	99.005	99.006	98.989	98.968	98.968*
Density [g/cm³]	2.391	2.391	2.390	2.388	2.388	2.388	2.389	2.389	2.389*
Weight fract. [%]	89.20	79.17	68.29	57.42	46.59	38.65	26.86	17.20	7.18
Molar fract. [%]	88.32	77.69	66.36	55.25	44.41	36.58	25.17	15.98	6.62
Lithium metatitanate									
a [Å]	5.075	5.073	5.073	5.073	5.072	5.072	5.072	5.072	5.072
b [Å]	8.777	8.771	8.771	8.771	8.771	8.771	8.774	8.775	8.774
c [Å]	9.746	9.745	9.747	9.747	9.747	9.746	9.748	9.747	9.749
ß [°]	99.847	99.854	99.855	99.818	99.816	99.822	99.866	99.869	99.825
Density [g/cm³]	3.408	3.413	3.412	3.412	3.412	3.413	3.411	3.411	3.411
Weight fract. [%]	10.80	20.83	31.71	42.58	53.41	61.35	73.14	82.80	92.82
March parameter	1.052	1.054	1.066	1.082	1.082	1.057	1.059	1.032	1.047
Molar fract. [%]	11.68	22.31	33.64	44.75	55.59	63.42	74.83	84.02	93.38
Refinement quality									
R_{wp}	6.365	5.886	6.365	4.554	4.889	6.122	5.443	6.526	6.064

Table C.10: Determined parameters of the biphasic lithium orthosilicate samples that contain lithium aluminate as second phase. The samples were fabricated by the emulsion method. In the table, the nominal LAO content is given.

EM-X-LAO	10 mol%	20 mol%	30 mol%
Lithium orthosilicate			
a [Å]	11.556	11.567	11.571
b [Å]	6.106	6.110	6.113
c [Å]	16.737	16.747	16.758
ß [°]	99.019	99.039	99.075
Density [g/cm³]	2.389	2.384	2.380
Weight fract.	92.12 %	86.22 %	77.70 %
Molar fract.	86.94 %	78.00 %	66.43 %
Lithium aluminate			
a [Å]	5.166	5.168	5.170
b [Å]	5.166	5.168	5.170
c [Å]	6.247	6.256	6.261
ß [°]	90	90	90
Density [g/cm³]	2.626	2.621	2.617
Weight fract.	7.61 %	13.38 %	21.60 %
Molar fract.	13.06 %	22.00 %	33.57 %
Refinement quality			
R_{wp}	6.244	5.549	5.680

Determined parameters of the thermal expansion coefficients

The determined coefficients of the empirical relationship of the thermal expansion of the lattice parameters as a function of temperature (see equation (3.11)) of lithium orthosilicate and β-lithium metatitanate are detailed in Table C.11.

Table C.11: Determined coefficients for describing the thermal expansion of LOS and LMT as a function of temperature according to equation (3.11).

Phase	a [Å]	b [Å]	c [Å]	β [°]
Li_4SiO_4, up to 725 °C				
A (x10^{-6})	17.571	19.995	16.131	-6.603
B (x10^{-8})	1.129	0.263	0.344	0.323
C (x10^{-10})	89.541	-43.281	-22.260	-0.006
D (x10^{-4})	-99.265	-118.414	-124.049	-233.583
Li_4SiO_4, above 725 °C				
A (x10^{-6})	13.221	6.683	20.112	-0.298
B (x10^{-8})	0.190	4.625	1.806	0.556
C (x10^{-10})	-260.681	319.811	196.792	244.104
D (x10^{-4})	-55.805	-59.778	-60.701	-58.366
β-Li_2TiO_3				
A (x10^{-6})	14.165	16.999	19.241	-4.372
B (x10^{-8})	0.451	0.189	0.758	0.592
C (x10^{-10})	0.200	0.643	0.986	319.720
D (x10^{-4})	-119.998	-90.343	-104.621	-48.595

Mean pebble diameters of biphasic emulsion method pebbles

The mean pebble diameters of the biphasic emulsion method pebbles are given in the following tables. The values were obtained by the optical method that is detailed in section 3.6. As a measure of the breadth of the pebble size distributions, also the standard deviation of these distributions is listed.

Table C.12: The mean pebble diameters as well as the respective standard deviations of the LMT containing emulsion method pebbles as obtained by optical method for green and sintered pebbles. The nominal LMT contents are given in the table.

EM-X-LMT	10 mol%	20 mol%	30 mol%	40 mol%	50 mol%	60 mol%	70 mol%	80 mol%	90 mol%
Mean dia., green [μm]	1534.8	1542.2	1548.1	1531.0	1504.3	1519.6	1564.6	1641.0	1414.1
Std. dev., green [μm]	120.2	27.9	37.2	41.7	40.7	26.6	39.2	56.1	47.1
Mean dia., sintered [μm]	1004.8	1030.0	1034.4	1039.5	1025.7	1033.1	1057.2	1129.9	1037.2
Std. dev., sintered [μm]	99.6	46.3	33.0	67.1	50.3	24.8	52.1	48.0	49.9

Table C.13: The mean pebble diameters as well as the respective standard deviations of the LAO containing emulsion method pebbles as obtained by optical method for green and sintered pebbles. The nominal LAO contents are given in the table.

EM-X-LAO	10 mol%	20 mol%	30 mol%
Mean dia., green [μm]	1549.5	1544.9	1528.0
Std. dev., green [μm]	36.2	39.6	55.3
Mean dia., sintered [μm]	1028.7	1031.7	1036.5
Std. dev., sintered [μm]	53.5	56.9	73.7

Additional results on the pebble surface

In Figure C.1 the surface structure of a nominally 30 mol% LMT containing KALOS pebble is shown. While considerable grain coarsening has taken place, the initially present lamellar structure is still noticeable.

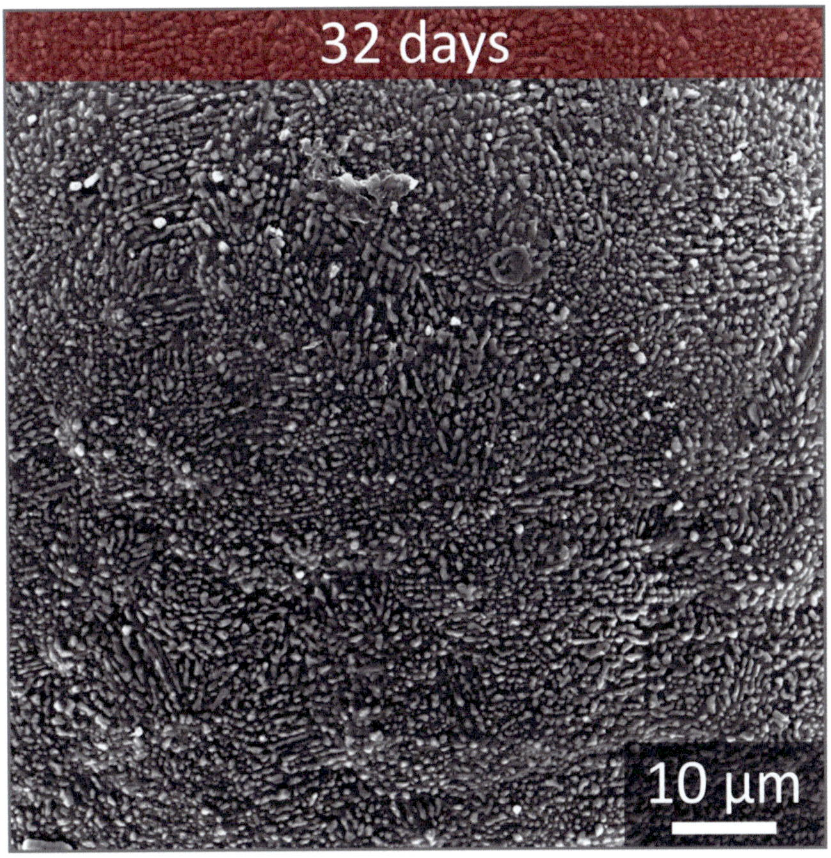

Figure C.1: SEM surface image of a nominally 30 mol% LMT containing KALOS pebble after 32 days of annealing. The lamellar surface structure is still clearly visible, although significant grain coarsening has taken place.

In Figure C.2 the surface of a nominally 30 mol% LMT containing KALOS pebble which was annealed for 128 days is captured by two different detectors. One shows the secondary electron image, and thus predominantely the relief of the sample, while the other one shows the backscattered electron image which provides a Z-contrast. From these images it can be concluded that the grains which accumulate between large LMT platelets are made of lithium orthosilicate, although and contrary to the expectations, these grains appear lighter than the surrounding material.

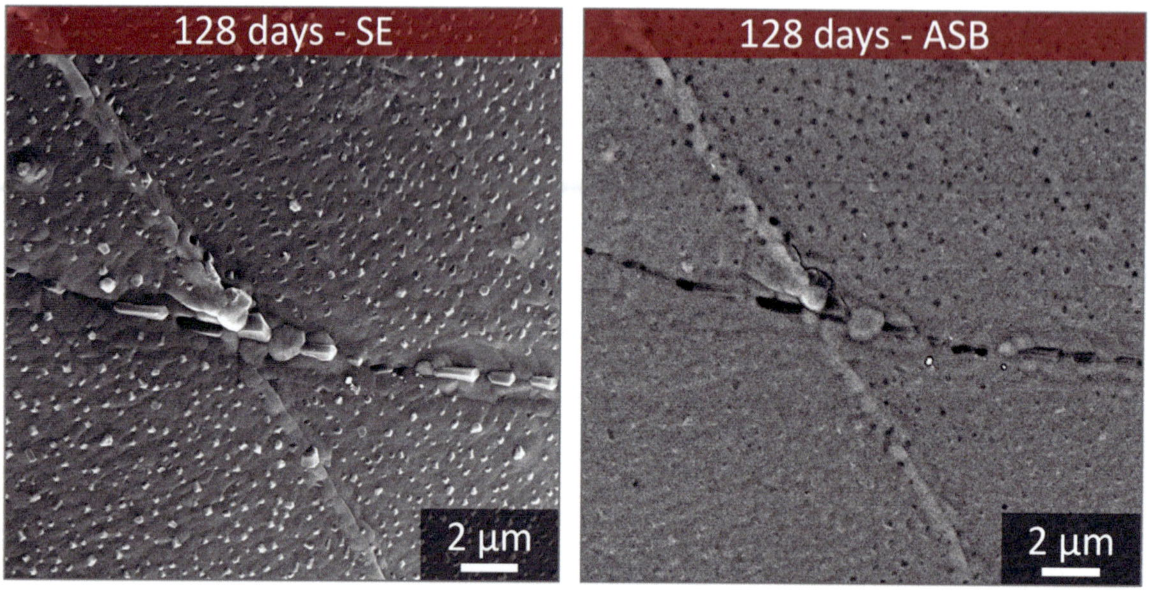

Figure C.2: SEM surface image of a nominally 30 mol% LMT containing KALOS pebble which was annealed for 128 days. The image on the left shows the secondary electron image, while the image on the right shows the Z-contrast by analyzing backscattered electrons (ASB detector). The dark grey in the ASB image reflects the lithium orthosilicate phase, while the light grey color reflects the lithium metatitanate phase.

Additional results on the pebble microstructure

In Figure C.3 overviews images of cross sections of nominally 20 mol% LMT containing KALOS pebbles are shown. These particular pebbles are of the type in which a fraction of the LMT content is dissolved within the LOS phase.

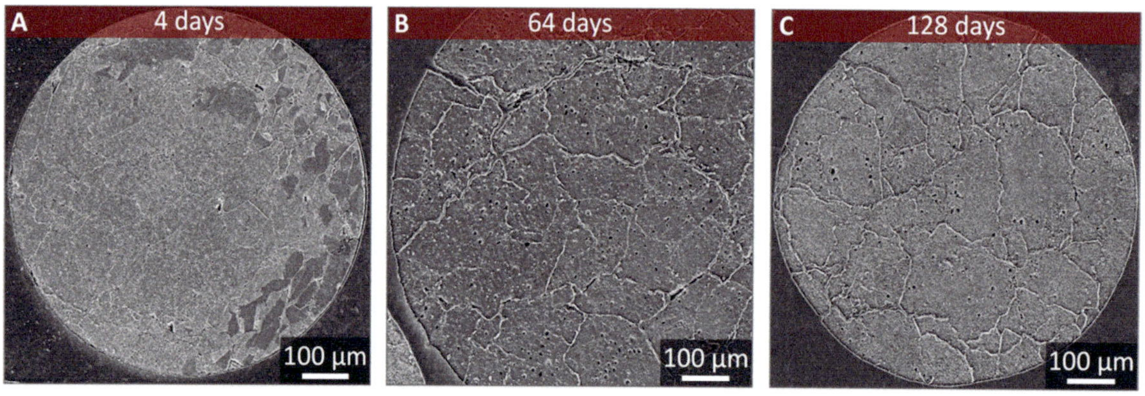

Figure C.3: Cross sections of nominally 20 mol% LMT containing pebbles that were fabricated by the KALOS process imaged by the SEM. These pebbles do not show the predominant microstructure which features eutectic domains and primary lithium orthosilicate. Along most cracks that the samples show after 64 days (B) and 128 days (C) of annealing epoxy resin infiltrated the pebbles, while the sample that was annealed for 4 days (A) does not show this feature.

The removal of a significant amount of material from as-received lithium excessive lithium metatitanate pebbles that were fabricated by the sol-gel method is observable in Figure C.4. The mechanical preparation of the cross sections in conjunction with the relatively little adherence between the individual lithium metatitanate grains lead to the observed removal of pebble material. Yet, only a fraction of the pebbles shows this behavior, as Figure C.4 also unambiguously shows.

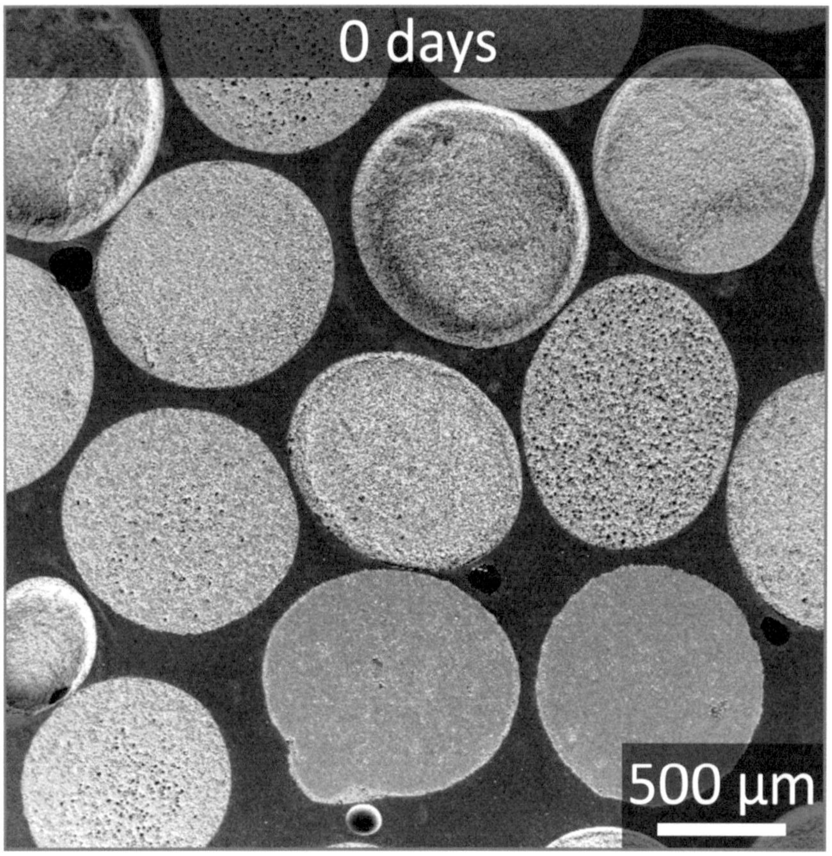

Figure C.4: SEM micrograph of a cross section of as-received lithium excessive lithium metatitanate pebbles that were fabricated by the sol-gel method. Especially the pebbles in the upper part of the micrograph show a significant removal of material after the mechanical preparation of the sample.

The elemental composition of the crust and the core of lithium excessive lithium metatitanate pebbles which were fabricated by the sol-gel method are exemplarily given in Table C.14. Figure C.5 shows the approximate areas from which the EDS signals were obtained. The signals of gold and palladium can easily be related to the coating to remove electrical charges from the surface, which was applied before the investigation by sputtering (see also section 3.6). The increased amount of carbon which is detected for the pebble crust may originate from the organic resin which has in part penetrated into the crust and also from the relatively large excitation volume of the electron beam which can reach the surrounding resin of the crust. It is clearly observable from the obtained values that the amount of titanium is reduced in the crust with respect to the core, also with respect to the oxygen signal.

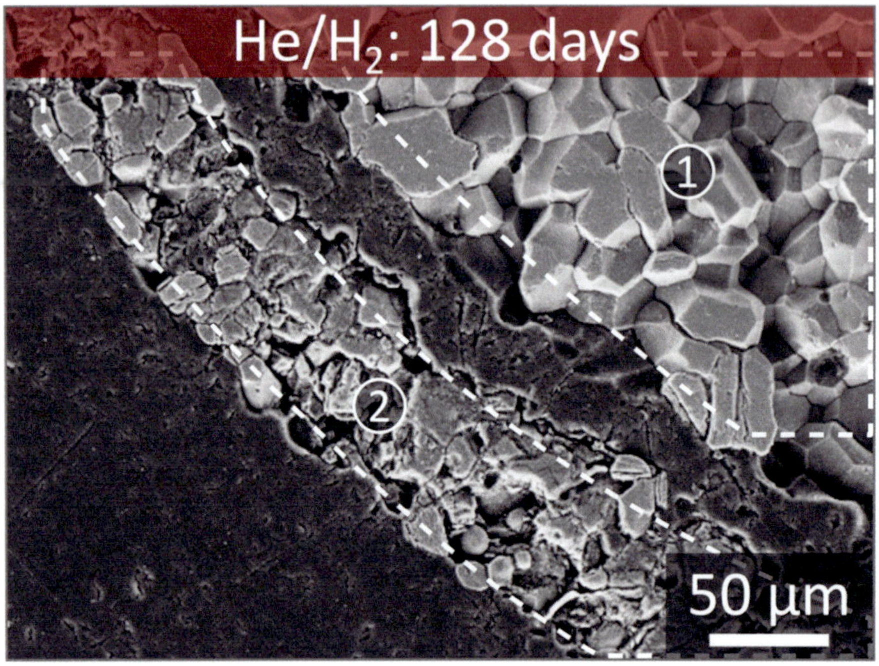

Figure C.5: SEM micrograph of the crust and the core of a lithium excessive lithium metatitanate pebble, fabricated by the sol-gel method, that was annealed for 128 days. The numbers and dashed lines indicate the approximate areas from which the EDS signal was acquired. The so-determined element concentrations are given in Table C.14.

Table C.14: Quantification of the EDS signals of the relevant elements for the marked areas in Figure C.5.

Area in Figure C.5	C Kα	O Kα	Ti Kα	Au Mα	Pd Lα
1: Pebble core	6.34 at.%	62.03 at.%	29.09 at.%	1.63 at.%	0.91 at.%
2: Pebble crust	23.01 at.%	61.53 at.%	12.99 at.%	1.63 at.%	0.84 at.%

Detailed results of the optical porosity determination

In the following, the determined total porosity is listed, which of course contains a certain amount of imaginary porosity as explained in section 3.6. The open porosity, as displayed in section 4.4, is also listed which results from the subtraction of the determined closed porosity from the optically determined total porosity. The so-derived open porosity thus also contains the imaginary porosity and does not represent the true open porosity.

Table C.15: The total porosity as well as the open porosity as determined by the optical method of the samples that were annealed in a He/H_2 environment as well as in their as-received state. All values are increased by an unknown amount of imaginary porosity.

	As-received	4 days	32 days	64 days	128 days
KA-20-LMT					
Total porosity	17.26 %	14.51 %	16.72 %	17.01 %	15.01 %
Open porosity	17.08 %	10.93 %	14.78 %	14.56 %	13.89 %
KA-30-LM					
Total porosity	29.02 %	22.08 %	23.93 %	26.81 %	26.58 %
Open porosity	28.33 %	20.11 %	23.42 %	26.25 %	24.91 %
EM-LMT					
Total porosity	23.94 %	22.35 %	21.55 %	20.03 %	19.44 %
Open porosity	19.80 %	12.85 %	19.32 %	19.84 %	18.65 %
SG-LMT					
Total porosity	33.81 %	36.98 %	36.74 %	37.50 %	37.63 %
Open porosity	28.04 %	29.71 %	32.07 %	35.03 %	34.51 %

Table C.16: The total porosity as well as the open porosity as determined by the optical method of the biphasic samples that were fabricated by the emulsion method. All values are increased by an unknown amount of imaginary porosity. The nominal concentrations of LMT and LAO are given in the table.

	10 mol%	20 mol%	30 mol%	40 mol%	50 mol%	60 mol%	70 mol%	80 mol%	90 mol%
EM-X-LMT									
Total porosity [%]	32.92	26.58	28.15	26.37	23.12	24.93	36.20	35.58	23.68
Open porosity [%]	26.88	24.53	25.40	24.12	21.48	21.99	33.77	31.77	19.36
EM-X-LAO									
Total porosity [%]	22.13	22.37	14.88	–	–	–	–	–	–
Open porosity [%]	14.76	13.65	3.76	–	–	–	–	–	–

Appendix D

This appendix chapter features additional analysis and details which supplement the discussion in section 5.

Modelling of multiple failure mechanisms

To model the failure by multiple failure mechanisms, i.e. multiple Weibull distributions, the inverse function of the resultant compound distribution has to be determined. This is done iteratively and by approximation by the following R code for the two relevant compound distributions for this work. The iterative approximation uses a simulated annealing approach in the first step. The so-determined values are then improved by a second step with uses the 'Brent' algorithm of R, which is conveniently fast and specialized on one-dimensional problems.

```
## Define compund functions

## Non-competing failure
NcomF  <-function(x, m1, c1, m2, c2, p)  {
              F1 <- pweibull(x, m1, c1)
              F2 <- pweibull(x, m2, c2)
              F  <- p*F1+(1-p)*F2
              value<-F
}

## Non-competing failure with three Weibull distributions
triNcomF      <-function(x, m1, c1, m2, c2, m3, c3, p, p2)  {
              F1 <- pweibull(x, m1, c1)
              F2 <- pweibull(x, m2, c2)
              F3 <- pweibull(x, m3, c3)
              F  <- p*F1+p2*F3+(1-p-p2)*F2
              value<-F
}

## Define compund functions to be minimized
NcomF0 <-function(x, m1, c1, m2, c2, p, rnk)  {
              value<- abs(NcomF(x, m1, c1, m2, c2, p) - rnk)
}

triNcomF0     <-function(x, m1, c1, m2, c2, m3, c3, p, p2, rnk)  {
              value<- abs(triNcomF(x, m1, c1, m2, c2, m3, c3, p, p2) - rnk)
}

## Define Weibull parameters
m1 <- 14.67
c1 <- 3.453

m2 <- 12.39
c2 <- 4.4

m3 <- 1
c3 <- 7

p <- 0.25
```

```
p2 <- 0.35

n <- 50

NcomF.val      <- c(NULL)
triNcomF.val   <- c(NULL)

lower0 <- 0.2
upper0 <- 15
x <- 1


## Noncompeting failure
        rankvec <- c(NULL)

        for (i in 1:n) {
        rankvec <- append(rankvec, qbeta(0.5,i,n-i+1))
        }

        for (i in 1:n) {
                if (i == 1) {
                lower.run <- lower0
                upper.run <- upper0
                x <- 1
                } else {
                lower.run <- NcomF.val[i-1]
                upper.run <- upper0
                x <- NcomF.val[i-1]
                }
        NcomF.run <- optim(par=x, fn=NcomF0, gr=NULL, method = "SANN", m1=m1, c1=c1, m2=m2,
c2=c2,   p=p,   rnk=rankvec[i],   control=list(abstol=1e-15,   reltol=1e-15,   ndeps=1e-12,
fnscale=1e3, maxit=10000), hessian = TRUE)$par
        NcomF.val <- append(NcomF.val, NcomF.run)
        }
        for (cycles in 1:100) {
        x.init <- NcomF.val
        NcomF.val <- c(NULL)
        for (i in 1:n) {
                lower.run <- x.init[i]*0.8
                upper.run <- x.init[i]*1.2
                x <- x.init[i]
        NcomF.run <- optim(par=x, fn=NcomF0, gr=NULL, method = "Brent", m1=m1, c1=c1, m2=m2,
c2=c2, p=p, rnk=rankvec[i], lower = lower.run, upper = upper.run, control=list(abstol=1e-
15, reltol=1e-15, ndeps=1e-9, fnscale=1e21, maxit=500000), hessian = TRUE)$par
        NcomF.val <- append(NcomF.val, NcomF.run)
        }
        }

F.val.NcomF <- data.frame(NcomF.val, rankvec)
colnames(F.val.NcomF) <- c("time", "rank.median")
wei.val.NcomF <- c(m1, c1, m2, c2)


## Non-competing failure with three Weibull distributions
        rankvec <- c(NULL)

        for (i in 1:n) {
        rankvec <- append(rankvec, qbeta(0.5,i,n-i+1))
        }

        for (i in 1:n) {
                if (i == 1) {
                lower.run <- lower0
                upper.run <- upper0
                x <- 1
                } else {
```

```
                lower.run <- triNcomF.val[i-1]
                upper.run <- upper0
                x <- triNcomF.val[i-1]*1.1
                }
        triNcomF.run <- optim(par=x, fn=triNcomF0, gr=NULL, method = "SANN", m1=m1, c1=c1,
m2=m2, c2=c2, m3=m3, c3=c3, p=p, p2=p2, rnk=rankvec[i], control=list(abstol=1e-15, rel-
tol=1e-15, ndeps=1e-12, fnscale=1e3, maxit=10000), hessian = TRUE)$par
        triNcomF.val <- append(triNcomF.val, triNcomF.run)
        }
        for (cycles in 1:100) {
        x.init <- triNcomF.val
        triNcomF.val <- c(NULL)
        for (i in 1:n) {
                lower.run <- x.init[i]*0.8
                upper.run <- x.init[i]*1.2
                x <- x.init[i]
        triNcomF.run <- optim(par=x, fn=triNcomF0, gr=NULL, method = "Brent", m1=m1, c1=c1,
m2=m2, c2=c2, m3=m3, c3=c3, p=p, p2=p2, rnk=rankvec[i], lower = lower.run, upper = up-
per.run, control=list(abstol=1e-15, reltol=1e-15, ndeps=1e-9, fnscale=1e21, maxit=500000),
hessian = TRUE)$par
        triNcomF.val <- append(triNcomF.val, triNcomF.run)
        }
        }

F.val.triNcomF <- data.frame(triNcomF.val, rankvec)
colnames(F.val.triNcomF) <- c("time", "rank.median")
wei.val.triNcomF <- c(m1, c1, m2, c2, m3, c3)
```

Contact pressures for lithium excessive LMT pebbles

In the following table the mean contact pressures according to equation (2.5) that were determined for 4 selected pebbles are shown. The fragments of these pebbles are shown in Figure 4.64 and Figure 4.66. From these images, the diameter of the flattened area at the points of loading can be determined. It must be noted, that the observed diameter is not necessarily the true diameter of the flattened region, because the pebbles does not have to fracture into hemispheres. Yet, it is possible to derive the withstood contact pressure p^* according to equation (D.1) by relating the failure load F to the (approximately) circular contact region at the loading points of the pebbles. By doing so, also the plastic deformation is considered.

$$p^* = \frac{F}{\pi\left(\frac{d}{2}\right)^2} \qquad \text{(D.1)}$$

Table D.1 shows the so calculated contact pressures as well as the ratio of p_m and p^* for the two pebbles of the same type. It is obvious that the determined values differ largely with respect to the absolute values, but the tendency that for stronger pebbles a higher contact pressure is assigned is true, regardless of the method that is used for the estimation.

Table D.1: For 4 pebbles that were examined in Figure 4.64 and Figure 4.66, the determined mean contact pressure according to equation (2.5), which is only valid for a fully elastic system is given, as well as the contact pressure which is determined from the failure load and the observed flattend area. Furthermore, the obtained pressures of the two pebbles of the same type are compared to each other.

	EM-LMT		SG-LMT	
	0 days		0 days	
p_m [GPa]	1.95	5.99	3.66	6.90
F [N]	2.16	62.81	9.90	66.30
d [μm]	447	300	201	243
p^* [GPa]	0.014	0.889	0.312	1.430
p_m-ratio	307 %		189 %	
p^*-ratio	6456 %		458 %	

Additional crystallographic data

In the following table additional crystallographic parameters of the published structures of lithium orthosilicate, lithium metasilicate and lithium carbonate are presented. Also the volume of the unit cell normalized for the number of Li-ions is given as a measure of the density of the structures.

Table D.2: Basic crystallographic parameters of lithium orthosilicate, lithium metasilicate and lithium carbonate at standard conditions as published by Deng et al. [56] (Li_4SiO_4), Völlenkle [173] (Li_2SiO_3) and Idemoto et al. [172] (Li_2CO_3). The volume V of the unit cell is normalized per Li-ion in the structure. The cell angles α and γ are identical for all three phases with 90°. The parameters of alumina according to Többens et al. [207] are also given. For this structure, α and β show a value of 90°.

Phase	Space group	a [Å]	b [Å]	c [Å]	β (γ for Al_2O_3) [°]	V [Å³/Li]
α-Li_4SiO_4	$P2_1/m$ (11)	11.5556(4)	6.0986(2)	16.7300(5)	99.0580(10)	20.79
Li_2SiO_3	$Cmc2_1$ (36)	9.396(1)	5.396(1)	4.661(1)	90	29.54
Li_2CO_3	C12/c1 (15)	8.35263(8)	4.97353(4)	6.18942(5)	114.677(1)	29.21
Al_2O_3	R-3c (167)	4.7597(1)	4.7597(1)	12.9935(3)	120	–

Raw powder weight for the fabrication of the biphasic emulsion method pebbles

In Table D.3 the initial weight of the raw material powders which were used for fabricating the biphasic emulsion method pebbles is detailed. Also the theoretically resultant amounts of the anticipated phases are given. As explained in section 3.2.2, the synthesis of lithium orthosilicate required a substantial surplus of lithium carbonate. The so-introduced surplus of lithium ions is also given as a fraction of the respective second phase content (either LMT or LAO).

Table D.3: Initial weight of the raw material powders that were introduced into the slurry of emulsion method, from which the biphasic pebbles were fabricated are given. Also the theoretically resultant weights of lithium orthosilicate and lithium metatitanate or lithium aluminate are given. Furthermore, the molar surplus of lithium as a fraction of the second phase content is given.

Nominal 2nd phase content	Initial weight Li_2CO_3 [g]	Initial weight Li_2SiO_3 [g]	Initial weight Li_2TiO_3/ Al_2O_3 [g]	Theoretical weight of Li_4SiO_4 [g]	Theoretical weight of Li_2TiO_3/ $LiAlO_2$ [g]	Molar lithium surplus per LMT/LAO
EM: LOS+LMT						
10 mol%	7.786	8.917	1.209	11.878	1.209	898%
20 mol%	7.181	8.223	2.508	10.954	2.508	399%
30 mol%	6.528	7.476	3.908	9.958	3.908	233%
40 mol%	5.822	6.667	5.422	8.882	5.422	150%
50 mol%	5.057	5.791	7.064	7.714	7.064	100%
60 mol%	4.224	4.837	8.851	6.444	8.851	67%
70 mol%	3.314	3.795	10.802	5.056	10.802	43%
80 mol%	2.316	2.653	12.943	3.534	12.943	25%
90 mol%	1.217	1.394	15.301	1.857	15.301	11%
EM: LOS+LAO						
10 mol%	8.548	8.810	0.555	11.736	0.717	1271%
20 mol%	8.755	8.021	1.136	10.685	1.469	636%
30 mol%	8.973	7.192	1.747	9.581	2.259	424%

Lithium metatitanate can in principle incorporate excessive lithium as shown in Figure 2.10, this number thus illustrates that it is increasingly unlikely with increasing LMT content, that the LMT phase of the fabricated pebbles bears a lithium excess. The values in Table D.3 of the theoretically produced amount lead to a total sample weight of approximately 11 g to 17 g, which is in contrast to the actually sintered amounts that are considerably lower as specified in Table 3.5. The discrepancy between both values is due to difficulties to reproduce the sintering conditions in QST at KIT, which impeded the sintering of the available green pebbles.

Cross sections of green LAO containing pebbles

Figure D.1 shows cross sections of green lithium aluminate containing lithium orthosilicate pebbles that were shaped by the emulsion method. It is clearly visible that toward the center of the pebbles, the green density substantially decreases. This decrease does not seem to be completely smooth as 'rings' appear in the image at which a relatively sharp transition of the green density is observable. The images do not reveal a significant amount of pores within the pebbles.

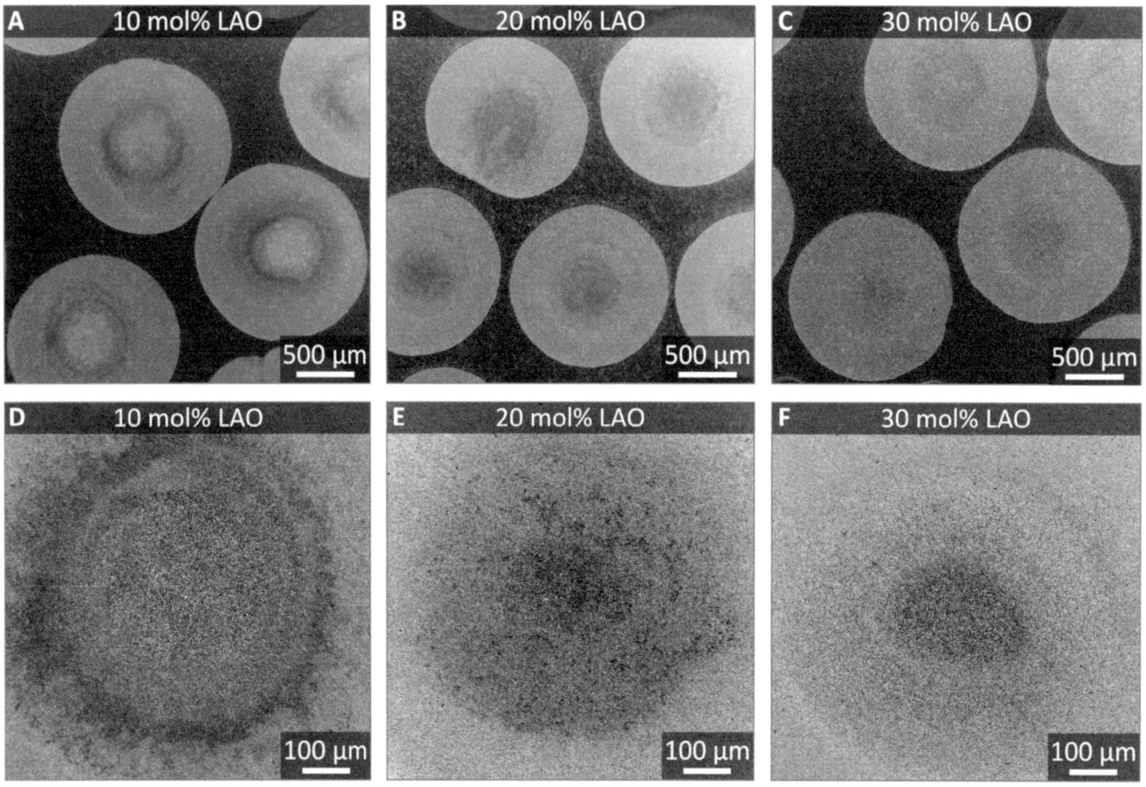

Figure D.1: SEM micrographs of cross sections of green LOS+LAO pebbles that were shaped by the emulsion method. In the image, the nominal LAO content is given.

Weibull size effect for the nominally 20 mol% LMT containing KALOS pebbles

Likelihood contour plots are an effective tool to compare the Weibull parameters of two sets of data also taking the confidence boundaries into account. For the different samples of the KA-20-LMT, the determined Weibull parameters are compared in this way in Figure D.2 while applying equation (5.2) to the 1000 μm pebbles to compensate for the Weibull size effect. The shift of the characteristic strength that originates from equation (5.2) with respect to the actually determined value is also relayed to the determined contour. Since the Weibull moduli of the larger and smaller pebbles substantially differ for all but the as-received samples, the mean value of the respective Weibull moduli is used as common Weibull modulus for the calculation of the size effect, yet the principal results do not change regardless of which experimentally justified Weibull modulus is chosen. In Figure D.2, the initially determined Weibull moduli of the data sets are used for the graph in order to display the substantial differences. From Figure D.2 it is also clear that just in case of the samples that were annealed for 32 days, the use of a common Weibull modulus would lead to an intersection of the likelihood contours. Since apart from this sample, the other samples remain statistically different, also the two sizes of pebbles that were annealed for 32 days are considered to show a statistically different mechanical failure behavior.

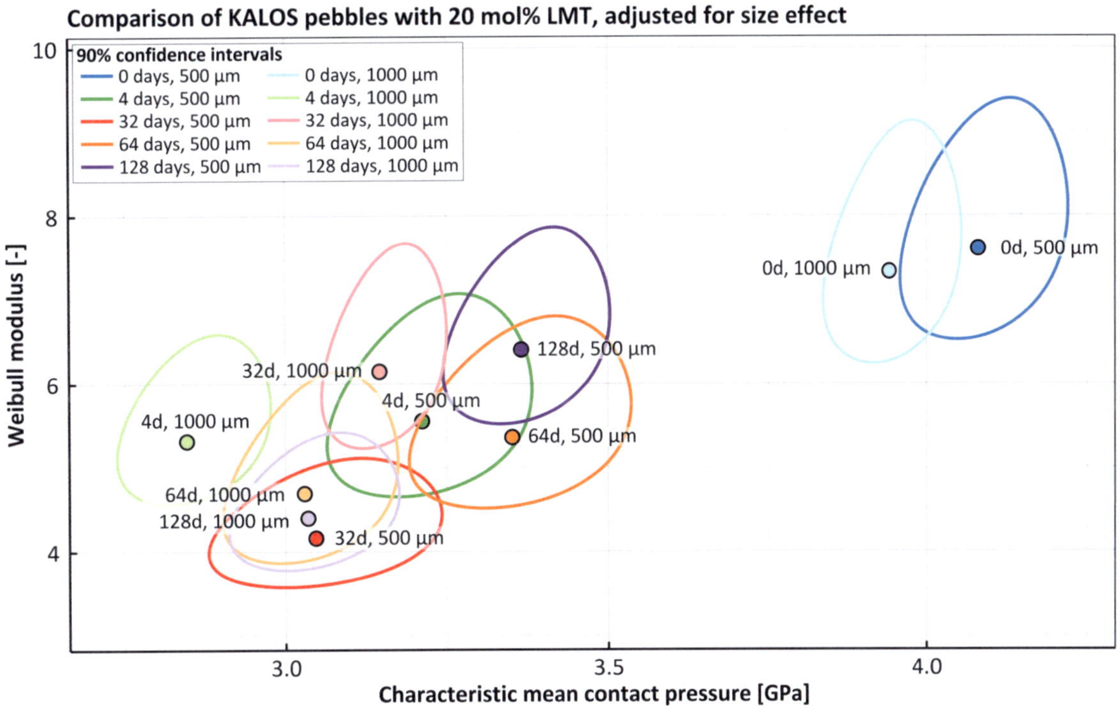

Figure D.2: Likelihood contour plot for the nominally 20 mol% LMT containing KALOS pebbles that were annealed for up to 128 days. The characteristic strength of the larger pebbles was corrected according to equation (5.2) to match the characteristic strength of the smaller pebbles.

Possible models of the failure behavior of KALOS pebbles

The Weibull plots of several samples can be reasonably be explained by the presence of two non-competitive failure modes. In the following figures, the resultant estimated compound failure models are compared to the actual experimental data. In all cases a relatively good agreement between model and experiment can be observed. Yet, in case of the nominally 30 mol% containing 1000 µm pebbles that were annealed for 128 days, considerable deviations between model and experiment can be observed in the low-strength part of the Weibull plot.

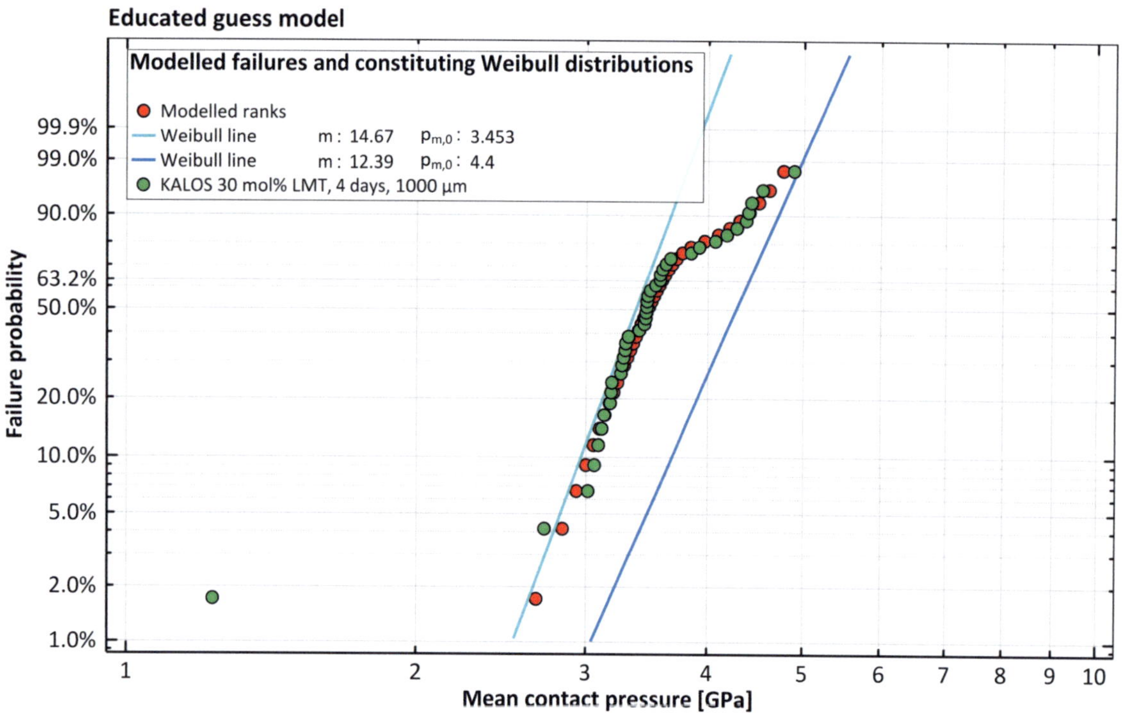

Figure D.3: Modelled failure behavior for two non-competing failure modes with Weibull moduli of 14.67 and 12.39. The probability of testing a sample of latter failure mechanism is 25 %. The model is compared to the experimental data of nominally 30 mol% LMT containing KALOS pebbles of 1000 µm diameter that were annealed for 4 days.

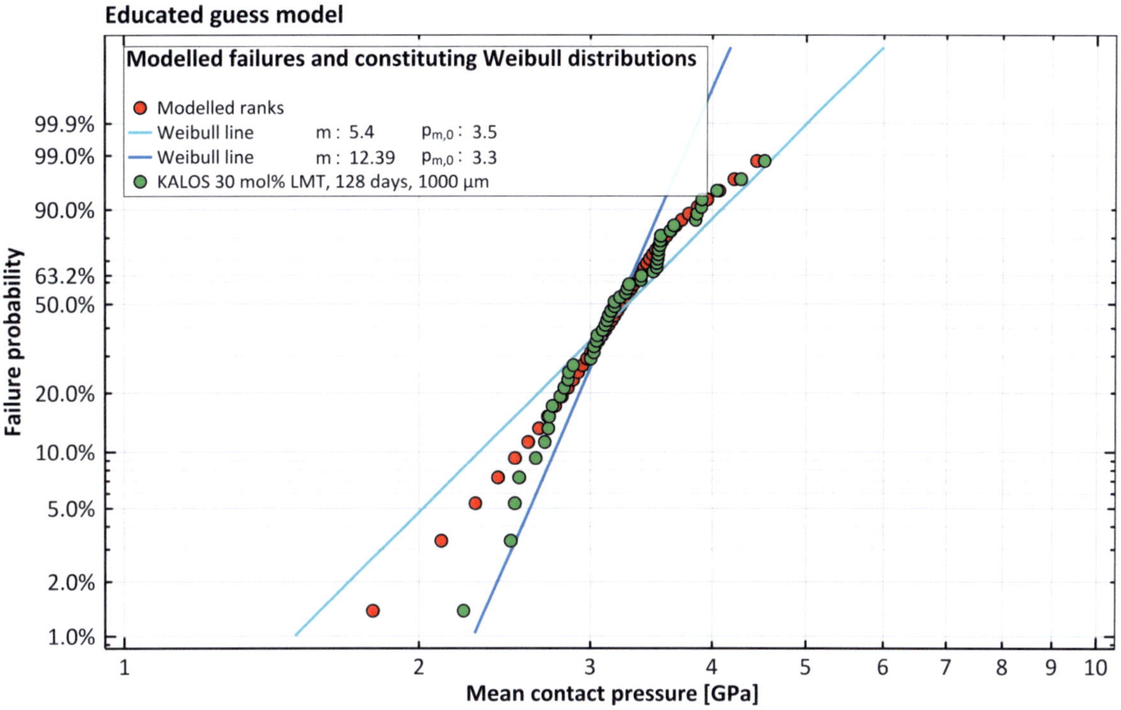

Figure D.4: Modelled failure behavior for two non-competing failure modes with Weibull moduli of 5.4 and 12.39. The probability of testing a sample of latter failure mechanism is 50 %. The model is compared to the experimental data of nominally 30 mol% LMT containing KALOS pebbles of 1000 µm diameter that were annealed for 128 days.

Correlations of the XRD results with other properties for the lithium excessive lithium metatitanate samples

The a/c-ratios of the β-lithium metatitanate unit cell as a function of the lithium ratio differences are detailed in Figure D.5. These results are virtually identical to the observations of the lattice strain in [100]-direction that are shown in Figure 5.19, but can readily compared to the results that were obtained for the biphasic emulsion method pebbles that contain lithium metatitanate (see Figure 5.11).

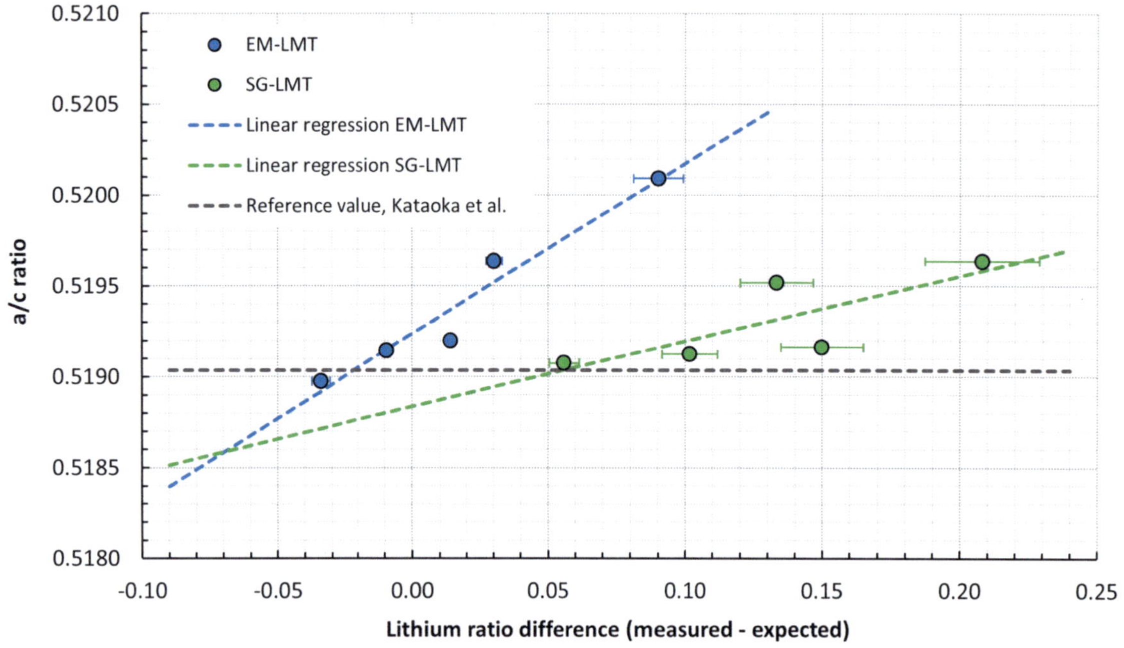

Figure D.5: The a/c ratio of the β-lithium metatitanate unit cell is shown as a function of the difference between the measured and expected (stoichiometric) lithium content. The error bars in x-direction show ±10 % on the obtained value, whereas the error that the refinement suggests is too low to be displayed in this graph. The linear regression shows a coefficient of determination of 0.944 and 0.632 for the EM-LMT and SG-LMT samples, respectively. Also the according reference value of β-lithium metatitanate, as reported by Kataoka et al. [78], is shown.

The preferred orientation in [001] direction of the lithium metatitanate crystallites of the lithium excessive lithium metatitanate pebble samples correlates excellently well with the optically determined grain size as Figure D.6 shows. Although the linear regression in Figure D.6 only uses the data points of the emulsion method pebbles, it is evident that also the data points of the sol-gel method pebbles are represented reasonably well.

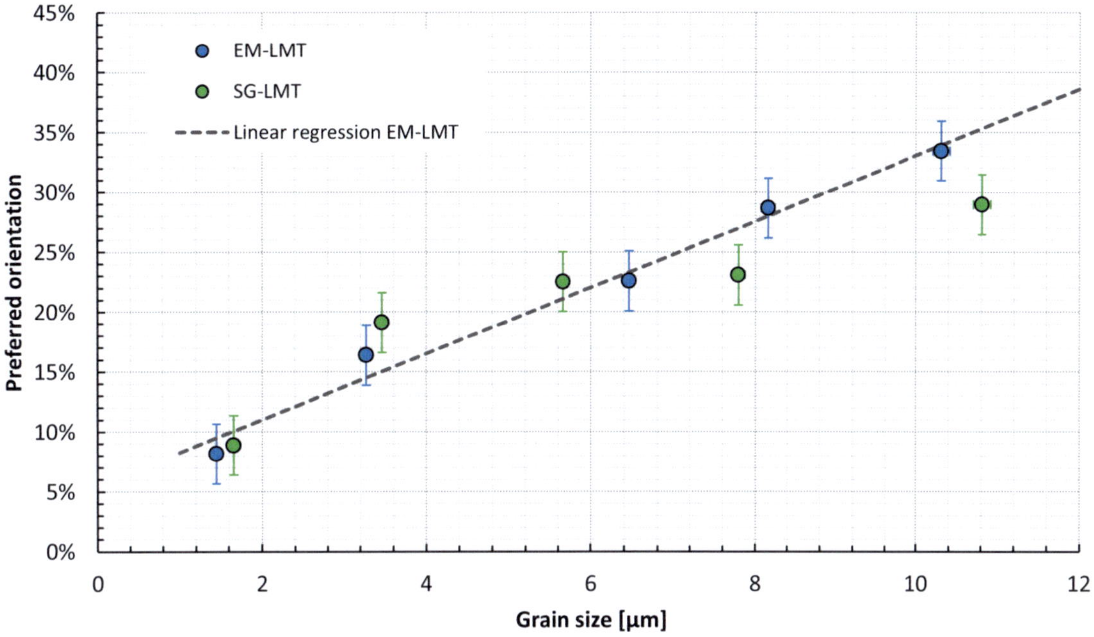

Figure D.6: The preferred orientation of the powdered lithium excessive lithium metatitanate samples as determined by XRD as a function of the optically determined grain size. In the graph, a linear regression of the data points of the emulsion method pebbles is given. The coefficient of determination of the regression is 0.98. For the experimental uncertainty in y-direction a constant value of ±2.5 % is assumed. The errorbars in x-direction vanish with the chosen scale.

Publications

In the following, the scientific publications of mine, until and including May 2018, are listed in chronological order. The publications that include results which are also presented here are formatted in bold.

F.A. Hernandez, F. Arbeiter, L.V. Boccaccini, E. Bubelis, V.P. Chakin, I. Cristescu, B.E. Ghidersa, M. Gonzalez, W. Hering, T. Hernandez, X.Z. Jin, M. Kamlah, B. Kiss, R. Knitter, M.H.H. Kolb, P. Kurinskiy, O. Leys, I.A. Maione, M. Moscardini, G. Nadasi, H. Neuberger, P. Pereslavtsev, S.Pupeschi, R. Rolli, S. Ruck, G.A. Spagnuolo, P.V. Vladimirov, C. Zeile, G. Zhou: Overview of the HCPB Research Activities in EUROfusion
IEEE Transactions on Plasma Science (2018) p.1-15
DOI: 10.1109/TPS.2018.2830813

J.M. Heuser, M.H.H. Kolb, T. Bergfeldt, R. Knitter: Long-term thermal stability of two-phased lithium orthosilicate/metatitanate ceramics
Journal of Nuclear Materials 489 (2018) in press
DOI: 10.1016/j.jnucmat.2018.05.010

M. Zmitko, P. Vladimirov, R. Knitter, M. Kolb, O. Leys, J. Heuser, H.-C. Schneider, R. Rolli, V. Chakin, S. Pupeschi, L. Magielsen, A. Fedorov, Y. Poitevon: Development and qualification of functional materials for the European HCPB TBM
Fusion Engineering and Design 124 (2018) in press
DOI: 10.1016/j.fusengdes.2018.05.014

M.H.H. Kolb, R. Rolli, R.Knitter: Tritium adsorption/release behaviour of advanced EU breeder pebbles
Journal of Nuclear Materials 489 (2017) p.229-235
DOI: 10.1016/j.jnucmat.2017.03.051

A. Zariņš, O. Valtenbergs, G. Ķizāne, A. Supe, S. Tamulevičius, M. Andrulevičius, E. Pajuste, L. Baumane, O. Leys, M.H.H. Kolb, R. Knitter: Characterisation and radiolysis of modified lithium orthosilicate pebbles with noble metal impurities
Fusion Engineering and Design 124 (2017) p.934-939
DOI: 10.1016/j.fusengdes.2017.01.008

M.H.H. Kolb, R.Knitter, T.Hoshino: Li_4SiO_4 based breeder ceramics with Li_2TiO_3, $LiAlO_2$ and $Li_xLa_yTiO_3$ additions, part II: Pebble properties
Fusion Engineering and Design 115 (2017) p.6-16
DOI: 10.1016/j.fusengdes.2016.12.038
identical to reference [96]

M.H.H. Kolb, K.Mukai, R.Knitter, T.Hoshino: Li_4SiO_4 BASED BREEDER CERAMICS WITH Li_2TiO_3, $LiAlO_2$ AND $Li_XLa_YTiO_3$ ADDITIONS, PART I: FABRICATION
Fusion Engineering and Design 115 (2017) p.39-48
DOI: 10.1016/j.fusengdes.2016.12.033
identical to reference [183]

O. Leys, T. Bergfeldt, M.H.H. Kolb, R. Knitter, A. A. Goraieb: THE REPROCESSING OF ADVANCED MIXED LITHIUM ORTHOSILICATE/METATITANATE TRITIUM BREEDER PEBBLES
Fusion Engineering and Design 107 (2016) p.70-74
DOI: 10.1016/j.fusengdes.2016.04.025

A. Zarins, O. Valtenbergs, G. Kizane, A. Supe, R. Knitter, M.H.H. Kolb, O. Leys, L. Baumane, D. Conka: FORMATION AND ACCUMULATION OF RADIATION-INDUCED DEFECTS AND RADIOLYSIS PRODUCTS IN MODIFIED LITHIUM ORTHOSILICATE PEBBLES WITH ADDITIONS OF TITANIUM DIOXIDE
Journal of Nuclear Materials 470 (2016) p.187-196
DOI: 10.1016/j.jnucmat.2015.12.027

M. González, E. Carella, A. Moroño, M.H.H. Kolb, R. Knitter: THERMALLY INDUCED OUTDIFFUSION STUDIES OF DEUTERIUM IN CERAMIC BREEDER BLANKET MATERIALS AFTER IRRADIATION
Fusion Engineering and Design 98-99 (2015) p.1771-1774
DOI: 10.1016/j.fusengdes.2015.03.027

D.A.H. Hanaor, M.H.H. Kolb, Y. Gan, M. Kamlah, R. Knitter: SOLUTION BASED SYNTHESIS OF MIXED-PHASE MATERIALS IN THE Li_2TiO_3–Li_4SiO_4 SYSTEM
Journal of Nuclear Materials 456 (2015) p.151-161
DOI: 10.1016/j.jnucmat.2014.09.028

R.K. Annabattula, M. Kolb, Y. Gan, R. Rolli, M. Kamlah: SIZE DEPENDENT CRUSH ANALYSIS OF LITHIUM ORTHOSILICATE PEBBLES
Fusion Science and Technology 66 (2014) p. 136-141
DOI: 10.13182/FST13-737

P. Waibel, J. Matthes, O. Leys, M.H.H. Kolb, H.B. Keller, R.Knitter: HIGH-SPEED CAMERA-BASED ANALYSIS OF THE LITHIUM CERAMIC PEBBLE FABRICATION PROCESS
Chemical Engineering & Technology 37 (2014) p. 1654-1662
DOI: 10.1002/ceat.201300769

A. Zarins, G. Kizane, A. Supe, M.H.H. Kolb, R. Knitter, J. Tiliks Jr., L. Baumane: INFLUENCE OF CHEMISORPTION PRODUCTS OF CARBON DIOXIDE AND WATER VAPOUR ON RADIOLYSIS OF TRITIUM BREEDER
Fusion Engineering and Design 89 (2014) p. 1426-1430
DOI: 10.1016/j.fusengdes.2014.01.005

V. Chakin, R. Rolli, A. Moeslang, M. Klimenkov, M.H.H. Kolb, P. Vladimirov, P. Kurinskiy, H.-C. Schneider, S. van Til, A.J. Magielsen, M. Zmitko: TRITIUM RELEASE AND RETENTION PROPERTIES OF HIGHLY NEUTRON-IRRADIATED BERYLLIUM PEBBLES FROM HIDOBE-01 EXPERIMENT
Journal of Nuclear Materials 442 (2013) p.483-489
DOI: 10.1016/j.jnucmat.2013.03.032

F. Hernández, M. Kolb, M. Ilic, A. Kunze, J. Németh, A. von der Weth: SET-UP OF A PRE-TEST MOCK-UP EXPERIMENT IN PREPARATION FOR THE HCPB BREEDER UNIT MOCK-UP EXPERIMENTAL CAMPAIGN
Fusion Engineering and Design 88 (2013) p. 2378-2383
DOI: 10.1016/j.fusengdes.2013.02.107

O. Leys, C. Odemer, U. Maciejewski, M.H.H. Kolb, R. Knitter: MICROSTRUCTURE ANALYSIS OF MELT-BASED LITHIUM ORTHOSILICATE/ METATITANATE PEBBLES – MIKROSTRUKTURANALYSE VON SCHMELZPROZESSIERTEN LITHIUMORTHOSILIKAT-/LITHIUMMETATITANAT-KUGELN
Praktische Metallographie – Practical Metallography, 2013/03 p.196-203
Link: http://www.practical-metallography.com/PM110194

J. Eymery, X. Chen, C. Durand, M. Kolb, G. Richter: SELF-ORGANIZED AND SELF-CATALYST GROWTH OF SEMICONDUCTOR AND METAL WIRES BY VAPOUR PHASE EPITAXY: GAN RODS VERSUS CU WHISKERS
Comptes Rendus Physique 14 (2013) p.221-227
DOI: 10.1016/j.crhy.2012.10.009

R. Knitter, M.H.H. Kolb, U. Kaufmann, A.A. Goraieb: FABRICATION OF MODIFIED LITHIUM ORTHOSILICATE PEBBLES BY ADDITION OF TITANIA
Journal of Nuclear Materials 442 (2012) p.433-436
DOI: 10.1016/j.jnucmat.2012.10.034

R. Knitter, M.H.H. Kolb, C. Odemer: SYNTHESIS OF TRITIUM BREEDER CERAMICS FROM METALLIC LITHIUM
Journal of Nuclear Materials, 420 (2012) p.268-272
DOI: 10.1016/j.jnucmat.2011.10.008

M.H.H. Kolb, M. Bruns, R. Knitter, S. van Til: LITHIUM ORTHOSILICATE SURFACES: CHARACTERIZATION AND EFFECT ON TRITIUM RELEASE
Journal of Nuclear Materials, 427 (2012) p.126-132
DOI: 10.1016/j.jnucmat.2012.04.024

M.H.H. Kolb, R. Knitter, U. Kaufmann, D. Mundt: ENHANCED FABRICATION PROCESS FOR LITHIUM ORTHOSILICATE PEBBLES AS BREEDING MATERIAL
Fusion Engineering and Design, 86 (2011) p.2148-2151
DOI: 10.1016/j.fusengdes.2011.01.104

M. Kolb, G. Richter: GROWTH OF SINGLE CRYSTALLINE COPPER NANOWHISKERS
AIP Conference Proceedings, 1300 (2010) p.98-105
DOI: 10.1063/1.3527143

U. Kaufmann, M. Kolb, R. Knitter: DIE CHEMISCH-THERMISCHE BEANSPRUCHUNG VON PLATIN-WERKSTOFFEN DURCH LITHIUMREICHE SILIKATSCHMELZEN
Kneissl, A. [Hrsg.], Fortschritte in der Metallographie : Berichte der 13.Internat.Metallographie-Tagung, Leoben, A, September 29 - October 1, 2010 – Frankfurt : MAT-INFO, Werkstoff-Informationsges., 2010 p.43-48 (Praktische Metallographie : Sonderbd. ; 42)

S. Van Til, M. Kolb, A.J. Magielsen, R. Knitter: IRRADIATION OF LITHIUM ORTHOSILICATE IN THE HIGH FLUX REACTOR IN PETTEN FOR THE FUSION FUEL CYCLE
1st Internat. Conf. on Materials for Energy, Karlsruhe, July 4-8, 2010 – Extended Abstracts – Book A p.272-274, Frankfurt a.M. : Dechema